VOLUME SEVEN

ADVANCES IN NEUROTOXICOLOGY

Occupational Neurotoxicology

VOLUME SEVEN

ADVANCES IN NEUROTOXICOLOGY

Occupational Neurotoxicology

Edited by

ROBERTO G. LUCCHINI
Florida International University, Miami, United States; University of Brescia, Brescia, Italy

MICHAEL ASCHNER
Albert Einstein College of Medicine, Bronx, NY, United States

LUCIO G. COSTA
University of Washington, Seattle, WA, United States; University of Parma, Parma PR, Italy

ELSEVIER

ACADEMIC PRESS
An imprint of Elsevier

Academic Press is an imprint of Elsevier
50 Hampshire Street, 5th Floor, Cambridge, MA 02139, United States
525 B Street, Suite 1650, San Diego, CA 92101, United States
The Boulevard, Langford Lane, Kidlington, Oxford OX5 1GB, United Kingdom
125 London Wall, London, EC2Y 5AS, United Kingdom

First edition 2022

ISBN: 978-0-12-819176-7
ISSN: 2468-7480

For information on all Academic Press publications
visit our website at https://www.elsevier.com/books-and-journals

Publisher: Zoe Kruze
Acquisitions Editor: Mariana Kuhl
Developmental Editor: Jhon Michael Peñano
Production Project Manager: Abdulla Sait
Cover Designer: Christian Bilbow

Typeset by STRAIVE, India

Contents

Contributors

Gaafar Abdel Rasoul
Community, Environmental, and Occupational Medicine Department, Faculty of Medicine, Menoufia University, Shebin El-Kom, Egypt

Claudia Ambrosi
University of Brescia; Spedali Civili of Brescia, Brescia, Italy

W. Kent Anger
Oregon Health & Science University, Oregon Institute of Occupational Health Sciences, Portland, OR, United States

Gabriela de Paula Arrifano
Laboratório de Farmacologia Molecular, Instituto de Ciências Biológicas, Universidade Federal do Pará, Belém, Brazil

Marcus Augusto-Oliveira
Laboratório de Farmacologia Molecular, Instituto de Ciências Biológicas, Universidade Federal do Pará, Belém, Brazil

Somaiyeh Azmoun
Florida International University, Miami, United States

Matthew R. Bonner
Department of Epidemiology and Environmental Health, State University of New York at Buffalo, Buffalo, NY, United States

Maryse F. Bouchard
Department of Environmental and Occupation Health, University of Montreal School of Public Health, QC, Canada

William K. Boyes
Neurological and Endocrine Toxicology Branch, Public Health and Integrated Toxicology Division, Center for Public Health and Environmental Assessment, Office of Research and Development, Environmental Protection Agency, Research Triangle Park, NC, United States

Evelyn J. Bromet
Stony Brook University, New York, United States

Jeremy W. Chambers
Florida International University, Miami, United States

Sean A.P. Clouston
Stony Brook University, New York, United States

Daniele Corbo
University of Brescia; Spedali Civili of Brescia, Brescia, Italy

Maria Elena Crespo-Lopez
Laboratório de Farmacologia Molecular, Instituto de Ciências Biológicas, Universidade Federal do Pará, Belém, Brazil

Felipe Farias da Rocha
Laboratório de Farmacologia Molecular, Instituto de Ciências Biológicas, Universidade Federal do Pará, Belém, Brazil

Yenny Fariñas Diaz
Florida International University, Miami, United States

Hana-May Eadeh
University of Iowa, Department of Psychological and Brain Sciences, G60 Psychological and Brain Sciences, Iowa City, IA, United States

Sam Gandy
Icahn School of Medicine at Mount Sinai; James J. Peters VA Medical Center, Bronx, New York, United States

Roberto Gasparotti
University of Brescia; Spedali Civili of Brescia, Brescia, Italy

G. Hageman
Medical Spectrum Twente, Hospital Enschede, The Netherlands

Olfat Hendy
National Liver Institute, Menoufia University, Shebin El-Kom, Egypt

Megan Horton
Icahn School of Medicine at Mount Sinai, New York, United States

Ahmed A. Ismail
Kansas Department of Health and Environment, Topeka, KS, United States; Community, Environmental, and Occupational Medicine Department, Faculty of Medicine, Menoufia University, Shebin El-Kom, Egypt

Amanda Lopes-Araújo
Laboratório de Farmacologia Molecular, Instituto de Ciências Biológicas, Universidade Federal do Pará, Belém, Brazil

Roberto G. Lucchini
Florida International University, Miami, United States; University of Brescia, Brescia, Italy

Ben J. Luft
Stony Brook University, New York, United States

S.J. Mackenzie Ross
Research Department of Clinical, Educational and Health Psychology, University College London, London, United Kingdom

Lorella Mascaro
Spedali Civili of Brescia, Brescia, Italy

J. Nihom
Medical Spectrum Twente, Hospital Enschede, The Netherlands

James R. Olson
Department of Epidemiology and Environmental Health; Department of Pharmacology and Toxicology, State University of New York at Buffalo, Buffalo, NY, United States

Barbara Paghera
University of Brescia; Spedali Civili of Brescia, Brescia, Italy

Donatella Placidi
University of Brescia; Spedali Civili of Brescia, Brescia, Italy

Marcia H. Ratner
Department of Pharmacology and Experimental Therapeutics, Boston University School of Medicine, Boston, MA, United States

Carlo Rodella
University of Brescia; Spedali Civili of Brescia, Brescia, Italy

Diane S. Rohlman
Department of Occupational and Environmental Health, University of Iowa, College of Public Health, Iowa City, IA, United States

Jonathan Rutchik
Division of Occupational & Environmental Medicine, University of California San Francisco, San Francisco; Neurology, Environmental and Occupational Medicine Associates, Mill Valley, CA, United States

Trisha Saha
Department of Environmental and Occupation Health, University of Montreal School of Public Health, QC, Canada

Letícia Santos-Sacramento
Laboratório de Farmacologia Molecular, Instituto de Ciências Biológicas, Universidade Federal do Pará, Belém, Brazil

José Rogério Souza-Monteiro
Faculdade de Medicina, Campus de Altamira, Universidade Federal do Pará, Altamira, Brazil

Cheuk Y. Tang
Icahn School of Medicine at Mount Sinai, New York, United States

Kim Tieu
Florida International University, Miami, United States

Carola Tuerk
Department of Psychology, Concordia University, Québec, Canada

G. van der Laan
Foundation Learning and Developing Occupational Health, Leusden, The Netherlands

Christoph van Thriel
Leibniz Research Centre for Working Environment and Human Factors, Department of Toxicology, Research Group: Neurotoxicology and Chemosensation, Dortmund, Germany

Preface

A variety of biological, chemical, and physical agents can affect the brain in the workplace. They can induce neurotoxic effects including any adverse structural or functional change in the central/peripheral nervous system. Many of these agents are known occupational neurotoxicants and can frequently interact with each other, producing mixture effects that can be more harmful than the individual components. Various gases, metals, solvents, and pesticides can yield different neurological phenotypes and behavioral modifications that can manifest acutely or in the long term, depending on the exposure intensity, timing, and potential vulnerabilities. From cognitive impacts to motor impairment and behavioral changes, there is a variety of manifestations that can be assessed with appropriate tools. While neurotoxicology generates relevant scientific data for individual neurotoxicants, bioinformatics and biostatistics allow a broader assessment of mixed exposure. New statistical models can provide evidence of mixture effects and the toxicological weight of each component (Bobb et al., 2015; Czarnota et al., 2015). The importance of early neurotoxicity has been questioned. To appreciate the relevance of adverse changes, it is important to distinguish between individual and population risks. Small shifts in the mean value of a health index within a study sample can have substantial implications for the health status of the population from which the study sample was drawn. Under such circumstances, a population-based approach to prevention might be as effective as a patient-based approach (Bellinger, 2007).

New emerging threats are reported in workplaces such as airplanes, and a specific chapter in this book provides updated evidence about the "aerotoxic syndrome." Further studies are needed to elucidate the causes and the potential interventions to reduce any risk of neurotoxicity in flight crew and passengers. Organic solvents are widely used in numerous occupations and can also imply exposure for customers, e.g., in nail salons. Pesticides are widely used worldwide in agriculture and also in homes. Carbon monoxide is a well-known occupational neurotoxicant but health effects are still caused by this gas in many types of occupations. Mercury is used extensively in gold mining and still impacts workers and communities through contamination of the ecosystem. This book includes the methodological aspects that are essential to assess neurotoxicity among workers: from brain imaging to neurobehavioral testing and clinical evaluation.

Updated information on all these old and newly emerging topics are provided by renowned scientists who are experts in the field. Focusing on occupational toxicology has a growing significance not only for what the exposure to neurotoxicants can cause during the working life, but also for the long-term effects that can increase the load of neurodegenerative diseases. Since occupational exposure is a truly modifiable factor, all efforts must be taken to control, minimize, and eliminate this risk.

Roberto G. Lucchini
Florida International University, Miami, USA
University of Brescia, Italy

References

Bellinger, D.C., 2007. Interpretation of small effect sizes in occupational and environmental neurotoxicology: individual versus population risk. Neurotoxicology 28 (2), 245–251.

Bobb, J.F., et al., 2015. Bayesian kernel machine regression for estimating the health effects of multi-pollutant mixtures. Biostatistics 16 (3), 493–508.

Czarnota, J., Gennings, C., Wheeler, D.C., 2015. Assessment of weighted quantile sum regression for modeling chemical mixtures and cancer risk. Cancer Inform. 14 (Suppl. 2), 159–171.

CHAPTER ONE

Current evolution of neurobehavioral methods

Maryse F. Bouchard[a,*], Carola Tuerk[c], Hana-May Eadeh[d], Diane S. Rohlman[b], Trisha Saha[e], and W. Kent Anger[f]

[a]Department of Environmental and Occupation Health, University of Montreal School of Public Health, QC, Canada
[b]Department of Occupational and Environmental Health, University of Iowa, College of Public Health, Iowa City, IA, United States
[c]Department of Psychology, Concordia University, Québec, Canada
[d]University of Iowa, Department of Psychological and Brain Sciences, G60 Psychological and Brain Sciences, Iowa City, IA, United States
[e]Department of Environmental and Occupation Health, University of Montreal School of Public Health, QC, Canada
[f]Oregon Health & Science University, Oregon Institute of Occupational Health Sciences, Portland, OR, United States
*Corresponding author: e-mail address: maryse.bouchard@umontreal.ca

Contents

1. A brief historical perspective on neurobehavioral evaluation in neurotoxicology assessment

Neurobehavioral tests have played a significant role in the growth and impact of the field of human epidemiological research aimed at identifying and characterizing the hazards of workplace chemicals. This chapter offers a brief review of the development of this field and an introduction to the need, methods and challenges of using neurobehavioral tests in human workplace studies.

Advances in Neurotoxicology, Volume 7
ISSN 2468-7480
https://doi.org/10.1016/bs.ant.2022.04.004

With industrialization, workplace exposures to never-before encountered chemicals grew rapidly and virtually exploded during World War II. This led to uncontrolled workplace exposures to many new chemicals in the rush to meet wartime needs (Davies et al., 1997). Nervous system effects were recognized as consistently associated with many occupational exposures by the 1950s, but the observations were not systematic. For example, case reports reflected consistent effects in workers such as tremors, ataxia, "general slowing of mental processes," emotional instability and other poorly defined symptoms in a significant proportion of workers exposed to tricholorethylene in degreasing operations (Grandjean et al., 1955). The "patients" who were exposed to carbon disulfide in the viscose rayon industry, which came into its' own in the 1940s, were typically and aptly described as "poisoned" and were seen in the clinics of neuropsychologists who used then-standard clinical measures to test their patients. Systematic testing for functional behavioral effects of exposures to workplace chemicals began when Helena Hänninen in Finland used standardized neuropsychological tests to assess workers exposed to carbon disulfide. As the full range of effects was not known, Hänninen and colleagues began with a broad range of tests of different functions, and then used multivariate techniques to categorize the tests to select a smaller number of tests to use to assess new patients exposed to existing and new chemicals. She began to call the resulting group of tests a "battery" (Hänninen, 1966). This systematizing of the process of evaluating patients with workplace exposures served to guide future research.

Charles Xintaras and Barry Johnson with colleagues from around the world convened an international meeting in Cincinnati, Ohio (USA) to describe the rapidly expanding field and let people working in the domain see demonstrations of the different of batteries available (Xintaras et al., 1974). Ten years later in 1984, the same group convened a meeting, again in Cincinnati, to consider the growing literature on human studies of behavioral effects of chemical exposures with the express goal of identifying a battery of neuropsychological tests that could be used in all populations and for all chemicals to establish their [what is now termed] neurotoxicity. The assembled group selected a "core" battery of tests, and some additional tests for select exposures or exposure concentrations. The "core" functions assessed (and the tests of those functions) were affect (Profile of Moods States), attention/response speed (Simple Reaction Time), auditory memory (Digit Span [WAIS]), manual dexterity (Santa Ana peg rotation test), perceptual motor speed (Digit Symbol [WAIS], visual perception/memory

(Benton Visual Retention), and motor steadiness (Pursuit Aiming II). They boldly decided to recommend a guide to the use of those tests written so that any scientist, with a modicum of training, could apply the tests consistently and identify their adverse effects on the nervous system (Johnson et al., 1987). Each goal was accomplished.

The group meeting in Cincinnati developed the recommendation under the auspices of the World Health Organization (WHO), so naming the resulting testing "system" the WHO-recommended Neurobehavioral Core Test Battery, or WHO NCTB. In the following years a user guide was developed (available from author Anger), and again with the help of WHO's then Office of Occupational Health, the test battery was administered to 2300 unexposed "control" working population adult participants to establish its feasibility for worldwide testing (Anger et al., 1993). While the too lofty goal of a universal testing system was not realized (e.g., people with limited education performed poorly on some of the tests among other limitations) (Anger et al., 1993; Anger and Cassitto, 1993), the feasibility study did demonstrate that the battery of tests could be used in epidemiological field studies when both an exposed and a comparable control group were available for cross-sectional epidemiologic studies. While a series of studies using the neuropsychological tests in the NCTB could establish norms for any population, such would require trained neuropsychologists with access and funding to assess the performance of the population and establish norms by age, sex and other variables in a "control" population for comparison, but such would not be feasible or practical and the use of a comparable control group was the recommendation. Moreover, the NCTB was economical to assemble and administer by people who were not trained neuropsychologists by using the Operational Guide. Data from the NCTB was published for scientists to compare with their own data, on a country-by-country basis (Cassitto et al., 1990). The battery was used extensively in some countries (e.g., Liang et al., 1990) and reported in meetings of the scientific committee on Neurotoxicology & Psychophysiology of the International Commission on Occupational Health (ICOH) (Liang et al., 1990).

Other test batteries were developed contemporaneously with the WHO NCTB; most used related neuropsychological tests. Hanninen and Lindstrom (1979) developed the Neuropsychological Psychological Screening Test Battery with varied optional groupings of tests to apply for the various chemicals under study, for use in Finland. Cherry et al. (1984) developed the British battery of neuropsychological tests, aimed at addressing toluene and paint solvent exposures. Scientists in Asia administered neuropsychological tests adapted to their countries (e.g., Jeyaratnam et al., 1986; Sakong et al., 2007), though they did not develop into testing systems or batteries available for use outside the developing lab or group. Maizlish et al. (1985) developed a battery that was drawn from tests used in experimental psychology research, assessing psychomotor speed, ability to ignore distracting visual and verbal stimuli, spatial reasoning or visual intelligence, short term memory scanning, memory span, and memory for digits after a lapse of time up to 40 s. The test format differed in that each was presented and measured electronically by various types of equipment.

The influence of the NCTB, or the publications drawn on to select the NCTB tests, was seen in all the testing systems except the Maizlish battery (1985). While the development of computers usable for administering such tests would quickly outstrip the benefits of the NCTB, it influenced most subsequently developed testing batteries or systems and has remained a tool useable by people in developing countries who could not afford more expensive systems, to this day.

The development of computer technology overtook each of these testing batteries due, arguably, to the consistency, ease, and economy of administration, once the tests were developed. This can be seen as the second wave of testing systems to assess neurotoxic effects in human populations. The most widely used was the Neurobehavioral Evaluation System (NES) developed by Richard Letz (1990). Letz developed a series of computer-administered tests in a single system that were based on the WHO NCTB recommended tests, and grew to include other neuropsychological tests several of which were recommended as additional tests by the NCTB group (Johnson et al., 1987). A strength of the NES system and other computer-based systems was that the user could not only select the tests to use in a study, but they could also select many of the tests' parameters such as test duration and difficulty (Letz, 1990). Several tests in the NES are also presented in Swedish by Iregren et al. (1996), the Swedish Performance Evaluation System) and they were also translated into other languages. The Milan Automated [computer-based] Neurobehavioral System (MANS), was also designed to present neuropsychological tests based on the WHO NCTB recommendations (Cassitto et al., 1989), as was the Behavioral Assessment and Research System (BARS) (Rohlman et al., 2003); it, too adopted other neuropsychological tests as it developed and the tests were translated into Spanish, Portuguese, and Arabic.

With more powerful computer systems and the availability of ancillary measurement equipment, a new (third) wave of computer-based testing systems developed. The first was the Cambridge Neuropsychological Test Automated Battery (CANTAB) which has the unique heritage of being developed based on neuropsychological tests but designed to be language-independent and culture-free. It has been used to study non-human primates as well as patients with many different types of neurodegenerative diseases; the developers have recommended the application to humans exposed to workplace chemicals (Fray and Robbins, 1996).

Of these early testing systems, the only ones currently available for use are the CANTAB and BARS, both of which can be licensed from the developers, and the WHO NCTB, as the user can purchase the tests from various sources and construct the Santa Ana pegboard test or find a commercially available alternative. The other test systems described above, and more, fell into disuse and/or the developers left the field of neurotoxicity research. Importantly, computer-based testing systems need updating; for this they need to be sponsored/supported by a professional group or society, or a commercial company to support their maintenance as is the case for the CANTAB (https://www.cambridgecognition.com/cantab/). This and other or advanced testing systems in this third wave of neurobehavioral testing are described next; they may be seen as the contemporary testing systems available to identify neurotoxic effects in human working or environmentally exposed populations.

2. Contemporary computerized testing

Viewed as an alternative to traditional paper-and-pencil tests for research and clinical practice, computerized testing allows for time efficiency, standardized quantitative data collection, and improved data analysis. Extensively trained personnel are not needed for data collection, recruitment of larger and specialized samples becomes more attainable, and coordinated data collection between multiple sites is facilitated, thereby boosting power Gosling (Feenstra et al., 2017; Gosling et al., 2004; Kraut et al., 2004). Measures that are difficult to record manually, such as reaction time, can be easily gathered using computerized tests.

Usually seated in front of a computer screen, test subjects are instructed to complete a series of cognitive tasks using a keyboard, touchscreen, mouse or joystick, following the presentation of visual or auditory stimuli.

Research regarding psychometric properties of computerized neurobehavioral tests have shown similar reliability and validity compared to traditional methods. As such, computerized testing has been a more accessible and cost-effective option for researchers and has been gradually applied in the neurotoxicology field to assess the possible effects of occupational exposure on neuropsychological functions. For example, computerized cognitive tests (e.g., N-back, Stroop, Finger Tapping and Reaction Time) were used to examine associations between cognition and solvent exposure in the automotive industry (Bates et al., 2016; Golbabaei et al., 2018).

Four well-established test batteries are summarized below (see Table 1 for more detailed information):

Cambridge Neuropsychological Test Automated Battery (CANTAB): First suggested to be used in the field of neurotoxicology in 1996 (Fray and Robbins, 1996), this neurobehavioral test battery can be self-administered using a computer with a touch-sensitive screen, a mouse or a keyboard. It captures various areas of cognitive function including visual memory, visual attention, learning and memory, working memory, planning, set-shifting, social cognition, fluid intelligence and sustained attention in subjects aged 6 to 90. In adults, web-based administration has been shown to be comparable to in-person lab-based assessments (Backx et al., 2020), and several studies have validated this battery among diverse clinical populations (e.g., O'Connell et al., 2004; Sterr et al., 2006) including those exposed to environmental contaminants (e.g., van Wijngaarden et al., 2017). Norms are available upon request (www.cambridgecognition.com).

National Institutes of Health (NIH) Toolbox: Intended for large-scale epidemiologic and longitudinal studies, the NIH Toolbox is a set of assessment tools measuring cognitive, emotional, motor and sensory function from ages 3 to 85. The cognitive portion of the Toolbox, called NIH Toolbox Cognition Battery (NIHTB-CB), specifically focusses on executive function, attention, episodic memory, language, processing speed and working memory using seven sub-tests. In adult populations, research has found the NIH Toolbox to be reliable and valid (Scott et al., 2019; Weintraub et al., 2013b), with norms available across the lifespan (Weintraub et al., 2014).

Cambridge Brain Sciences (CBS): CBS is a web-based self-administered battery comprised of 12 tasks measuring short-term memory, reasoning, concentration and verbal ability for subjects ages six and older. Mostly used in adult studies, no significant effect of at home verses in laboratory

Table 1 Description of the CANTAB, NIH-CB, CBS and BARS in terms of administration, subtests, and psychometric properties.

Battery	Domains assessed	Subtests	Data on psychometric properties
Cambridge Neuropsychological Test Automated Battery (CANTAB) Age: 4-90y	– Memory	– Delayed Matching to Sample – Paired Associates Leaning – Pattern Recognition Memory – Spatial Span – Verbal Recognition Memory	In adults, web-based administration is comparable to in-person lab-based assessments (Backx et al., 2020). Norms available upon request (www.cambridgecognition.com).
Hardware: PC, Laptop, iPad Administration time: 40–45 min (according to number of tests)	– Executive functions	– Cambridge Gambling Task – Intra-extra dimensional set shift – Multitasking Test – One Touch Stockings of Cambridge – Spatial Working Memory – Stockings of Cambridge – Stop Signal Task	Test-retest reliabilities $r=0.75$ for 3 tests, but mostly between 0.39 and 0.73, in line with previous studies (3 months) (Karlsen et al., 2022).
Mode: Lab- and web-based (self-) administration	– Attention/ Psychomotor speed	– Reaction Time – Rapid Visual Information Processing – Motor Screening Task – Match to Sample Visual Search	
	– Social/Emotional Cognition	– Emotional Bias Task – Emotion Recognition Task	

Continued

Table 1 Description of the CANTAB, NIH-CB, CBS and BARS in terms of administration, subtests, and psychometric properties.—cont'd

Battery	Domains assessed	Subtests	Data on psychometric properties
NIH Toolbox Cognition Battery (NIHTB-CB) Age: 3-85y Hardware: iPad, computer, tablet Administration time: 30–35 min Mode: Lab-based	– Executive Function	– Dimensional Change Card Sort Test – Flanker Inhibitory Control & Attention Test	In adult populations, research has found NIH Toolbox to be reliable and valid, with norms available across the lifespan.
	– Attention	– Flanker Inhibitory Control & Attention Test	Excellent psychometric properties (e.g., Weintraub et al., 2013a, 2013b).
	– Episodic Memory	– Picture Sequence Memory Test	Demographically adjusted norms available (age, education, sex, race/ethnicity).
	– Language	– Picture Vocabulary Test – Oral Reading Recognition Test	
	– Processing Speed	– Pattern Comparison Processing Speed Test	
	– Working Memory	– List Sorting Working Memory Test	
Cambridge Brain Sciences (CBS) Age: 6y+ Hardware: Laptop, PC, tablets (web-browser compatible)	– Short-term memory	– Spatial Span – Paired associates – Token Search – Monkey Ladder	Measurement properties (e.g., reliability) have not been extensively investigated. Test-retest reliabilities of the tasks average = 0.69.
	– Reasoning	– Odd One Out – Polygons – Spatial Planning – Rotations	A normative database of 75 k participants ages 6 to 75 + is available from developers.

Administration time: 35–45 min Mode: Lab- or web-based (self-administration)	– Concentration – Verbal ability	– Feature Match – Double Trouble (Stroop Task) – Digit Span – Grammatical Reasoning	
Behavioral Assessment and Research System (BARS) Age: 4y+ Hardware: Laptop Administration time: varies depending on test selection and parameter settings Mode: Lab-based	– Attention – Memory – Motor Function – Complex Functions	– Continuous Performance – Selective Attention – Digit-Span – Match-to-Sample – Oregon Dual Task Procedure – Progressive Ratio – Simple Reaction Time – Finger Tapping – Reversal Learning – Serial Digit Learning – Symbol Digit	Developed specifically to detect neurobehavioral deficits due to neurotoxic exposures (Rohlman et al., 2003). Test-retest reliability has been adequate within a day and a week (Farahat et al., 2003).

testing has been found (Sternin et al., 2019). Measurement properties (e.g., reliability) have not been extensively investigated. A normative database of 75,000 participants ages 6 to 75 years and older is available (www.cambridgebrainsciences.com).

Behavioral Assessment and Research System (BARS): Developed in the 1990s, BARS was specifically designed for population research to examine adverse nervous system effects associated with occupational exposures (Rohlman et al., 1996) and has been widely used around the world. It includes tests of memory, attention, learning motivation, executive function, response speed, and coordination and instructions are available in English, Spanish, Arabic, Thai, and Portuguese. The spoken language instructions as well as modifiable parameters (e.g., number of trials, inter-stimulus intervals, stimulus shape, instruction language, test difficulty, delays, alternate forms) allows it to be tailored for different populations, including children, or for individuals with limited education (Rohlman et al., 2001a), and to be used in repeated assessments. Reliability and stability have been assessed in both adults and children (Campbell et al., 1999; Farahat et al., 2003; Rohlman et al., 2000a).

Despite the many benefits of computerized testing, several aspects are to be kept in mind when choosing this approach. Relevant performance measures, such as free recall memory, cannot be collected (Howieson, 2019). Furthermore, if unsupervised, observations affecting the test subjects' performance, such as distraction and fatigue, can be missed and invalid assessment of performance can ensue if instructions are not understood. Potential errors may arise and influence stimulus presentation, participant behavior, input processing, and recorded performance due to interactions between software, internet speed (for online tests), processing power, and external hardware (Feenstra et al., 2017). Overall, what is generally missing from these contemporary testing systems is a strategy or guidance for selecting the tests and test parameters for a particular exposed population, especially for one exposed to a chemical or mixture suspected to be causing neurotoxic effects but for which there is no prior evidence in human or animal studies. Symptoms that are assumed to occur in the exposed population based on individual cases or reports in the literature remain the guide in these cases.

3. Assessment of cognitive function

Cognition refers to a wide range intellectual processes performed by the central nervous system, often with the input of sensory information.

Wide variations in cognitive performance and functioning are naturally present in the population, and originate both from hereditary and environmental factors, starting in utero and extending throughout life (Bouchard Jr. and McGue, 2003). Neuropsychologists often refer to the different cognitive processes involved in a particular function as cognitive domains. In most instances, be it for research or clinical purposes, the assessment of cognition in workers exposed to neurotoxic substances is conducted using several tests in order to draw a portrait of the deficits.

It is challenging to describe the broad range of cognitive processes occurring in the human brain, as they are numerous and overlapping with one another. Furthermore, performance on cognitive tests can be influenced by several factors, as cognitive domains targeted for assessment are dependent on sensory input, which might also be impaired by exposure to neurotoxic substances. There is also an overlap with mood and/or mental health, as symptoms of depression and anxiety, for example, can affect performance on tests designed to assess cognition, through changes in energy level (fatigue), motivation, and arousal state. With these caveats in mind, cognitive evaluation is an important aspect of any neurobehavioral assessment. Some of the domains most frequently assessed in relation with exposure to neurotoxicants in the workplace are listed:

- Attention
- Memory (short term, long term, recall learning)
- Executive function
- Reaction time
- Concentration
- Visuo-spatial processing
- Code-shifting
- Processing speed
- Reasoning
- Language (e.g., naming, fluency)

The selection of cognitive domains to be assessed in a specific workplace is directly linked with the nature of the suspected or known neurotoxicant and its effects on the nervous system, often using reported symptoms as a guide. Well over a hundred different tests have been used in the field of neurotoxicology, and it is beyond the scope of this chapter to cover this, but they can be found in several reference publications (e.g., Sherman et al., 2020).

Most cognitive assessments conducted in the context of occupational and environmental exposure to neurotoxicants are composed of several tests, each of which might generate multiple scores. For instance, the California Verbal Learning Test (CVLT) can generate over 15 different scores, such as total number of correct responses during the five learning trials, spontaneous recall after an interference list (short recall), and spontaneous recall of the initial list 20 minlater (long delay), and the number correctly recognized as being in the initial list. Outside the context of neurotoxicology, these multiple scores are useful for clinicians who aim to obtain a precise portrait of deficits of their patient, necessary for a precise diagnosis or devise a readaptation program. But caution must be exercised when the results from neurobehavioral assessments aim to determine whether workplace exposure is linked with deficits, because the numerous outcome variables might lead to a problem (real or perceived) of multiple testing. An elegant solution is to use advanced statistical techniques (i.e., structural equation models Bollen, 1989) to create fewer, more robust, indicators of cognitive performance by simultaneously including multiple outcome variables in the statistical models (Budtz-Jorgensen et al., 2002; Oulhote et al., 2014).

Exposure to certain neurotoxic substances can cause widespread, diffuse disruptions of nervous system function, such as those caused by changes in neurotransmitter activity. This differs from the focal brain damage such as those following strokes, trauma, or brain tumors. Hence, assessments originally designed to screen for dementia are often included in test batteries to detect impairment of general cognition, in addition to cognitive domain-specific tests. The Montreal Cognitive Assessment (MoCA) is one of the most frequently used screeners for cognitive impairment. It evaluates the following domains: delayed recall, visuospatial skills, executive function, calculation, abstraction, language, and orientation. The test has a sensitivity of 90% and specificity of 87% for mild cognitive impairment (Smith et al., 2007).

Another focus has emerged from a life course approach, such as investigations to determine whether exposure to neurotoxic substances in the workplace might accelerate loss of cognitive function due to aging (Bouchard et al., 2005), and/or increase the risk of developing neurodegenerative diseases (Arab and Mostafalou, 2021). For example, occupational exposure to pesticides is a well-documented risk factor for the development of Parkinsonian syndrome (Gunnarsson and Bodin, 2019). Furthermore, emerging evidence shows that several airborne pollutants may also disrupt the brain through molecular and cellular pathways that either directly damage brain structures or increase the predisposition to neurological diseases. For example, experiments in animal models revealed that air pollutants induce neuroinflammation, oxidative stress, microglial activation, cerebrovascular dysfunction, and alterations in the blood-brain barrier (Block and Calderon-Garciduenas, 2009; Genc et al., 2012; Kilian and Kitazawa, 2018).

In the last decade, the new developments in the field of cognitive effects of neurotoxic exposures were geared towards greater refinement in the interpretation of these assessments, and the exploration of factors that can modify the effects of neurotoxicants. For instance, a study showed that personality (viz., traits of neuroticism) could magnify the effects of exposure to noise on cognitive performance (Khajehnasiri et al., 2022).

4. Assessment of mood, mental state

Occupational exposure to neurotoxicants can cause disruption of mental and psychological well-being, either by interfering directly with neurochemical homeostasis, or indirectly by adversely impacting cognition and motor function. Several neurotoxicants are known to induce mood disorders, including symptoms of depression, irritability, anxiety, sleep problems, hyperactivity, aggressiveness, and confusion. At high levels of exposure, more severe displays of behavioral problems can occur, like extreme emotional lability (uncontrolled laughing and crying) and psychotic symptoms, such as paranoid ideation and hallucinations. These neuropsychiatric manifestations have been reported as occurring with several occupational neurotoxicants, including inorganic mercury (Jackson, 2018) and manganese (Emara et al., 1971).

For obvious reasons, mood is most often assessed with a self-administered questionnaire. For instance, the Profile of Mood States (POMS) is an instrument composed of six mood scales: anger, confusion, depression, fatigue, tension, and vigor (McNair et al., 1971). However, since these subscales were developed using clinical populations, the structure may not be appropriate for nonclinical populations, such as those comprised of workers (e.g., lead exposed workers Lindgren et al., 1999). Several questionnaires were initially designed to assess neurotoxic symptoms related to solvents, given their potency and their widespread use in many workplaces. Some of the most often used questionnaires are EUROQUEST for solvents (Chouaniere et al., 1997), the Symptom Checklist (SCL-90, Karlson et al., 2000), and the Q16 (Lundberg et al., 1997). Symptom ascertainment is non-invasive and inexpensive, hence, it can be used to assess neurotoxicity in the context of interventions to reduce exposure in the workplace. For example, interventions aimed at better use of personal protective equipment and good workplace hygiene by solvent-exposed spray painters were found to reduce self-reported symptoms of neurotoxicity, ascertained with the EUROQUEST (Keer et al., 2018). Despite the limitations of self-reported symptoms, subjective complaints are an important part of the assessment of brain function since they relate directly to perceived health status and day-to-day functioning.

5. Assessment of neuromotor function

The control of movement requires conscious and unconscious processing involving a complex ensemble of neurological structures. Voluntary movement necessitates transmission of an influx message from the motor strip of the cerebral cortex to the appropriate muscle on the opposite side of the body, and the ability to achieve smooth and coordinated movements involves the cerebellum and the basal ganglia (Groenewegen, 2003). Some neurotoxicants can impair the integrity of sensorimotor neurons, causing a clinical portrait of neuropathy and distal weakness, as reported among solvent-exposed workers (Puri et al., 2007). A wide range of neuromotor outcomes can be assessed within a neurobehavioral evaluation in the context of occupational exposures. Neuromotor deficits following neurotoxicant exposure include decreased gross and fine motor ability, weakness, poor eye-hand coordination, slowing of movements, tremor, postural abnormalities, rigidity, and decreased facial expression (Costa and Manzo, 2020).

Neuromotor deficits caused by neurotoxicant exposures can be assessed with standardized clinical evaluations, such as the Luria Nebraska Neuropsychological Battery (Lucchini et al., 2014; Mergler et al., 1994) and the Unified Parkinson's Disease Rating Scale (Criswell et al., 2021; Lewis et al., 2016). However, these clinical scales are based on observations and have a subjective component. The results of these assessments are semi-quantitative and might not be sensitive enough to detect subclinical deficits (Lewis et al., 2016). Hence, other methods of assessment were developed to provide quantitative and objective measures. For instance, relatively simple devices have been used extensively in neurotoxicity studies, such as the dynamometer for grip strength, Finger Tapping for motor speed, pegs and board for fine dexterity. Normative values are available for some of the most common tests, such as the Finger Tapping and Grooved Pegboard tests (Ruff and Parker, 1993). Motor ability of the upper limbs is often evaluated for the dominant and non-dominant hand, which can be analyzed separately or averaged. Furthermore, neuromotor tests often involve repetition of the task to obtain several measures, which can be summed or averaged, to calculate a performance score that reliably represent the level of function of the examinee.

Some specific neuromotor outcomes are better assessed with more complex instruments, such as alterations of hand tremor frequency parameters, which requires measurement with accelerometers and processing of data to calculate different parameters (Despres et al., 2000). Another sophisticated instrument used to assess neuromotor function in relation to neurotoxicants is the diadochokinesimeter, which is designed to measure rapid alternating rotation of the forearms, a movement often impaired in patients with extrapyramidal dysfunction (Beuter et al., 1994). More recently, the use of smartphones has been suggested to assess finger tapping performance (Bohannon and Wang, 2021), tremors (Calvo and Ferrara, 2021), gait and balance (Abou et al., 2021). Smartphones have also been explored as means to detect early signs of Parkinson's disease by using artificial intelligence to analyze touchscreen typing patterns (Iakovakis et al., 2020). These portable devices, now owned by most people, could also be used to collect data allowing assessment of several neuromotor outcomes in the context of occupational exposures.

The most extensive assessments of neuromotor function have been conducted in the context of exposure to manganese, such among welders (Bowler et al., 2006), workers in the production of manganese oxide and salt production (Roels et al., 1987), and workers from the metallurgical industry (Mergler et al., 1994). Manganese in the brain causes severe dysfunctions in the central nervous system known as manganism, an extrapyramidal disorder characterized by motor disturbances associated with neuropsychiatric and cognitive disabilities (Rodier, 1955). Upon overexposure, this metal accumulates preferentially in the basal ganglia, which are brain structures known to be involved in the pathophysiology of extrapyramidal disorders. Manganese-induced neuromotor deficits are similar to those displayed by patients diagnosed with Parkinson's disease, although there are some differences in pathophysiological characteristics (Andruska and Racette, 2015). Manganese can markedly impact voluntary movements requiring coordination, induce tremors, bradykinesia, rigidity, lack of facial expression, and postural instability. Hence, several studies used the Coordination Ability Test System (CATSYS, Danish Product Development Ltd., Denmark), a computerized system to assess neuromotor function in manganese-exposed individuals (e.g., Gyntelberg et al., 1990; Rolle-McFarland et al., 2022). The CATSYS is a portable, Windows-based system consisting of a data logger and four different sensors (tremor pen, touch recording plate, reaction time handle, and force plate for balance recording) for quantification of neuromotor functions. Efforts have been made to generate reference population norms in order to standardize scores from the CATSYS system (Despres et al., 2000), allowing the discrimination of normal from abnormal scores. The CATSYS has also been used in groups of workers exposed to lead (Blond and Netterstrom, 2007), mercury (Schutzmeier et al., 2018), and pyrethroid insecticides (Hansen et al., 2017).

Additional considerations are relevant when assessing neuromotor function on workers. For instance, welders can have better than normal fine motor abilities since excellent fine motor skills are necessary in this trade (Bowler et al., 2006). Hence, the comparison of motor performance with a "general" reference group might not allow the detection of diminished abilities among welders, even if neuromotor deficits are indeed present.

6. Sensitivity and reliability issues in neurobehavioral assessment

To evaluate the impact of environmental exposures on neurobehavioral outcomes it is necessary to have effective methods that will allow accurate conclusions to be drawn (Williamson, 1990). Neurotoxic effects of environmental chemicals can vary along a continuum from minor subclinical deficits in sensory memory, motor, or cognitive functioning to clinical diseases. Several neurobehavioral assessment tools are available for use across both research and clinical settings (Anger, 1989, 2003). Over time, several tests have been identified as being the most sensitive to detect neurotoxic effects (i.e., symbol-digit, digit-span, continuous performance task, simple reaction time, and tapping; Anger, 2003). As such, it is imperative that neurobehavioral assessment batteries used to assess deficits in individuals with environmental exposures are sensitive to these deficits, even if exposures are subthreshold or low-level (Anger, 1989; Williamson, 1990). Moreover, to best understand neurobehavioral deficits in this population over time, there needs to be repeated measurements which raises questions of methodological design concerns (Bowler and Lezak, 2015), such as practice effects, test re-test reliability, and stability of measurements over time.

Another consideration is that many populations being assessed for deficits related to pesticide and other environmental toxic exposures come from marginalized, immigrant, and/or low-income backgrounds which are frequently underrepresented in normative data or from countries that do not share the same cultural background where tests were developed. These factors can lead to misestimating the true level of neurobehavioral performance which can have large detrimental impacts on how regulatory agencies set threshold limits for exposures (Anger, 1989; Rohlman et al., 2008a; Williamson, 1990). Together, these concerns suggest the need for 1) pilot testing of neurobehavioral batteries to ensure they are sensitive to neurotoxicant effects of pesticides and other environmental toxins, 2) appropriate use of comparison groups based on education level, cultural and other demographic factors, exposure histories and differences in language, and 3) the ability to use neurobehavioral test batteries in repeated measures designs, with multiple test forms.

Environmental or occupational exposures can occur in any population, and neurobehavioral batteries need to be appropriate for the target audience and demonstrate reliability and sensitivity (White et al., 1994). Several neurobehavioral batteries were developed specifically for use in agricultural workers or with environmentally toxin exposed populations (Anger and Cassnrot, 1994). Work has been completed to validate or develop neurobehavioral test batteries in, for example, Spanish-speaking populations (Casaletto et al., 2016; Gershon et al., 2020; Rohlman et al., 2001a), China (Liang et al., 1997; Zhou et al., 2002), Thailand (Rohitrattana et al., 2014), Korea (Chung et al., 2007; Sakong et al., 2007), Sweden (Iregren et al., 1996), South Africa (London et al., 1997), Bangladesh (Khan et al., 2021), and several Asian populations (Chung et al., 2003). Broadly speaking, there is a great need to ensure neurobehavioral batteries are culturally sensitive (Bowler and Lezak, 2015). In addition to culturally relevant variables, performance on these batteries may still be sensitive to and impacted by typical demographic variables such as sex, handedness, and age (Chen et al., 1990). There are also important considerations when assessing the impact of neurotoxic exposure among children (Amler et al., 1996; Krasnegor et al., 1994; Seppalainen, 1988). For instance the Pediatric Environmental Neurobehavioral Test Battery (Amler et al., 1996) was developed specifically for use with children and adolescent populations that have been exposed to toxins. This child and adolescent version was developed from an adult version that was already demonstrated to detect neurotoxic effects of environmental toxicants. Other batteries have also adapted existing tests for use among younger populations. This includes the addition of new tests and the use of flexible parameter settings to allow use among participants of all ages (Khan et al., 2021; Lee et al., 2012; Rohitrattana et al., 2014; Rohlman et al., 2001a, 2007, 2008b; Sizemore and Amler, 1996; Zeitz et al., 2002) For instance, the BARS has flexible test parameters, simple instructions, and the ability to be used in populations that do not speak English or may have limited education or literacy, including children (Rohlman et al., 2000b, 2003). Fortunately, research investigating the various neurobehavioral assessment batteries described in this paragraph has been conducted across many countries with varying cultural backgrounds.

The accepted study design to understand neurobehavioral deficits has been to compare performance from exposed populations at one time-point to data from an unexposed population (Anger, 1989). This is necessary because these test batteries are not clinical tools with established norms applicable to populations who happen to be exposed to a neurotoxic substance (White and Proctor, 1992). It is essential that the control or reference group has the same characteristics of the exposed group (e.g., age, education, occupation, socioeconomic and cultural variables) in order to accurately determine the impact of the exposure on the exposed population. More recently, there has been a move towards examining within-subject changes using longitudinal study designs (as opposed to between subject designs described prior). This allows for a comparison of individual-level change, although it introduces potential confounds of test-retest reliability and practice effects. Importantly, this has been evaluated for several neurobehavioral batteries (Arcia and Otto, 1992; Campbell et al., 1999; Chung et al., 2007; Farahat et al., 2003; Heaton et al., 2014; Lee et al., 2012; Lowe and Rabbitt, 1998; Lucero et al., 2019; Rohlman et al., 2000a; Sakong et al., 2007; Syvaoja et al., 2015; Taylor et al., 2020). Some samples have shown improved performance over time, for example the BARS (Rohlman et al., 2001b) or the Swedish Performance Evaluation System (Chung et al., 2007). The CANTAB showed practice effects that may inhibit its use in repeated measures designs (Lowe and Rabbitt, 1998) in an elderly population and the reliability could not be determined for all measures in a sample of children over a year-long period (Syvaoja et al., 2015). Similarly, recent work examining the reliability of the NIH Toolbox cognitive battery urged researchers to interpret data with caution given limited tests met standards of research reliability in a sample of children and adolescents across a 3-year longitudinal study (Taylor et al., 2020). Of note, this is contrary to work supporting strong test re-test reliability of the

NIH Toolbox cognitive battery in the short term for both children and adults (e.g., days to weeks; Heaton et al., 2014; Weintraub et al., 2013a).

7. Training of neurobehavioral ascertainers

Given the range of sensitivity and reliability issues outlined, there are also several concerns and suggestions for ensuring efficient and accurate neurobehavioral data collection. First, psychologists experienced in human testing should be employed during the development of test batteries, including when training test administrators, and neurobehavioral test batteries must be appropriately standardized and rigorously tested themselves (Iregren and Letz, 1994). One way of rigorously testing neurobehavioral batteries is through field epidemiological and pilot studies. Importantly, these should be conducted with the target population; this is especially important when the target population is a particular cultural group which may differ from where the tests were developed (Anger, 2003), given the cultural sensitivity concerns identified above. It is also important to include psychosocial measures and other potential covariates as part of the overall assessment to gather information about potential confounders (Bowler and Lezak, 2015; Letz, 1993). Indeed, research has shown that confounders can lead to misinterpretation of effects that appear as neurobehavioral deficits when only controlled for in statistical analyses, as opposed to during the study design phase (Anger et al., 1997).

Traditional paper/pencil neurobehavioral task administration requires extensive training to ensure tests are administered reliably and accurately, however computerized versions are easier to administer and less error prone (McCauley et al., 2006). Importantly, they may help with reducing time it takes to completing testing, which can be limited in occupational settings (Baker and Letz, 1986). Additionally, comparisons of computerized tests to typical paper/pencil administration have found the computerized versions were reliable for normative samples (Campbell et al., 1999; Cassitto et al., 1989). Computerized tests also allow for a greater number of participants to engage in neurobehavioral testing at once, depending upon age and developmental ability (McCauley et al., 2006) and provide a more flexible interface including easily adaptable test items and task parameters (Rohlman et al., 1996). A review of the WHO Neurobehavioral Core Test Battery (NCTB) as used to detect neurotoxicity across the world indicated significant problems in administrator drift (Anger et al., 2000) or straying from ideal testing procedures over time. This speaks to the need for consistent quality control checks throughout the study period to ensure examiners are meeting testing standards (McCauley et al., 2006). Data should also be reviewed throughout the data collection period to check for any systematic errors (e.g., technological issues or administrator differences).

Research has shown there are consistent neurobehavioral deficits associated with low-level and chronic exposure to pesticides, however the pattern of deficits has varied across samples (Rohlman et al., 2011). One reason for the differences may be that researchers often report results of neurobehavioral testing in different ways including at the individual test level, sum scores, latent scores, and other types of combined data (e.g., Heyer et al., 1996). Other differences across results may be due to study designs and methodologies, participant exposure histories and other individual-level differences, as well as the use of culturally appropriate comparison groups (Anger et al., 2000; Rohlman et al., 2011).

8. Future directions

Neurobehavioral test batteries have demonstrated sensitivity and reliability in detecting the adverse effects of occupational exposures, with some variability across specific tests and test batteries. They have largely been used in populations exposed to a wide array of toxic chemicals. However, given that they were selected to detect and characterize sub-clinical deficits, they may have an even wider range of applicability (e.g., infectious diseases, COVID-19, changing hazards related to climate change) which should be explored. Further, additional research may elucidate how performance on these batteries map onto more traditional neurological exams, such as neuroimaging. Combining such approaches could yield important information about localization of deficits within the brain, structural changes or impairments, and how these relate to the assessed neurobehavioral performance due to toxin exposures.

Additionally, research has shown that exposures to toxicants are linked with other types of impairments, beyond neurobehavioral deficits. Specifically, individuals may be at higher risk for psychiatric diagnoses such as attention deficit/hyperactivity disorder, autism spectrum disorders, anxiety, or depression. Importantly, risk for developing these and related concerns may vary depending upon the type of exposure and point in development at which exposure occurred (e.g., prenatal versus adolescence).

Further research is needed to understand how exposures are linked with such mental health problems and how to best identify those at greatest risk. Importantly, there may be other physical health risks of concern as well. Identifying how to protect individuals at risk of exposure and communicating that risk may be an important step for future work. Perhaps regularly screening for neurobehavioral deficits and then communicating feedback from the neurobehavioral test batteries is one way of communicating this information to individuals at risk. For example, work has been completed evaluating safety and educational interventions (Rohlman et al., 2020) which may be an effective way to limit and reduce exposure while increasing knowledge about these various mental and physical health risks. However, further research is needed to identify the most effective ways to implement and disseminate these types of interventions, and to communicate feedback about neurobehavioral performance.

Lastly, given many populations at greatest risk for exposure are from lower-income and/or more rural areas, there can be accessibility concerns. That is, individuals in most need of support may be unable to access treatment or have neurobehavioral performance regularly screened for deficits. One potential avenue to boost accessibility is through telemedicine. However, to date, neurobehavioral batteries have not been tested or validated for delivery through a telemedicine format. This will require additional testing in the target populations and may be an important direction for future work.

References

Abou, L., Peters, J., Wong, E., Akers, R., Dossou, M.S., Sosnoff, J.J., Rice, L.A., 2021. Gait and balance assessments using smartphone applications in parkinson's disease: a systematic review. J. Med. Syst. 45, 87.

Amler, R.W., Gibertini, M., Lybarger, J.A., Hall, A., Kakolewski, K., Phifer, B.L., Olsen, K.L., 1996. Selective approaches to basic neurobehavioral testing of children in environmental health studies. Neurotoxicol. Teratol. 18, 429–434.

Andruska, K.M., Racette, A.B., 2015. Neuromythology of manganism. Curr. Epidemiol. Rep 2, 143–148.

Anger, W.K., 1989. Human neurobehavioral toxicology testing: current perspectives. Toxicol. Ind. Health 5, 165–180.

Anger, W.K., 2003. Neurobehavioural tests and systems to assess neurotoxic exposures in the workplace and community. Occup. Environ. Med. 60 (531–538), 474.

Anger, W.K., Cassitto, M.G., 1993. Report on workshop: who nctb and other neurobehavioral test batteries. Environ. Res. 60, 84–86.

Anger, W.K., Cassnrot, M.G., 1994. Individual-administered human behavioral test batteries to identify neurotoxic chemicals. In: Araki, S. (Ed.), Neurobehavioral Methods and Effects in Occupational and Environmental Health. Academic Press, pp. 101–114.

Anger, W.K., Cassitto, M.G., Liang, Y.X., Amador, R., Hooisma, J., Chrislip, D.W., Mergler, D., Keifer, M., Hortnagl, J., Fournier, L., et al., 1993. Comparison of performance from three continents on the who-recommended neurobehavioral core test battery. Environ. Res. 62, 125–147.
Anger, W.K., Sizemore, O.J., Grossmann, S.J., Glasser, J.A., Letz, R., Bowler, R., 1997. Human neurobehavioral research methods: impact of subject variables. Environ. Res. 73, 18–41.
Anger, W.K., Liang, Y.X., Nell, V., Kang, S.K., Cole, D., Bazylewicz-Walczak, B., Rohlman, D.S., Sizemore, O.J., 2000. Lessons learned- -15 years of the who-nctb: A review. Neurotoxicology 21, 837–846.
Arab, A., Mostafalou, S., 2021. Neurotoxicity of pesticides in the context of cns chronic diseases. Int. J. Environ. Health Res., 1–38.
Arcia, E., Otto, D.A., 1992. Reliability of selected tests from the neurobehavioral evaluation system. Neurotoxicol. Teratol. 14, 103–110.
Backx, R., Skirrow, C., Dente, P., Barnett, J.H., Cormack, F.K., 2020. Comparing web-based and lab-based cognitive assessment using the Cambridge neuropsychological test automated battery: a within-subjects counterbalanced study. J. Med. Internet Res. 22, e16792.
Baker, E.L., Letz, R., 1986. Neurobehavioral testing in monitoring hazardous workplace exposures. J. Occup. Med. 28, 987–990.
Bates, M.N., Reed, B.R., Liu, S., Eisen, E.A., Hammond, S.K., 2016. Solvent exposure and cognitive function in automotive technicians. Neurotoxicology 57, 22–30.
Beuter, A., Mergler, D., de Geoffroy, A., Carriere, L., Belanger, S., Varghese, L., Sreekumar, J., Gauthier, S., 1994. Diadochokinesimetry: a study of patients with parkinson's disease and manganese exposed workers. Neurotoxicology 15, 655–664.
Block, M.L., Calderon-Garciduenas, L., 2009. Air pollution: mechanisms of neuroinflammation and cns disease. Trends Neurosci. 32, 506–516.
Blond, M., Netterstrom, B., 2007. Neuromotor function in a cohort of danish steel workers. Neurotoxicology 28, 336–344.
Bohannon, R.W., Wang, I., 2021. Measurement of finger tapping performance using a smartphone application: a pilot study. J. Phys. Ther. Sci. 33, 618–620.
Bollen, K.A., 1989. Structural Equations with Latent Variables. John Wiley & Sons.
Bouchard Jr., T.J., McGue, M., 2003. Genetic and environmental influences on human psychological differences. J. Neurobiol. 54, 4–45.
Bouchard, M., Mergler, D., Baldwin, M., 2005. Manganese exposure and age: neurobehavioral performance among alloy production workers. Environ. Toxicol. Pharmacol. 19, 687–694.
Bowler, R.M., Lezak, M.D., 2015. Chapter 3 - neuropsychologic evaluation and exposure to neurotoxicants. In: Lotti, M., Bleecker, M.L. (Eds.), Handbook of Clinical Neurology. vol. 131. Elsevier, pp. 23–45.
Bowler, R.M., Gysens, S., Diamond, E., Nakagawa, S., Drezgic, M., Roels, H.A., 2006. Manganese exposure: neuropsychological and neurological symptoms and effects in welders. Neurotoxicology 27, 315–326.
Budtz-Jorgensen, E., Keiding, N., Grandjean, P., Weihe, P., 2002. Estimation of health effects of prenatal methylmercury exposure using structural equation models. Environ. Health. 1, 2.
Calvo, N.E., Ferrara, J.M., 2021. Diagnosis of orthostatic tremor using smartphone accelerometry. BMC Neurol. 21, 457.
Campbell, K.A., Rohlman, D.S., Storzbach, D., Binder, L.M., Anger, W.K., Kovera, C.A., Davis, K.L., Grossmann, S.J., 1999. Test-retest reliability of psychological and neurobehavioral tests self-administered by computer. Assessment 6, 21–32.

Cassitto, M.G., Gilioli, R., Camerino, D., 1989. Experiences with the Milan automated neurobehavioral system (mans) in occupational neurotoxic exposure. Neurotoxicol. Teratol. 11, 571–574.

Casaletto, K.B., Umlauf, A., Marquine, M., Beaumont, J.L., Mungas, D., Gershon, R., Slotkin, J., Akshoomoff, N., Heaton, R.K., 2016. Demographically corrected normative standards for the spanish language version of the NIH Toolbox cognition battery. J. Int. Neuropsychol. Soc. 22, 364–374.

Cassitto, M., Camerino, D., Hanninen, H., Anger, W., 1990. International collaboration to evaluate the who neurobehavioral core test battery. In: Advances in neurobehavioral toxicology: Applications in environmental and occupational health, pp. 203–223.

Chen, Z.Q., Yu, J.H., Cao, S.H., 1990. Reference values of indicators for who neurobehavioral core test battery. Chin Med J (Engl) 103, 61–65.

Cherry, N., Venables, H., Waldron, H.A., 1984. British studies on the neuropsychological effects of solvent exposure. Scand. J. Work Environ. Health 10 (Suppl 1), 10–12.

Chouaniere, D., Cassitto, M.G., Spurgeon, A., Verdier, A., Gilioli, R., 1997. An international questionnaire to explore neurotoxic symptoms. Environ. Res. 73, 70–72.

Chung, J.H., Sakong, J., Kang, P.S., Kim, C.Y., Lee, K.S., Jeon, M.J., Sung, N.J., Ahn, S.H., Won, K.C., 2003. Cross-cultural comparison of neurobehavioral performance in asian workers. Neurotoxicology 24, 533–540.

Chung, S.T., Jeon, M.J., Sakong, J., 2007. Assessment of changes in computerized neurobehavioral performance across multiple test sessions. Korean J Occup Environ Med 19, 251–258.

Costa, L.G., Manzo, L., 2020. Occupational Neurotoxicology. CRC Press.

Criswell, S.R., Searles Nielsen, S., Dlamini, W.W., Warden, M.N., Perlmutter, J.S., Sheppard, L., Moerlein, S.M., Lenox-Krug, J., Checkoway, H., Racette, B.A., 2021. Principal component analysis of striatal and extrastriatal d2 dopamine receptor positron emission tomography in manganese-exposed workers. Toxicol. Sci. 182, 132–141.

Davies, D.S., Stammers, J.R., Jones, R.V., 1997. The effect of world war ii on industrial science. Proc. R. Soc. A: Math. Phys. Eng. Sci. 342, 505–518.

Despres, C., Lamoureux, D., Beuter, A., 2000. Standardization of a neuromotor test battery: the CATSYS system. Neurotoxicology 21, 725–735.

Emara, A.M., el-Ghawabi, S.H., Madkour, O.I., el-Samra, G.H., 1971. Chronic manganese poisoning in the dry battery industry. Br. J. Ind. Med. 28, 78–82.

Farahat, F.M., Rohlman, D.S., Storzbach, D., Ammerman, T., Anger, W.K., 2003. Measures of short-term test-retest reliability of computerized neurobehavioral tests. Neurotoxicology 24, 513–521.

Feenstra, H.E., Vermeulen, I.E., Murre, J.M., Schagen, S.B., 2017. Online cognition: factors facilitating reliable online neuropsychological test results. Clin. Neuropsychol. 31, 59–84.

Fray, P.J., Robbins, T.W., 1996. Cantab battery: proposed utility in neurotoxicology. Neurotoxicol. Teratol. 18, 499–504.

Genc, S., Zadeoglulari, Z., Fuss, S.H., Genc, K., 2012. The adverse effects of air pollution on the nervous system. J. Toxicol 2012, 782462.

Gershon, R.C., Fox, R.S., Manly, J.J., Mungas, D.M., Nowinski, C.J., Roney, E.M., Slotkin, J., 2020. The NIH Toolbox: Overview of development for use with hispanic populations. J. Int. Neuropsychol. Soc. 26, 567–575.

Golbabaei, F., Dehghani, F., Saatchi, M., Zakerian, S.A., 2018. Evaluation of occupational exposure to different levels of mixed organic solvents and cognitive function in the painting unit of an automotive industry. Health Promot. Perspect. 8, 296–302.

Gosling, S.D., Vazire, S., Srivastava, S., John, O.P., 2004. Should we trust web-based studies? A comparative analysis of six preconceptions about internet questionnaires. Am. Psychol. 59, 93–104.

Grandjean, E., Munchinger, R., Turrian, V., Hass, P.A., Knoepfel, H.K., Rosenmund, H., 1955. Investigations into the effects of exposure to trichlorethylene in mechanical engineering. Br. J. Ind. Med. 12, 131–142.

Groenewegen, H.J., 2003. The basal ganglia and motor control. Neural Plast. 10, 107–120.

Gunnarsson, L.G., Bodin, L., 2019. Occupational exposures and neurodegenerative diseases-a systematic literature review and meta-analyses. Int. J. Environ. Res. Public Health 16.

Gyntelberg, F., Flarup, M., Mikkelsen, S., Palm, T., Ryom, C., Suadicani, P., 1990. Computerized coordination ability testing. Acta Neurol. Scand. 82, 39–42.

Hänninen, H., 1966. Psychological tests in the diagnosis of carbon disulphide poisoning. Work Environ. Health 2, 16–20.

Hanninen, H., Lindstrom, K., 1979. Behavioral Test Battery for Toxicopsychological Studies: Used at the Institute of Occupational Health in Helsinki. Institute of Occupational Health, Helsinki.

Hansen, M.R.H., Jors, E., Lander, F., Condarco, G., Debes, F., Bustillos, N.T., Schlunssen, V., 2017. Neurological deficits after long-term pyrethroid exposure. Environ Health Insights 11. 1178630217700628.

Heaton, R.K., Akshoomoff, N., Tulsky, D., Mungas, D., Weintraub, S., Dikmen, S., Beaumont, J., Casaletto, K.B., Conway, K., Slotkin, J., Gershon, R., 2014. Reliability and validity of composite scores from the NIH Toolbox cognition battery in adults. J. Int. Neuropsychol. Soc. 20, 588–598.

Heyer, N.J., Bittner Jr., A.C., Echeverria, D., 1996. Analyzing multivariate neurobehavioral outcomes in occupational studies: a comparison of approaches. Neurotoxicol. Teratol. 18, 401–406.

Howieson, D., 2019. Current limitations of neuropsychological tests and assessment procedures. Clin. Neuropsychol. 33, 200–208.

Iakovakis, D., Chaudhuri, K.R., Klingelhoefer, L., Bostantjopoulou, S., Katsarou, Z., Trivedi, D., Reichmann, H., Hadjidimitriou, S., Charisis, V., Hadjileontiadis, L.J., 2020. Screening of parkinsonian subtle fine-motor impairment from touchscreen typing via deep learning. Sci. Rep. 10, 12623.

Iregren, A., Letz, R., 1994. Computerized test batteries. In: Araki, S. (Ed.), Neurobehavioral Methods and Effects in Occupational and Environmental Health. Academic Press, pp. 973–974.

Iregren, A., Gamberale, F., Kjellberg, A., 1996. Spes: a psychological test system to diagnose environmental hazards. Swedish performance evaluation system. Neurotoxicol Teratol 18, 485–491.

Jackson, A.C., 2018. Chronic neurological disease due to methylmercury poisoning. Can. J. Neurol. Sci. 45, 620–623.

Jeyaratnam, J., Boey, K.W., Ong, C.N., Chia, C.B., Phoon, W.O., 1986. Neuropsychological studies on lead workers in Singapore. Br. J. Ind. Med. 43, 626–629.

Johnson, B.L., Baker, E.L., El Batawi, M., Gilioli, R., Hanninen, H., Seppalainen, A., 1987. Prevention of Neurotoxic Illness in Working Populations.

Karlsen, R.H., Karr, J.E., Saksvik, S.B., Lundervold, A.J., Hjemdal, O., Olsen, A., Iverson, G.L., Skandsen, T., 2022. Examining 3-month test-retest reliability and reliable change using the Cambridge neuropsychological test automated battery. Appl. Neuropsychol. Adult 29, 146–154.

Karlson, B., Osterberg, K., Orbaek, P., 2000. Euroquest: the validity of a new symptom questionnaire. Neurotoxicology 21, 783–789.

Keer, S., McLean, D., Glass, B., Douwes, J., 2018. Effects of personal protective equipment use and good workplace hygiene on symptoms of neurotoxicity in solvent-exposed vehicle spray painters. Ann Work Expo Health 62, 307–320.

Khajehnasiri, F., Samaei, S.E., Mehrzad, S., Taban, E., Ghalenoei, M., Salimi, F., 2022. The effect of traffic noise on cognitive performance with regard to personality traits: a laboratory experiment study. Work. https://doi.org/10.3233/WOR-210002.

Khan, K.M., Hatch, L.C., Akhter, S., Eunus, M., Zhou, Z., Parvez, F., Rohlman, D., 2021. Reliability of a computer-based neurobehavioral assessment test battery for Bangladeshi adolescent children. Neurotoxicology 85, 47–53.

Kilian, J., Kitazawa, M., 2018. The emerging risk of exposure to air pollution on cognitive decline and alzheimer's disease - evidence from epidemiological and animal studies. Biom. J. 41, 141–162.

Krasnegor, N.A., Otto, D.A., Bernstein, J.H., Burke, R., Chappell, W., Eckerman, D.A., Needleman, H.L., Oakley, G., Rogan, W., Terracciano, G., et al., 1994. Neurobehavioral test strategies for environmental exposures in pediatric populations. Neurotoxicol. Teratol. 16, 499–509.

Kraut, R., Olson, J., Banaji, M., Bruckman, A., Cohen, J., Couper, M., 2004. Psychological research online: report of board of scientific affairs' advisory group on the conduct of research on the internet. Am. Psychol. 59, 105–117.

Lee, K.S., Jeon, M.J., Hwang, T.Y., Kim, C.Y., Sakong, J., 2012. Evaluation of reliability of computerized neurobehavioral tests in Korean children. Neurotoxicology 33, 1362–1367.

Letz, R., 1990. The neurobehavioral evaluation system: an international effort. In: Johnson, B.L. (Ed.), Advances in neurobehavioral toxicology: applications in environmental and occupational health. Lewis Publishers, Chelsea, MI, pp. 189–201.

Letz, R., 1993. Covariates of computerized neurobehavioral test performance in epidemiologic investigations. Environ. Res. 61, 124–132.

Lewis, M.M., Lee, E.Y., Jo, H.J., Du, G., Park, J., Flynn, M.R., Kong, L., Latash, M.L., Huang, X., 2016. Synergy as a new and sensitive marker of basal ganglia dysfunction: a study of asymptomatic welders. Neurotoxicology 56, 76–85.

Liang, Y., Chen, Z., Sun, R., Fang, Y., Yu, J., 1990. Application of the WHO neurobehavioral core test battery and other neurobehavioral screening methods. In: Johnson, B.L., Anger, W.K., Durao, A., Xintaras, C. (Eds.), Advances in Neurobehavioral Toxicology: Applications in Environmental and Occupational Health. Lewis Publishers, Chelsea, MI, pp. 225–243.

Liang, Y.X., Chen, Z.Q., Wang, C., Tang, H.W., Yang, H.G., 1997. The development of neurobehavioral toxicology in China: the 1994 Hanninen lecture. Environ. Res. 73, 9–17.

Lindgren, K.N., Masten, V.L., Tiburzi, M.J., Ford, D.P., Bleecker, M.L., 1999. The factor structure of the profile of mood states (POMS) and its relationship to occupational lead exposure. J. Occup. Environ. Med. 41, 3–10.

London, L., Myers, J.E., Nell, V., Taylor, T., Thompson, M.L., 1997. An investigation into neurologic and neurobehavioral effects of long-term agrichemical use among deciduous fruit farm workers in the western cape, South Africa. Environ. Res. 73, 132–145.

Lowe, C., Rabbitt, P., 1998. Test/re-test reliability of the cantab and ispocd neuropsychological batteries: theoretical and practical issues. Cambridge neuropsychological test automated battery. International study of post-operative cognitive dysfunction. Neuropsychologia 36, 915–923.

Lucchini, R.G., Guazzetti, S., Zoni, S., Benedetti, C., Fedrighi, C., Peli, M., Donna, F., Bontempi, E., Borgese, L., Micheletti, S., Ferri, R., Marchetti, S., Smith, D.R., 2014. Neurofunctional dopaminergic impairment in elderly after lifetime exposure to manganese. Neurotoxicology 45, 309–317.

Lucero, B., Ceballos, P.A., Munoz-Quezada, M.T., Reynaldos, C., Saracini, C., Baumert, B.O., 2019. Validity and reliability of an assessment tool for the screening of neurotoxic effects in agricultural workers in Chile. Biomed. Res. Int. 2019, 7901760.

Lundberg, I., Hogberg, M., Michelsen, H., Nise, G., Hogstedt, C., 1997. Evaluation of the q16 questionnaire on neurotoxic symptoms and a review of its use. Occup. Environ. Med. 54, 343–350.

Maizlish, N.A., Langolf, G.D., Whitehead, L.W., Fine, L.J., Albers, J.W., Goldberg, J., Smith, P., 1985. Behavioural evaluation of workers exposed to mixtures of organic solvents. Br. J. Ind. Med. 42, 579–590.

McCauley, L.A., Anger, W.K., Keifer, M., Langley, R., Robson, M.G., Rohlman, D., 2006. Studying health outcomes in farmworker populations exposed to pesticides. Environ. Health Perspect. 114, 953–960.

McNair, D.M., Lorr, M., Droppleman, L.F., 1971. Manual Profile of Mood States. Educational and Industrial Testing Service, San Diego, CA.

Mergler, D., Huel, G., Bowler, R., Iregren, A., Belanger, S., Baldwin, M., Tardif, R., Smargiassi, A., Martin, L., 1994. Nervous system dysfunction among workers with long-term exposure to manganese. Environ. Res. 64, 151–180.

O'Connell, H., Coen, R., Kidd, N., Warsi, M., Chin, A.V., Lawlor, B.A., 2004. Early detection of alzheimer's disease (AD) using the CANTAB paired associates learning test. Int. J. Geriatr. Psychiatry 19, 1207–1208.

Oulhote, Y., Mergler, D., Barbeau, B., Bellinger, D.C., Bouffard, T., Brodeur, M.E., Saint-Amour, D., Legrand, M., Sauve, S., Bouchard, M.F., 2014. Neurobehavioral function in school-age children exposed to manganese in drinking water. Environ. Health Perspect. 122, 1343–1350.

Puri, V., Chaudhry, N., Tatke, M., 2007. N-hexane neuropathy in screen printers. Electromyogr. Clin. Neurophysiol. 47, 145–152.

Rodier, J., 1955. Manganese poisoning in Moroccan miners. Br. J. Ind. Med. 12, 21–35.

Roels, H., Lauwerys, R., Buchet, J.P., Genet, P., Sarhan, M.J., Hanotiau, I., de Fays, M., Bernard, A., Stanescu, D., 1987. Epidemiological survey among workers exposed to manganese: effects on lung, central nervous system, and some biological indices. Am. J. Ind. Med. 11, 307–327.

Rohitrattana, J., Siriwong, W., Suittiwan, P., Robson, M.G., Strickland, P.O., Rohlman, D.S., Fiedler, N., 2014. Adaptation of a neurobehavioral test battery for Thai children. Rocz. Panstw. Zakl. Hig. 65, 205–212.

Rohlman, D.S., Sizemore, O.J., Anger, W.K., Kovera, C.A., 1996. Computerized neurobehavioral testing: techniques for improving test instructions. Neurotoxicol. Teratol. 18, 407–412.

Rohlman, D.S., Bailey, S.R., Brown, M., Blanock, M., Anger, W.K., McCauley, L., 2000a. Establishing stable test performance in tests from the behavioral assessment and research system (BARS). Neurotoxicology 21, 715–723.

Rohlman, D.S., Davis, J.W., Ismail, A., Abdel Rasoul, G.M., Hendy, O., Olson, J.R., Bonner, M.R., 2020. Risk perception and behavior in Egyptian adolescent pesticide applicators: an intervention study. BMC Public Health 20, 679.

Rohlman, D.S., Gimenes, L.S., Ebbert, C., Anger, W.K., Bailey, S.R., McCauley, L., 2000b. Smiling faces and other rewards: using the behavioral assessment and research system (BARS) with unique populations. Neurotoxicology 21, 973–978.

Rohlman, D.S., Anger, W.K., Tamulinas, A., Phillips, J., Bailey, S.R., McCauley, L., 2001a. Development of a neurobehavioral battery for children exposed to neurotoxic chemicals. Neurotoxicology 22, 657–665.

Rohlman, D.S., Bailey, S.R., Anger, W.K., McCauley, L., 2001b. Assessment of neurobehavioral function with computerized tests in a population of hispanic adolescents working in agriculture. Environ. Res. 85, 14–24.

Rohlman, D.S., Gimenes, L.S., Eckerman, D.A., Kang, S.K., Farahat, F.M., Anger, W.K., 2003. Development of the behavioral assessment and research system (BARS) to detect and characterize neurotoxicity in humans. Neurotoxicology 24, 523–531.
Rohlman, D.S., Bodner, T., Arcury, T.A., Quandt, S.A., McCauley, L., 2007. Developing methods for assessing neurotoxic effects in hispanic non-english speaking children. Neurotoxicology 28, 240–244.
Rohlman, D.S., Lucchini, R., Anger, W.K., Bellinger, D.C., van Thriel, C., 2008a. Neurobehavioral testing in human risk assessment. Neurotoxicology 29, 556–567.
Rohlman, D.S., Villanueva-Uy, E., Ramos, E.A., Mateo, P.C., Bielawski, D.M., Chiodo, L.M., Delaney-Black, V., McCauley, L., Ostrea Jr., E.M., 2008b. Adaptation of the behavioral assessment and research system (BARS) for evaluating neurobehavioral performance in filipino children. Neurotoxicology 29, 143–151.
Rohlman, D.S., Anger, W.K., Lein, P.J., 2011. Correlating neurobehavioral performance with biomarkers of organophosphorous pesticide exposure. Neurotoxicology 32, 268–276.
Rolle-McFarland, D., Liu, Y., Mostafaei, F., Zauber, S.E., Zhou, Y., Li, Y., Fan, Q., Zheng, W., Nie, L.H., Wells, E.M., 2022. The association of bone and blood manganese with motor function in chinese workers. Neurotoxicology 88, 224–230.
Ruff, R.M., Parker, S.B., 1993. Gender- and age-specific changes in motor speed and eye-hand coordination in adults: normative values for the finger tapping and grooved pegboard tests. Percept. Mot. Skills 76, 1219–1230.
Sakong, J., Kang, P.S., Kim, C.Y., Hwang, T.Y., Jeon, M.J., Park, S.Y., Lee, S.J., Won, K.C., Lee, S.B., Chung, J.H., 2007. Evaluation of reliability of traditional and computerized neurobehavioral tests. Neurotoxicology 28, 235–239.
Schutzmeier, P., Focil Baquerizo, A., Castillo-Tandazo, W., Focil, N., Bose-O'Reilly, S., 2018. Efficacy of n,n'bis-(2-mercaptoethyl) isophthalamide on mercury intoxication: A randomized controlled trial. Environ. Health 17, 15.
Scott, E.P., Sorrell, A., Benitez, A., 2019. Psychometric properties of the NIH Toolbox cognition battery in healthy older adults: reliability, validity, and agreement with standard neuropsychological tests. J. Int. Neuropsychol. Soc. 25, 857–867.
Seppalainen, A.M., 1988. Neurophysiological approaches to the detection of early neurotoxicity in humans. Crit. Rev. Toxicol. 18, 245–298.
Sherman, E.M.S., Tan, J.E., Hrabok, M., 2020. A Compendium of Neuropsychological Tests : Administration, Tests, Norms, and Commentary, Fourth edition. ed. Oxford University Press, New York.
Sizemore, O.J., Amler, R.W., 1996. Characteristics of ATSDR's adult and pediatric environmental neurobehavioral test batteries. Neurotoxicology 17, 229–236.
Smith, T., Gildeh, N., Holmes, C., 2007. The Montreal cognitive assessment: validity and utility in a memory clinic setting. Can. J. Psychiatry 52, 329–332.
Sternin, A., Burns, A., Owen, A.M., 2019. Thirty-five years of computerized cognitive assessment of aging-where are we now? Diagnostics (Basel) 9, 114.
Sterr, A., Herron, K.A., Hayward, C., Montaldi, D., 2006. Are mild head injuries as mild as we think? Neurobehavioral concomitants of chronic post-concussion syndrome. BMC Neurol. 6, 7.
Syvaoja, H.J., Tammelin, T.H., Ahonen, T., Rasanen, P., Tolvanen, A., Kankaanpaa, A., Kantomaa, M.T., 2015. Internal consistency and stability of the cantab neuropsychological test battery in children. Psychol. Assess. 27, 698–709.
Taylor, B.K., Frenzel, M.R., Eastman, J.A., Wiesman, A.I., Wang, Y.P., Calhoun, V.D., Stephen, J.M., Wilson, T.W., 2020. Reliability of the NIH TOOLBOX cognitive battery in children and adolescents: a 3-year longitudinal examination. Psychol. Med., 1–10.
van Wijngaarden, E., Thurston, S.W., Myers, G.J., Harrington, D., Cory-Slechta, D.A., Strain, J.J., Watson, G.E., Zareba, G., Love, T., Henderson, J., Shamlaye, C.F.,

Davidson, P.W., 2017. Methyl mercury exposure and neurodevelopmental outcomes in the Seychelles child development study main cohort at age 22 and 24years. Neurotoxicol. Teratol. 59, 35–42.
Weintraub, S., Bauer, P.J., Zelazo, P.D., Wallner-Allen, K., Dikmen, S.S., Heaton, R.K., Tulsky, D.S., Slotkin, J., Blitz, D.L., Carlozzi, N.E., Havlik, R.J., Beaumont, J.L., Mungas, D., Manly, J.J., Borosh, B.G., Nowinski, C.J., Gershon, R.C., 2013a. I. NIH TOOLBOX cognition battery (cb): introduction and pediatric data. Monogr. Soc. Res. Child Dev. 78, 1–15.
Weintraub, S., Dikmen, S.S., Heaton, R.K., Tulsky, D.S., Zelazo, P.D., Bauer, P.J., Carlozzi, N.E., Slotkin, J., Blitz, D., Wallner-Allen, K., Fox, N.A., Beaumont, J.L., Mungas, D., Nowinski, C.J., Richler, J., Deocampo, J.A., Anderson, J.E., Manly, J.J., Borosh, B., Havlik, R., Conway, K., Edwards, E., Freund, L., King, J.W., Moy, C., Witt, E., Gershon, R.C., 2013b. Cognition assessment using the NIH TOOLBOX. Neurology 80, S54–S64.
Weintraub, S., Dikmen, S.S., Heaton, R.K., Tulsky, D.S., Zelazo, P.D., Slotkin, J., Carlozzi, N.E., Bauer, P.J., Wallner-Allen, K., Fox, N., Havlik, R., Beaumont, J.L., Mungas, D., Manly, J.J., Moy, C., Conway, K., Edwards, E., Nowinski, C.J., Gershon, R., 2014. The cognition battery of the NIH TOOLBOX for assessment of neurological and behavioral function: validation in an adult sample. J. Int. Neuropsychol. Soc. 20, 567–578.
White, R.F., Proctor, S.P., 1992. Research and clinical criteria for development of neurobehavioral test batteries. J. Occup. Med. 34, 140–148.
White, R.F., Gerr, F., Cohen, R.F., Green, R., Lezak, M.D., Lybarger, J., Mack, J., Silbergeld, E., Valciukas, J., Chappell, W., et al., 1994. Criteria for progressive modification of neurobehavioral batteries. Neurotoxicol. Teratol. 16, 511–524.
Williamson, A.M., 1990. The development of a neurobehavioral test battery for use in hazard evaluations in occupational settings. Neurotoxicol. Teratol. 12, 509–514.
Xintaras, C., Johnson, B., De Groot, I., 1974. Behavior Toxicology Early Detection of Occupational Hazards.
Zeitz, P., Kakolewski, K., Imtiaz, R., Kaye, W., 2002. Methods of assessing neurobehavioral development in children exposed to methyl parathion in Mississippi and Ohio. Environ. Health Perspect. 110 (Suppl 6), 1079–1083.
Zhou, W., Liang, Y., Christiani, D.C., 2002. Utility of the who neurobehavioral core test battery in chinese workers-a meta-analysis. Environ. Res. 88, 94–102.

CHAPTER TWO

Cognitive impact of exposure to airborne particles captured by brain imaging

Somaiyeh Azmoun[a], Yenny Fariñas Diaz[a], Cheuk Y. Tang[b], Megan Horton[b], Sean A.P. Clouston[c], Ben J. Luft[c], Evelyn J. Bromet[c], Sam Gandy[b,d], Donatella Placidi[e,f], Claudia Ambrosi[e,f], Lorella Mascaro[f], Carlo Rodella[e,f], Barbara Paghera[e,f], Roberto Gasparotti[e,f], Jeremy W. Chambers[a], Kim Tieu[a], Daniele Corbo[e,f], and Roberto G. Lucchini[a,e,*]

[a]Florida International University, Miami, United States
[b]Icahn School of Medicine at Mount Sinai, New York, United States
[c]Stony Brook University, New York, United States
[d]James J. Peters VA Medical Center, Bronx, New York, United States
[e]University of Brescia, Brescia, Italy
[f]Spedali Civili of Brescia, Brescia, Italy
*Corresponding author: e-mail address: roberto.lucchini@unibs.it

Contents

1. Introduction

Moderate to severe cognitive impairment, meaning pathological dysfunction in fluid cognition, including memory, executive functioning, and verbal and category fluency can make it difficult to interact or navigate daily

Advances in Neurotoxicology, Volume 7
ISSN 2468-7480
https://doi.org/10.1016/bs.ant.2022.05.002

activities (Horn and Cattell, 1967). Such burden and loss of functioning is a major public health problem because, at present, the course of cognitive impairment (CI) is unpredictable, seemingly irreversible, and remains incurable (Adams, 2012). Cognitive impairment often represents significant losses in capabilities that are individually meaningful and are indicative of later clinical levels of pathology (Plassman et al., 2010, 2011; Ravaglia et al., 2008; Valenzuela and Sachdev, 2005). Indeed, a large body of evidence suggests that in its earliest stages, CI is commonly indicated by rapid losses in cognitive capability in the years preceding clinical pathology (Piccinin et al., 2011) and diagnosis with dementia (Jack et al., 2010). Occupational exposure to a variety of neurotoxicants causes cognitive impacts. From the early finding on carbon disulfide damage, observed in Finland by Helena Hänninen in the 1960s (Hänninen, 1966), the knowledge of neurotoxicity in the workplace has progressively been revealed throughout an impressive number of studies. Solvent-related encephalopathy, neurotoxic effects from multiple metals, pesticides, and organic compounds like PCB, dioxins, and more recently PFAS and phthalates, have been largely demonstrated as caused or exacerbated by workplace exposure. Airborne particles in the ultrafine range can impact the brain through olfactory absorption, posing the "nose-brain" connection as a relevant absorption route that can target different brain regions. The COVID-19 pandemic has further underlined the importance of the olfactory pathway in relation to neurotoxic impact from biological agents.

Mounting evidence is accumulating that exposure to air pollutants, especially fine particulate matter, contributes to an increased risk of CI. For example, epidemiologic work by Lucchini et al. (Lucchini et al., 2012a) found memory decline and CI in children, adults and the elderly exposed to various airborne contaminants. Furthermore, imaging studies using structural magnetic resonance imaging (MRI) reported reductions in total brain volume and white matter structures after lifetime exposure (Chen et al., 2015; Wilker et al., 2015), while Aβ increases were found in animal models exposed to air pollution (Kim et al., 2012). The investigators leading these studies theorized that airborne particles may interrupt brain functioning in two ways: (1) via the circulation, through increased distribution of pro-inflammatory cytokines (Block and Calderón-Garcidueñas, 2009) resulting in neuroinflammation, blood–brain barrier dysfunction, and neural degeneration (Calderón-Garcidueñas et al., 2002); or (2) intranasally by direct translocation through the olfactory bulb (Bench et al., 2001). Several animal studies have shown increased brain inflammation in response to air particulate exposures (Campbell et al., 2005), signs of blood–brain barrier dysfunction, neural degeneration, cerebrovascular pathological signs, and apoptosis in glial cells (Calderón-Garcidueñas et al., 2002).

Brain imaging is rapidly improving how we understand the mechanisms of brain toxicity through a variety of anatomical and functional modalities. Functional MRI, resting states, and other techniques generate information on brain connectivity and response that can be altered by the exposure to multiple hazards and by their mixture. In this chapter, we will consider the new methodological development of brain imaging, the suitability, and the advantages of its use in occupational health when targeting the cognitive impacts of exposure to neurotoxic agents. Specific cases will be described, related to (i) the exposure to a mixture of neurotoxic agents and psychological trauma caused by the 9/11 terrorist attack, and (ii) the long-term consequences of manganese exposure among ferroalloy workers.

1.1 Brain imaging techniques

Brain imaging is used to determine the impact on the brain from exposure to neurotoxins. Functional imaging with positron emission tomography (PET) and single-photon emission computed tomography (SPECT) are tools used to determine the damage done to the brain and the functional deficits to the brain from exposure to a neurotoxin. They also aid in evaluating the mechanisms in the brain that are involved in intoxication (de Vries et al., 2021). A brain PET scan uses a radioactive substance (a tracer) to look for disease or injury in the brain and how brain cells are working. The tracer is given intravenously (IV) or as a breathable radioactive gas. Then the tracer travels through the blood and collects in the organs and tissues helping assess specific areas or diseases. MRI as well as computed tomography (CT) scans focus on assessing the structure of the brain and on showing the blood flow to and from organs (Dougherty et al., 2004; Meyer et al., 2016).

These imaging technologies are important in evaluating chemically induced brain damage. Similarly, MR imaging and PET scans are proven effective in diagnosing Alzheimer's disease. Through MRI, we canlook at soft tissue contrast with tissue characterization from specific sections of the brain. This technology provides a three-dimensional structural distinction between cerebral gray and white matter of the brain, while PET scans can provide functional metabolic and molecular information as well as changes and lesions in the brain. Therefore, both imaging techniques provide a more accurate diagnostic technique when used together (Barthel et al., n.d.).

2. Case study 1: Cognitive decline among the World Trade Center responders

The WTC disaster exposed tens of thousands of individuals including employees at or near the WTC site; residents of the area; and responders to a host of traumatic experiences and, at the same time, exposed many to the toxic detritus of the Towers after they collapsed. Current estimates suggest that, at midlife, a growing number of WTC responders are experiencing early CI and dementia (Clouston et al., 2016). Exposure to psychological trauma and fine particulate matter may increase the risk of dementia in traumatized populations including military veterans and WTC responders. WTC responders were exposed to neurotoxicants including metals, PCBs, dioxins, and PAHs. Ultrafine particles can reach the brain through the olfactory pathway causing oxidative stress and chronic inflammation leading to neurodegeneration. A decrement in the ability to identify odors was observed in a group of 99 WTC exposed subjects compared to 99 controls (Altman et al., 2011). Carbon nanotubes (CNT) were found in the lung biopsies of WTC responders and in samples of WTC dust (Wu et al., 2010). Furthermore, research on the role of exposures at the WTC has found elevated levels of systemic inflammation (serum C-reactive protein), indicative of accelerated systemic aging (Kazeros et al., 2015). Different exposure-related potential mechanisms of neuroinflammation leading to neurodegeneration among the 9/11 responders need further investigation (Kritikos et al., 2020).

In the Stony Brook WTC cohort, 20% of responders developed PTSD since 9/11. In 2400 responders screened with the Montreal Cognitive Assessment (MoCA), the rate of cognitive impairment (CI) (scores <20) was 2.6%, and the association with PTSD was highly significant. Given the complex exposures leading to PTSD, the etiology needs to be fully understood. Since the NIOSH CDC is mandated to pay for care for WTC-related diseases, there is a pressing need for a better understanding of the nature of this impairment, including its presentation in brain tissue, in highly exposed and traumatized WTC responders.

2.1 PTSD and CI

The first question posed by the observation of cognitive decline among the 9/11 responders is whether this is potentially a consequence of mental health disorders. In fact, PTSD has been broadly associated with a reduced cognitive functioning and an increased risk of dementia (Schuitevoerder et al., 2013; Veitch et al., 2013). PTSD involves complex memory, emotional, and behavioral processes (Lawrence-Wood et al., 2016), and encompasses distinct domains including re-experiencing, effortful avoidance, emotional numbing, and hyperarousal resulting from a traumatic event (King et al., 1998). The exact mechanisms for this association remain unclear (Pitman, 2010). Theories alternatively suggest that: posttraumatic stress may be a unique part of the causal pathway leading to cognitive aging (Greenberg et al., 2014); the association may be confounded by comorbid features that are independently associated with PTSD and dementia, such as traumatic brain injury (McKee and Robinson, 2014). Another formulation is that symptoms indicative of PTSD, which is commonly comorbid with major depressive disorder (MDD) (Shalev et al., 2014), are also manifestations of dementia or related pathology (van Achterberg and Southwick, 2001). Moreover, consistent with these findings, structural and functional MRI have been successful in detecting associations between PTSD and neurological function. For example, structural MRI studies have highlighted hippocampus and amygdala volume (O'Doherty et al., 2015). Functional MRI studies have further noted the consistent associations found between PTSD and white matter lesions (Pitman et al., 2001) and PTSD, white matter integrity, diffusivity, and anisotropy (Daniels et al., 2013; Davenport et al., 2015). Finally, studies of neuronal function highlight the role of PTSD in modulating glucose uptake in the amygdala (Buchsbaum et al., 2015; Zhu et al., 2016) as well as generally elevated neuroinflammation (Acosta et al., 2013) and even reduced microglial activation (Seibyl, 2012).

Several other potential risk factors for CI need to be considered in addition to PTSD and massive exposure to air pollutants. They include family history, apolipoprotein ε4 allele, other early life and adult-onset extreme stress, head injury, educational attainment and childhood cognition, mid-life obesity and hypertension, smoking, lack of exercise, and excessive use of alcohol and drugs. As WTC responders age, they are increasingly faced with an aging-related disease burden, especially CI. It is, therefore, critical to understanding the role of WTC-PTSD and extreme exposure to toxic pollutants in accelerating the onset of CI in this population.

An imaging study using PET/MRI has been conducted by our group from the Stony Brook University and the Icahn School of Medicine at Mount Sinai, New York. The methodology developed in this study has included the following protocols.

2.2 MRI assessments

All imaging was performed on a Siemens 3T PET/MR. This instrument simultaneously acquires PET and MRI. Before administering the research PET/MRI protocol, a dual echo sequence was also obtained to screen for incidental abnormalities. The protocol included the following modalities: (i) anatomical T1-weighted MPRage (TR/TE = 1900/2.5 ms, FOV = 23 cm, Matrix 256 × 256, slice thickness 1.0 mm); (ii) Diffusion Tensor Imaging using a Pulsed-Gradient Spin Echo sequence with 33 gradient directions (TR/TE = 7800 ms/101 ms, FOV = 23 cm, Matrix = 128 × 128, slice thickness 3 mm, b-value = 1200 s/mm^2, 33 directions); (iii) Resting state fMRI scan (10 min); (iv) Task driven fMRI performed using a working memory task (N-back letter sequences).

Images were acquired using a Gradient-Echo-EPI sequence (TR/TE = 2000 ms/27 ms, FOV = 23 cm, Matrix 64 × 64, slice thickness 2.5 mm). All visual stimuli were presented using a high-definition goggle system by Resonance Technology Inc., Northridge, CA. Stimulus presentation and subject responses were acquired using Eprime and a fiber optic response glove by Psychology Software Tools (PST Inc. Pittsburgh PA).

2.3 PET image acquisition (PET/MR)

PET/MR (3T) Siemens mMR. The 3T MR/PET is a fully integrated and capable of simultaneous whole-body PET and MRI scanning. This allows more precisely coregistered functional and structural acquisition while reducing the radiation dose in PET imaging by replacing the CT scan with an MRI scan. True simultaneous acquisition of MR and PET data by the hybrid system merges the highly sensitive PET metabolic information with the highly specific MR anatomical and functional information. Attenuation data in PET/MRI are derived from the MRI scan. For each bed position, the MRI sequence for attenuation correction (AC) purposes is acquired first. The sequence used in the integrated PET/MRI system is a 2-point Dixon volume-interpolated breath-hold examination (VIBE). This sequence is preceded by scanning preparations that include shimming to optimize the homogeneity of the magnetic field (~40 s). The MRI data are then segmented to identify air, brain tissue, fatty tissue, and watery tissue as required for AC.

[^{18}F] fluorodeoxyglucose ([^{18}F]FDG) brain imaging was performed on cognitively impaired subjects in the SBU WTC cohort. To administer the [^{18}F]2-deoxyglucose, one catheter is placed in an intravenous line in an antecubital vein. Ten mCi (370 MBq) of [^{18}F]2-deoxyglucose are injected. ^{18}FDG is used according to its FDA approved indication and labeling. Thirty minutes after radiotracer injection, the subject is then positioned supine in the mMR PET scanner, image acquisition period lasts 15 min. Amyloid deposition was also studied using the Amyvid tracer.

2.4 [^{18}F] amyvid PET imaging

To administer the [^{18}F] florbetapir (Amyvid™), one catheter is placed in an intravenous line in an antecubital vein. Ten mCi (370 MBq) of [^{18}F] florbetapir are injected. Florbetapir is used according to its FDA approved indication and labeling. Sixty minutes after radiotracer injection, the subject is then positioned supine in the mMR-PET scanner. The imaging data acquisition lasts 15 min.

2.5 A WTC-CI neuro-phenotype

Unique brain characteristic patterns of CI among WTC responders emerged, compared to other signatures, such as AD. Hippocampal subfield volume analysis suggests reductions in specific subregions are associated with the duration of WTC exposure (Deri et al., 2021). Cortical thickness measurement from MRI provides a powerful non-invasive method for quantifying neurodegenerative disease risk including AD (Dickerson et al., 2009). Among 99 WTC responders aged 55.8 (SD = 0.52) years, 48 had CI. Compared to unimpaired responders, global means cortical thickness was reduced in CI subjects and across 21/34 cortical subregions. Surface-based analyses revealed reduced cortical thickness across frontal, temporal, and parietal lobes when adjusting for multiple comparisons (Clouston et al., 2020). Both CI and unimpaired WTC groups showed reduced cortical thickness in the entorhinal and temporal cortices compared to published normative data. An artificial neural network was also applied to accurately identify WTC responders who may be at higher risk for cognitive issues on MRI (Clouston et al., 2021). Cortical complexity can also be measured as Fractal Dimension (FD) from T1 MRI brain images and has been reported to be reduced in a variety of psychiatric and neurological conditions. Vertex clusters of complexity were altered in WTC responders with PTSD, with marked reductions in regions within the frontal, parietal, and temporal cortices, in addition to whole-brain absolute bilateral and unilateral complexity. The magnitude of changes in regional FD severity was associated with increased PTSD symptoms (reexperiencing, avoidance, hyperarousal, negative affect) severity (Kritikos et al., 2021).

Bilateral cortical thickness resulted also in 12 cerebellar lobules of these WTC responders. Mean cerebellar cortical thickness was reduced by 0.17 mm in responders with CI. Decrements in cerebellar cortical thickness were symmetric and located in the Cerebellar Crus (I and II), and in Lobules IV, VI, VIIb, VIIIa, VIIIb, and IX. Cerebellar cortical thickness was associated with episodic memory, response speed, and tandem balance. WTC responders with CI had evidence of reduced cerebellar cortical thickness that was present across lobules in a pattern unique to this cohort (Clouston et al., 2022).

Approximately 23% of World Trade Center (WTC) responders are experiencing chronic posttraumatic stress disorder (PTSD) associated with their exposures at the WTC following the terrorist attacks of 9/11/2001, which has been demonstrated to be a risk factor for CI raising concerns regarding their brain health (Kritikos et al., 2021). Diffuse brain atrophy, reduced cortical thickness, and hippocampal subfield volume analysis suggest that reductions in specific subregions are associated with the duration of WTC exposure. These findings support the hypothesis that WTC exposure to neurotoxicants is causing a long-term neurodegenerative impact. The neuro-phenotype of this impairment is different from the AD and inconsistent with signatures developed for known neurodegenerative diseases. The WTC-CI may be a WTC-specific encephalopathy with an unknown etiology characterized by widespread cortical atrophy.

3. Case study 2: Cognitive impact of occupational exposure to manganese

Manganese (Mn), one of the most abundant elements on the earth, is an essential trace element required for critical enzyme-mediated biological processes involved in development, growth, and neuronal function (Martins et al., 2019). Although necessary for life, Mn can cause health issues both in deficiency and excessive levels. Occupational settings such as mining, steel, and alloy production, welding, smelting and dry-cell battery manufacturing are among the main sources of environmental risk factors for chronic exposure to various metals (Briffa et al., 2020). Welders and smelters are among the most vulnerable groups exposed to fumes or inhaled dust types of manganese (Mn) at different levels. Different forms of Mn oxides including Mn_3O_4 and MnO_2 have been found in both welding and smelting aerosols

(Cowan et al., 2009; Jiang et al., 2007; Keane et al., 2010; Long et al., 2014). A great part of respirable Mn particles is transmitted through the circulation to the brain directly and is not being involved in the process of the Mn homeostasis control of the liver. This makes occupational workers more susceptible to the neurotoxic Mn effect compared to individuals with oral exposure to Mn (Roels et al., 2012). Mn accumulates selectively in the specific parts of the brain such as globus pallidus, putamen, caudate nucleus, and cerebral cortex. The neural system and cognitive functioning are the most relevant targets of exposure to neurotoxicants from early life to old age (Balachandran et al., 2020).

3.1 Manganese-induced cognitive impactive deficits

It is well documented that increased manganese (Mn) exposure is associated with various neurologic deficits that occur in multiple stages. Psychiatric, motor, and olfactory dysfunction and cognitive deficits (including working or visual memory and verbal disorders) have been reported in the various studies of environmental exposure to Mn in adults (Bowler et al., 2015; Lucchini et al., 1995, 2014, 2017; Zoni and Lucchini, 2013). Additionally, significant improvement in cognitive but not motor test scores has been recorded by ceasing or decreasing Mn exposure (Roels et al., 2012).

Cognitive dysfunction and structural changes in the brain regions could be implicated as important factors in contributing to Mn toxicity. Functional imaging and Mn exposure brain morphometric measurement allow analyzing cognitive function and the deleterious effects of Mn on the size, volume, and shape characteristics of different brain structures using sensitive methods such as MRI and PET scan.

A potential mechanism for Mn-induced neurotoxicity caused cognitive decline is amyloidogenesis, an increased formation of amyloid-beta (Aβ) deposition, a biomarker of cognitive dysfunction, and the main hallmark pathology in Alzheimer's disease (AD), the most common cause of dementia (DeTure and Dickson, 2019). Chronic manganese exposure may increase the risk of cognitive decline and induces amyloidogenesis mediated through whether alteration in the level of proinflammatory markers such as cytokines IL-1β and TNF-α or Amyloid Precursor Protein (APP) production. It is also shown that a high level of Mn leads to disruption Aβ degradation through down-regulation of major enzymes involved in this mechanism such as neprilysin (NEP) and insulin degrading enzyme (IDE) (Lin et al., 2020; Tong et al., 2014).

We investigated the presence of Aβ aggregation and cognitive function from a cohort of Mn-exposed workers, employed in ferroalloy plants located in different areas of the province of Brescia, Italy and historical unexposed control workers (Lucchini et al., 2022). This sample was randomly selected from the occupational cohort initiated in 1990 within the PHIME (Public Health Impact of the Metal Exposure) project, a multiage cohrt funded initially by the EY 6thFP and subsequently by NIEHS. This occupational cohort has provided numerous observations about the impact of manganese exposure on motor coordination, olfactory discrimination and cognitive decline (Lucchini et al., 1997, 1999, 2000). Workers of this cohort were originally enrolled from the company health surveillance program. They were asked to participate on a voluntary base, and were consented through a procedure approved by the Ethical Committee of Brescia. Exclusion criteria included being diagnosed with a neurological or psychiatric disorder.

Dose–response relationships were observed with cumulative exposure metrics based on the annual measurements of manganese in respirable/inhalable airborne particles in personal and stationary monitoring throughout the industrial operation (Lucchini et al., 1995). Structural MRI showed manganese deposition in the globus pallidus (Lucchini et al., 2000a,b), providing potential interpretation of a long-term diffuse deposition of manganese in the basal ganglia after prolonged exposure to low levels throughout the working life (Lucchini et al., 2009; Lucchini and Zimmerman, 2009).

In this recent pilot study, we examined differences in the cognitive functioning and β-amyloid brain deposition in 6 ferroalloy workers (average age 64 and average Mn exposure duration 31 years) and 5 historical sex- and age-matched control workers (average age 63), not exposed to metals. The cognitive function was assessed with a battery of neuropsychological tests including the Montreal Cognitive Assessment (MOCA). Mn exposure was based on the 25-year annual assessment of workplace air monitoring and biological monitoring. A similar PET/MRI brain imaging protocol to the one utilized for the WTC responders in NYC was applied to the Italian manganese workers. In this case, the presence of β-amyloid deposition was assessed with a General Electric Discovery 690 PET-CT scanner after the injection of 185 MBq of [^{18}F] flutematol (Vizamyl, GE Healthcare, Marlborough, MA, USA) through a catheter placed in an intravenous line in an antecubital vein (Vandenberghe et al., 2010). PET acquisition was carried out for 20 min, starting 90 min after injection.

We performed *t*-tests to compare the Mn exposed workers and the controls. β-amyloid deposition in the ferroalloy workers was more diffuse than in controls (*P*, 0.05), as shown in Fig. 1. The major regions with increased beta-amyloid uptake were the anterior and posterior cingulate, entorhinal cortex and part of hippocampus, dorso-lateral prefrontal cortices. The cognitive function testing did not differ between the two groups as shown in Table 1.

To our knowledge, this is the first study showing increased β-amyloid brain deposition in manganese exposed individuals. Further research is warranted to test the hypothesis of β-amyloid as a predictor of Mn-induced cognitive decline. Additional investigation including the analysis of blood biomarkers of signaling proteins and plasma metabolites would be effective to identify prognostic or promising diagnostic characteristics of AD and cognitive dysfunction in the Mn-exposed group. The significance of the study is the identification of the putative mechanisms contributed to chronic Mn toxicity provides new insights into the understanding of etiological, therapeutic, and preventive approaches to neurodegenerative diseases.

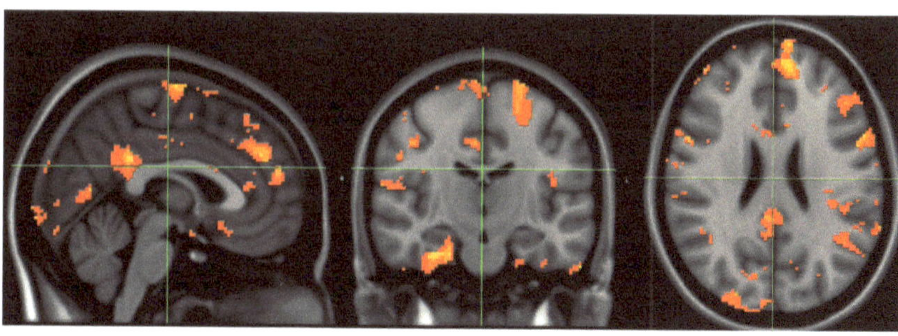

Fig. 1 Brain β-amyloid PET scanning showed widely diffused higher β-amyloid accumulation in the ferroalloy workers. Significant voxels ($P < 0.05$) are displayed in red superimposed on a standard anatomical MRI scan.

Table 1 The cognitive function testing did not differ between the two Mn exposed and control (non-exposed) groups

	Exposed ($N = 6$)	Not Exposed ($N = 5$)	Total ($N = 11$)	*P* value
Digit Span Standard Score				0.745
Mean (SD)	9.8 (1.9)	9.5 (1.3)	9.7 (1.6)	
Median (Q1, Q3)	9.5 (9.0, 11.5)	9.5 (8.8, 10.2)	9.5 (9.0, 10.8)	

Continued

Table 1 The cognitive function testing did not differ between the two Mn exposed and control (non-exposed) groups—cont'd

	Exposed ($N = 6$)	Not Exposed ($N = 5$)	Total ($N = 11$)	*P* value
Semantic Fluency Standard Score				0.927
Mean (SD)	47.0 (8.0)	46.6 (8.7)	46.8 (7.9)	
Median (Q1, Q3)	46.8 (41.2, 49.7)	44.3 (40.3, 46.8)	44.3 (40.3, 49.6)	
Phonemic Fluency Standard Score				0.584
Mean (SD)	26.4 (8.1)	24.2 (5.0)	25.4 (6.6)	
Median (Q1, Q3)	26.4 (19.8, 32.1)	26.0 (24.0, 26.5)	26.0 (20.6, 29.7)	
Total time dominant hand				0.361
Mean (SD)	69.4 (5.3)	72.9 (4.4)	71.0 (5.0)	
Median (Q1, Q3)	70.2 (66.2, 71.7)	73.0 (69.0, 75.3)	70.7 (68.7, 74.2)	
Total time dominant hand				0.465
Mean (SD)	77.8 (8.2)	76.0 (8.5)	76.9 (7.9)	
Median (Q1, Q3)	80.4 (74.4, 83.9)	72.0 (71.6, 74.2)	74.3 (71.7, 83.0)	
MOCA score adjusted				0.454
Mean (SD)	25.0 (2.8)	24.0 (2.2)	24.5 (2.5)	
Median (Q1, Q3)	26.5 (23.0, 27.0)	24.0 (23.0, 25.0)	25.0 (22.5, 27.0)	
Standard copy score				0.522
Mean (SD)	31.9 (3.7)	32.8 (5.1)	32.3 (4.2)	
Median (Q1, Q3)	32.5 (28.7, 34.9)	34.2 (31.5, 36.2)	34.2 (29.3, 35.5)	
Standard recall score				0.584
Mean (SD)	13.5 (3.5)	13.9 (8.1)	13.7 (5.7)	
Median (Q1, Q3)	13.8 (11.9, 16.0)	10.2 (9.5, 13.4)	12.8 (9.9, 15.6)	

Table 1 The cognitive function testing did not differ between the two Mn exposed and control (non-exposed) groups—cont'd

	Exposed (*N* = 6)	Not Exposed (*N* = 5)	Total (*N* = 11)	*P* value
Standard score words immediate retrieval				1
Mean (SD)	39.2 (10.8)	38.2 (14.4)	38.8 (11.6)	
Median (Q1, Q3)	39.3 (33.8, 43.3)	35.5 (26.4, 47.3)	39.3 (27.9, 44.6)	
Standard score words delay retrieval				0.67
Mean (SD)	7.7 (3.1)	8.5 (5.3)	8.0 (3.8)	
Median (Q1, Q3)	8.5 (5.2, 10.0)	7.5 (4.3, 11.8)	8.5 (4.3, 10.6)	
Words recognition test				0.426
Mean (SD)	14.2 (1.2)	13.5 (1.7)	13.9 (1.4)	
Median (Q1, Q3)	14.5 (14.0, 15.0)	14.0 (13.2, 14.2)	14.0 (14.0, 15.0)	
Short story standard score				0.855
Mean (SD)	7.4 (2.9)	6.9 (4.6)	7.2 (3.6)	
Median (Q1, Q3)	7.5 (5.1, 9.4)	6.6 (5.2, 6.7)	6.6 (4.9, 9.1)	
Sniffin Identification Score				0.665
Mean (SD)	12.0 (2.1)	12.4 (1.9)	12.2 (1.9)	
Median (Q1, Q3)	12.0 (10.0, 13.0)	13.0 (13.0, 13.0)	13.0 (10.5, 13.0)	
TMT Standard score time part a				0.715
Mean (SD)	24.1 (7.9)	23.6 (10.4)	23.9 (8.6)	
Median (Q1, Q3)	25.1 (17.9, 30.3)	20.1 (19.8, 20.5)	20.5 (17.8, 28.9)	
TMT Standard score time part b				1
Mean (SD)	71.4 (60.0)	58.0 (41.7)	66.9 (52.3)	
Median (Q1, Q3)	69.5 (18.2, 115.6)	52.3 (35.9, 77.3)	52.3 (19.4, 102.2)	

Continued

Table 1 The cognitive function testing did not differ between the two Mn exposed and control (non-exposed) groups—cont'd

	Exposed ($N = 6$)	Not Exposed ($N = 5$)	Total ($N = 11$)	*P* value
TMT Standard score time part a and b				1
Mean (SD)	47.6 (55.2)	32.0 (31.0)	42.4 (47.0)	
Median (Q1, Q3)	45.6 (3.9, 84.3)	36.7 (17.8, 48.6)	36.7 (−1.1, 60.4)	

4. Conclusions

The two case studies of the WTC responders and the ferromanganese workers outline the utility of modern brain imaging modalities in the understanding of neurotoxicity induced by exposure to airborne pollutants. These findings underline the importance of preventive action aimed to exposure reduction in the workplaces. Given the long-term projection of cognitive impairment, protection of the brain from workplace hazards is of fundamental importance. Accelerated cognitive decline can lead to neurodegeneration, and it is imperative to act on modifiable factors to minimize the potential risk of dementia and parkinsonian disturbances in heavily exposed populations. Further studies are needed to understand the potential underlying mechanisms leading to brain amyloidosis and neurodegeneration induced by neurotoxicants, especially when carried by small or ultrafine particles able to access the brain through the olfactory pathway (Lucchini et al., 2012b).

References

Acosta, S.A., et al., 2013. Influence of post-traumatic stress disorder on neuroinflammation and cell proliferation in a rat model of traumatic brain injury. PLoS One 8 (12), e81585.

Adams, S., 2012. Dementia Screening 'will Scare Patients from Seeing their GP'. The Telegraph, London, UK.

Altman, K.W., et al., 2011. Odor identification ability and self-reported upper respiratory symptoms in workers at the post-9/11 world trade center site. Int. Arch. Occup. Environ. Health 84 (2), 131–137.

Balachandran, R.C., et al., 2020. Brain manganese and the balance between essential roles and neurotoxicity. J. Biol. Chem. 295 (19), 6312–6329.

Barthel, H., et al., 2015. PET/MR in dementia and other neurodegenerative diseases. Semin. Nucl. Med. 45 (3), 224–233 (1558-4623 (Electronic)).

Bench, G., et al., 2001. Olfactory bulb uptake and determination of biotransfer factors in the California ground squirrel (Spermophilus beecheyi) exposed to manganese and cadmium in environmental habitats. Environ. Sci. Technol. 35 (2), 270–277.

Block, M.L., Calderón-Garcidueñas, L., 2009. Air pollution: mechanisms of neuroinflammation and CNS disease. Trends Neurosci. 32 (9), 506–516.
Bowler, R.M., et al., 2015. Environmental exposure to manganese in air: associations with cognitive functions. Neurotoxicology 49, 139–148.
Briffa, J., Sinagra, E., Blundell, R., 2020. Heavy metal pollution in the environment and their toxicological effects on humans. Heliyon 6 (9), e04691.
Buchsbaum, M.S., et al., 2015. Clusters of low 18F-fluorodeoxyglucose uptake voxels in combat veterans with traumatic brain injury and post-traumatic stress disorder. J. Neurotrauma 32 (22), 1736–1750.
Calderón-Garcidueñas, L., et al., 2002. Air pollution and brain damage. Toxicol. Pathol. 30 (3), 373–389.
Campbell, A., et al., 2005. Particulate matter in polluted air may increase biomarkers of inflammation in mouse brain. Neurotoxicology 26 (1), 133–140.
Chen, R., et al., 2015. Parallel comparative studies on mouse toxicity of oxide nanoparticle-and gadolinium-based T1 MRI contrast agents. ACS Nano 9 (12), 12425–12435.
Clouston, S.A., et al., 2016. Cognitive impairment among world trade center responders: Long-term implications of re-experiencing the 9/11 terrorist attacks. Alzheimers Dement (Amst) 4, 67–75.
Clouston, S.A.P., et al., 2020. Reduced cortical thickness in world trade center responders with cognitive impairment. Alzheimers Dement (Amst) 12 (1), e12059.
Clouston, S.A.P., et al., 2021. A cortical thinning signature to identify world trade center responders with possible dementia. Intell. Based Med., 100032.
Clouston, S.A.P., et al., 2022. Reduced cerebellar cortical thickness in world trade center responders with cognitive impairment. Transl. Psychiatry 12 (1), 107.
Cowan, D.M., et al., 2009. Manganese exposure among smelting workers: blood manganese-iron ratio as a novel tool for manganese exposure assessment. Biomarkers 14 (1), 3–16.
Daniels, J.K., et al., 2013. White matter integrity and its relationship to PTSD and childhood trauma—a systematic review and meta-analysis. Depress. Anxiety 30 (3), 207–216.
Davenport, N.D., Lim, K.O., Sponheim, S.R., 2015. White matter abnormalities associated with military PTSD in the context of blast TBI. Hum. Brain Mapp. 36 (3), 1053–1064.
de Vries, E.F.J., Dierckx, R.A.J.O., de Groot, D.M.G., 2021. PET and SPECT imaging of neurotoxicity. In: Dierckx, R.A.J.O., et al. (Eds.), PET and SPECT in Neurology. Springer International Publishing, Cham, pp. 917–946.
Deri, Y., et al., 2021. Selective hippocampal subfield volume reductions in world trade center responders with cognitive impairment. Alzheimers Dement (Amst) 13 (1), e12165.
DeTure, M.A., Dickson, D.W., 2019. The neuropathological diagnosis of Alzheimer's disease. Mol. Neurodegener 14 (1), 32.
Dickerson, B.C., et al., 2009. The cortical signature of Alzheimer's disease: regionally specific cortical thinning relates to symptom severity in very mild to mild AD dementia and is detectable in asymptomatic amyloid-positive individuals. Cereb. Cortex (New York, N.Y.: 1991) 19 (3), 497–510.
Dougherty, D.D., et al., 2004. Positron emission tomography and single photon emission computed tomography. In: Essentials of Neuroimaging for Clinical Practice. vol. 13(10).
Greenberg, M.S., et al., 2014. Stress, PTSD, and dementia. Alzheimers Dement. 10 (3, Supplement), S155–S165.
Hänninen, H., 1966. Psychological tests in the diagnosis of carbon disulphide poisoning. Work Environ. Health 2, 16–20.
Horn, J., Cattell, R., 1967. Age differences in fluid and crystallized intelligence. Acta Psychol. (Amst) 26, 107–129.
Jack Jr., C.R., et al., 2010. Hypothetical model of dynamic biomarkers of the Alzheimer's pathological cascade. Lancet Neurol. 9 (1), 119–128.

Jiang, Y., et al., 2007. Brain magnetic resonance imaging and manganese concentrations in red blood cells of smelting workers: search for biomarkers of manganese exposure. Neurotoxicology 28 (1), 126–135.
Kazeros, A., et al., 2015. Systemic inflammation associated with world trade center dust exposures and airway abnormalities in the local community. J. Occup. Environ. Med. 57 (6), 610–616.
Keane, M., Stone, S., Chen, B., 2010. Welding fumes from stainless steel gas metal arc processes contain multiple manganese chemical species. J. Environ. Monit. 12 (5), 1133–1140.
Kim, S.H., et al., 2012. Rapid doubling of Alzheimer's amyloid-β40 and 42 levels in brains of mice exposed to a nickel nanoparticle model of air pollution. F1000Res. 1.
King, D.W., et al., 1998. Confirmatory factor analysis of the clinician-administered PTSD scale: evidence for the dimensionality of posttraumatic stress disorder. Psychol. Assess. 10 (2), 90.
Kritikos, M., et al., 2020. Acute versus chronic exposures to inhaled particulate matter and neurocognitive dysfunction: pathways to Alzheimer's disease or a related dementia. J. Alzheimers Dis. 78 (3), 871–886.
Kritikos, M., et al., 2021. Cortical complexity in world trade center responders with chronic posttraumatic stress disorder. Transl. Psychiatry 11 (1), 597.
Lawrence-Wood, E., et al., 2016. Re-experiencing phenomena following a disaster: the long-term predictive role of intrusion symptoms in the development of post-trauma depression and anxiety. J. Affect. Disord. 190, 278–281.
Lin, G., et al., 2020. Manganese exposure aggravates beta-amyloid pathology by microglial activation. Front. Aging Neurosci. 12, 556008.
Long, Z., et al., 2014. Vulnerability of welders to manganese exposure- -a neuroimaging study. Neurotoxicology 45, 285–292.
Lucchini, R., Zimmerman, N., 2009. Lifetime cumulative exposure as a threat for neurodegeneration: need for prevention strategies on a global scale. Neurotoxicology 30 (6), 1144–1148.
Lucchini, R., et al., 1995. Neurobehavioral effects of manganese in workers from a ferroalloy plant after temporary cessation of exposure. Scand. J. Work Environ. Health 21 (2), 143–149.
Lucchini, R., et al., 1997. Motor function, olfactory threshold, and hematological indices in manganese-exposed ferroalloy workers. Environ. Res. 73 (1–2), 175–180.
Lucchini, R., et al., 1999. Long-term exposure to "low levels" of manganese oxides and neurofunctional changes in ferroalloy workers. Neurotoxicology 20 (2–3), 287–297.
Lucchini, R., et al., 2000a. Brain MRI in manganese exposed workers and hepatopatic patients. J. Alzheimers Dis. 2 (3–4), 344.
Lucchini, R., et al., 2000b. Brain magnetic resonance imaging and manganese exposure. Neurotoxicology 21 (5), 769–775.
Lucchini, R.G., Martin, C.J., Doney, B.C., 2009. From manganism to manganese-induced parkinsonism: a conceptual model based on the evolution of exposure. Neuromolecular Med. 11 (4), 311–321.
Lucchini, R.G., et al., 2012a. Tremor, olfactory and motor changes in Italian adolescents exposed to historical ferro-manganese emission. Neurotoxicology 33 (4), 687–696.
Lucchini, R.G., et al., 2012b. Neurological impacts from inhalation of pollutants and the nose-brain connection. Neurotoxicology 33 (4), 838–841.
Lucchini, R.G., et al., 2014. Neurofunctional dopaminergic impairment in elderly after lifetime exposure to manganese. Neurotoxicology 45, 309–317.
Lucchini, R., et al., 2017. Manganese and developmental neurotoxicity. Adv. Neurobiol. 18, 13–34.
Lucchini, R., et al., 2022. Diffuse brain deposition of beta-amyloid among italian ferroalloy workers. Saf. Health Work 13, S57–S58.

Martins Jr., A.C., et al., 2019. New insights on the role of manganese in Alzheimer's disease and Parkinson's disease. Int. J. Environ. Res. Public Health 16 (19).

McKee, A.C., Robinson, M.E., 2014. Military-related traumatic brain injury and neurodegeneration. Alzheimers Dement. 10 (3), S242–S253.

Meyer, P.T., et al., 2016. Functional neuroimaging: functional magnetic resonance imaging, positron emission tomography, and single-photon emission computed tomography. In: Daroff, R.B., et al. (Eds.), Bradley's Neurology in Clinical Practice. Elsevier, Philadelphia, PA.

O'Doherty, D.C., et al., 2015. A systematic review and meta-analysis of magnetic resonance imaging measurement of structural volumes in posttraumatic stress disorder. Psychiatry Res. Neuroimaging 232 (1), 1–33.

Piccinin, A., et al., 2011. Terminal decline from within-and between-person perspectives, accounting for incident dementia. J. Gerontol. B Psychol. Sci. Soc. Sci. 66 (4), 391–401.

Pitman, R.K., 2010. POsttraumatic stress disorder and dementiaposttraumatic stress disorder and dementia. JAMA 303 (22), 2287–2288.

Pitman, R.K., Shin, L.M., Rauch, S.L., 2001. Investigating the pathogenesis of posttraumatic stress disorder with neuroimaging. J. Clin. Psychiatry.

Plassman, B.L., et al., 2010. Systematic review: factors associated with risk for and possible prevention of cognitive decline in later life. Ann. Intern. Med. 153 (3), 182–193.

Plassman, B.L., et al., 2011. Incidence of dementia and cognitive impairment, not dementia in the United States. Ann. Neurol. 70 (3), 418–426.

Ravaglia, G., et al., 2008. Mild cognitive impairment: epidemiology and dementia risk in an elderly Italian population. J. Am. Geriatr. Soc. 56 (1), 51–58.

Roels, H.A., et al., 2012. Manganese exposure and cognitive deficits: a growing concern for manganese neurotoxicity. Neurotoxicology 33 (4), 872–880.

Schuitevoerder, S., et al., 2013. A meta-analysis of cognitive functioning in older adults with PTSD. J. Anxiety Disord. 27 (6), 550–558.

Seibyl, J., 2012. Imaging Neuroinflammation in Post Traumatic Stress Disorder. (DTIC Document).

Shalev, A.Y., et al., 2014. Prospective study of posttraumatic stress disorder and depression following trauma. Am. J. Psychiatry.

Tong, Y., et al., 2014. High manganese, a risk for Alzheimer's disease: high manganese induces amyloid-beta related cognitive impairment. J. Alzheimers Dis. 42 (3), 865–878.

Valenzuela, M.J., Sachdev, P., 2005. Brain reserve and dementia: a systematic review. Psychol. Med. 36, 441–454.

van Achterberg, M.E., Southwick, S.M., 2001. Emergence of PTSD in trauma survivors with dementia. J. Clin. Psychiatry 62 (3), 206–207.

Vandenberghe, R., et al., 2010. 18F-flutemetamol amyloid imaging in Alzheimer disease and mild cognitive impairment: a phase 2 trial. Ann. Neurol. 68 (3), 319–329.

Veitch, D.P., Friedl, K.E., Weiner, M.W., 2013. Military risk factors for cognitive decline, dementia and Alzheimer's disease. Curr. Alzheimer Res. 10 (9), 907–930.

Wilker, E.H., et al., 2015. Long-term exposure to fine particulate matter, residential proximity to major roads and measures of brain structure. Stroke 46 (5), 1161–1166.

Wu, M., et al., 2010. Case report: lung disease in world trade center responders exposed to dust and smoke: carbon nanotubes found in the lungs of world trade center patients and dust samples. Environ. Health Perspect. 118 (4), 499.

Zhu, Y., et al., 2016. PET mapping of neurofunctional changes in a post-traumatic stress disorder model. J. Nucl. Med. (p. jnumed. 116.173443).

Zoni, S., Lucchini, R.G., 2013. Manganese exposure: cognitive, motor and behavioral effects on children: a review of recent findings. Curr. Opin. Pediatr. 25 (2), 255–260.

CHAPTER THREE

Clinical evaluation and differential diagnosis of neurotoxic disease

Jonathan Rutchik[a,b,*,†] and Marcia H. Ratner[c,†]
[a]Division of Occupational & Environmental Medicine, University of California San Francisco, San Francisco, CA, United States
[b]Neurology, Environmental and Occupational Medicine Associates, Mill Valley, CA, United States
[c]Department of Pharmacology and Experimental Therapeutics, Boston University School of Medicine, Boston, MA, United States
*Corresponding author: e-mail address: jsrutch@neoma.com

Contents

1. Introduction

The diagnosis of neurotoxicant-induced nervous system injury is similar to the diagnosis of other types of neurological disease in that it is necessary for the occupational and environmental medicine (OEM) specialist to

† Drs. Rutchik and Ratner contributed equally to this work.

Advances in Neurotoxicology, Volume 7
ISSN 2468-7480
https://doi.org/10.1016/bs.ant.2022.05.003

establish a case specific cause-effect relationship between the observed signs and symptoms, neuropathology and/or results of objective testing, and the onset, progression, and resolution of the patient's symptoms and signs. In the idealized clinical case, the patient is currently exposed to the offending neurotoxicant(s) and, the requisite biological exposure data necessary to establish a cause-effect relationship is readily obtained from blood and urine samples. More likely an evaluation is performed after an exposure occurs either in or outside the workplace. The diagnosis of neurotoxic disease is often a clinical one which must be made based entirely on limited retrospective historical exposure data and the chronological relationship between this and the onset and progression of the patient's symptoms which serve as de facto biological markers of effect. Patients with a history of occupational and environmental exposure are also often seen in an outpatient office setting where they typically present with non-life threatening subtle neurological effects of chronic neurotoxicant exposure. As a result, the OEM specialist cannot turn to the well-known toxidromes used to diagnose and treat acute poisonings and overdoses. Due to these limitations, it is very important to effectively rule out all other possible etiologies of the observed neurological signs and symptoms even in those patients who present with a well-documented history of occupational and/or environmental exposure to neurotoxicants. The possibility that the exposure may have worsened or accelerated an idiopathic condition must be considered.

The well-trained OEM specialist is able recognize the acute and chronic effects of specific neurotoxicant exposures as well as possible points of toxicokinetic and toxicodynamic interaction between these chemicals and the mechanisms of action implicated in idiopathic neurodegenerative diseases such as oxidative stress and glutamatergic excitotoxicity which can further complicate the differential diagnosis (see Table S1 from Supplementary material in the online version at https://doi.org/10.1016/bs.ant.2022.05.003), the symptoms of Parkinson's disease (PD) can be unmasked by acute ingestion of antidopaminergic drugs while the age at onset of PD is often younger in workers exposed to metals and pesticides (Gamache et al., 2019; Ratner et al., 2014; Shin and Chung, 2012; Wilk et al., 2006).

In this chapter, we introduce and review the best clinical practices for effectively diagnosing neurotoxic disease. This process requires a team of specialists in addition to the OEM specialist which often includes an industrial hygienist, risk assessment professional, neurologist, neurophysiologist, neuroradiologist, neuropsychologist, and toxicologist or if available neurotoxicologist. The findings of the tests performed by these experts are presented in terms of a best practices approach and related to specific case examples of common neurotoxicants found in the workplace and environment. Diagnostic pearls are also interwoven into the chapter which is written to serve as a practical guide that can be effectively applied to the unique cases an OEM specialist will see in their own clinical practice.

2. Neurotoxicology patient intake questionnaire and interview

Standard patient intake questionnaires are not well-suited for collecting occupational and environmental neurotoxicant exposure data. Therefore, it is necessary to use a comprehensive occupational and environmental exposure and medical history questionnaire specifically designed for collecting data about neurotoxicant exposures (Feldman, 1999; Feldman et al., 1999; Ratner et al., 2020). We have developed a questionnaire specifically for this purpose (see Neurotoxicology patient intake questionnaire in Supplementary material in the online version at https://doi.org/10.1016/bs.ant.2022.05.003). The questionnaire is designed to collect data about all current and past occupational exposures to neurotoxicants in chronological order with duration of exposure. Relevant details about the exposure circumstances including the industry in which the exposure(s) occurred, the patient's job title(s) and duties during that time, as well as any health problems they experienced while working in this setting are recorded in the questionnaire. The questionnaire also collects data about environmental sources of exposure such as proximity of residences to mines, smelters and super fund sites which create additional non-occupational exposure risks. Because *chronic traumatic encephalopathy* must be differentiated from *chronic toxic encephalopathy* the questionnaire also should inquire about the patient's history of head injury and participation in contact sports. The patient is queried about their dietary habits including consumption of fish which may contain high concentrations of organic mercury (Hightower and Moore, 2003). Because neurotoxic disease can resemble hereditary neurodegenerative diseases such as Amyotrophic Lateral Sclerosis (ALS), Parkinson's disease (PD) and Alzheimer's disease (AD), it is important to obtain the medical histories of the patient's family members as well. When genetic testing is available a copy of this data should be obtained and reviewed and attached to the questionnaire as an addendum (see Neurotoxicology patient intake questionnaire in Supplementary material in the online version at https://doi.org/10.1016/bs.ant.2022.05.003). While this information does not control for nor exclude the interactions between these hereditary neurodegenerative diseases and patient's own medical and neurotoxicant exposure histories, it is nevertheless informative to the differential diagnostic process. Finally, the questionnaire collects information about the patient's education and professional achievements which can be used to estimate their premorbid intellectual functioning.

Upon completion of the questionnaire, the OEM specialist should conduct an in person or virtual interview of the patient to confirm and clarify their responses. The personal interview often reveals additional relevant information such as how specific chemicals were used and stored in the workplace (Feldman et al., 1999; Ratner et al., 2020). Patients may not be aware of the adverse effects of all chemicals they encounter and thus, it is incumbent open the OEM specialist to recognize neurotoxic exposures that the patient may overlook (Ratner et al., 2020). Because questionnaires and interviews inherently rely on self-reporting of retrospective exposure data by the patient, this information should be verified to the full extent possible. Coworkers, spouses and children are excellent sources of corroborative information. Because the field of neurotoxicology is a dynamic one, the intake questionnaire should be updated periodically to ensure it collects the most relevant and contemporary occupational and environmental exposure and medical history data.

Clinical pearl

Most important information from a questionnaire is to clarify chronology of possible exposure, and to review activities of daily living, before during and after an exposure and to consider whether any other activities or events impact the patients present level of function.

3. The neurological examination

The neurological examination serves to document the symptoms of cognitive, sensorimotor and autonomic nervous system dysfunction experienced by the patient which may be due to neurotoxicant exposure and to differentiate these clinical findings from those which may be due to other non-toxic etiologies such as neurodegenerative disease, diabetes, peripheral vascular disease, physical injury, autoimmune disease or stroke (Table 1). The differential diagnosis of neurotoxic disease follows a standard diagnostic

Table 1 10 Steps in evaluating neurological symptoms and chemical exposures.

1.	Establish medical problem exists
2.	Establish medical diagnosis
3.	Consider common conditions
4.	Is there objective evidence for exposure?
5.	Obtain biological exposure data
6.	Neuroimaging, neuropsychology, toxicology?
7.	What regulations or government agency recommendations exist?
8.	Are follow up Information consistent with diagnosis?
9.	Consult literature
10.	Reassessment of causation, conclusions

algorithm similar to that used in the differential diagnosis of other diseases (Feldman et al., 1999) (Fig. 1). However, this diagnostic process is unique it that it depends first on the documentation of neurotoxicant exposure and second, on establishing the chronological relationship between the exposure history and the onset of patient's neurological symptoms as outlined in this chapter. The exposure data and the patient's history must also be related to objective measures of neurologic dysfunction. Only conventional measures of neurologic function should be used for this purpose. Only the results of formal neuropsychological testing, neurophysiological testing (e.g., nerve conduction studies and electromyography) and neuroimaging reports based on tests performed by other experts board-certified in their respective fields are suitable for this purpose. Results of unique and unvalidated tests whether performed by the OEM specialist or other specialists should not be relied upon for this purpose.

Exposures to neurotoxicants is often associated with non-specific symptoms such as headache, dizziness and fatigue. These symptoms often emerge in close chronologic relationship to known acute exposures and, to resolve or improve with cessation of exposure. Although non-specific symptoms do not carry as much diagnostic information as specific symptoms, when these symptoms are consistent with the known effects of a particular neurotoxicant and/or occur in close chronological relationship to the onset of other known specific symptoms of the suspected neurotoxicant such as neuropathy these can be of great diagnostic value provided of course that the exposure is well-documented and other non-neurotoxic etiologies of these symptoms can be excluded.

Specific symptoms and signs, such as motor weakness, sensory loss, gate disturbances, or impaired cognitive functions reflect damage to the central

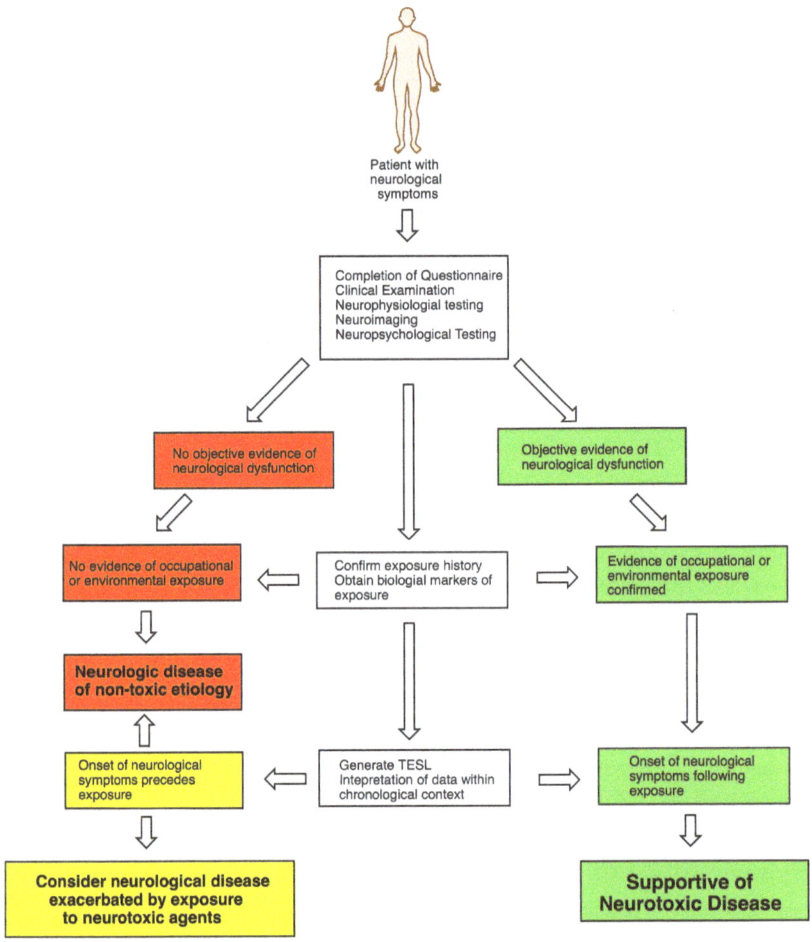

Fig. 1 Diagnostic algorithm for neurotoxic disease. Neurodegenerative diseases of non-toxic etiologies may be exacerbated by exposures to neurotoxicants.

and peripheral nervous systems. Neurologic signs and symptoms are physiological expressions of damage to specific regions of the nervous system irrespective of the etiology. Therefore, the OEM specialist must differentiate symptoms due to neurotoxicant exposure from those due to primary neurologic illness. Clinical manifestations are typically consistent with known effects of specific neurotoxicants. For example, the most common symptoms associated with exposure to the solvent toluene are toxic encephalopathy and tremor (Feldman et al., 1999) while the most common symptoms seen in patients exposed to *n*-hexane is peripheral neuropathy (Ruff et al., 1981).

It is not uncommon for neurotoxicants to produce constellations of symptoms that resemble those due to neurodegenerative disease (Feldman et al., 1999; Ratner and Fitzgerald, 2017; Rutchik and Ratner, 2018). It is also not uncommon for neurotoxicants to share mechanism of action in common with those implicated in neurodegenerative disease which can lead to unmasking and worsening of neurological disease (Gamache et al., 2019; Ratner, 2016; Ratner and Feldman, 2005; Ratner and Fitzgerald, 2017; Ratner et al., 2014, 2018; Rutchik and Ratner, 2019). Diseases, such as diabetes which is associated with neuropathy, as well as renal and hepatic disorders which reduce biotransformation and clearance of toxicants can also contribute to the clinical effects of exposure (Feldman and Pransky, 1988). For example, impaired hepatic function can result in the bioaccumulation manganese within the globus pallidus resulting in bilateral hyperintensities on T1-weighted MR images similar to that which is seen in welders (Ratner and Fitzgerald, 2017). Therefore, laboratory evidence of diabetes and liver or kidney disease or any other condition which may produce similar symptoms should be carefully related to the patient's symptoms, signs and test results.

Some patients exposed to neurotoxicants who do not develop objective neurological signs and symptoms may nevertheless experience anxiety or Post Traumatic Stress Disorder (PTSD) which are also chronologically related to the exposure circumstances. The risk for PTSD is greater when the exposure is perceived as life-threatening (Schottenfeld and Cullen, 1985). These patients may seek consultation with an OEM specialist due to their history of exposure, their concerns about persistent adverse effects of neurotoxicants, and/or their need for reassurance that they will fully recover with cessation of exposure. In such cases, the lack of objective measures of neurological dysfunction or cognitive deficits at the time of examination does not negate the connection between the exposure circumstances and the onset of the patient's symptoms and complaints of increased anxiety or PTSD.

4. Evaluation of cranial nerve functions for neurotoxicity

The cranial nerves are vulnerable to the adverse effects of neurotoxicants. Ingestion of ethylene glycol has been associated with bilateral dysfunction of cranial nerves II, V, VII, VIII, IX, X and XII (Lewis et al., 1997). Damage to the olfactory nerve is associated with a reduced sense of smell and taste. The olfactory system is vulnerable to the effect of neurotoxicants because chemoreceptors are directly exposed to the environment (Doty, 2015; Ratner et al., 2005). Chronic exposure to neurotoxic aromatic solvents and gases such as hydrogen sulfide is associated with olfactory adaptation which may persist after cessation of exposure (Hirsch and Zavala, 1999; Ratner et al., 2005; Tvedt et al., 1991). Olfactory dysfunction due to neurotoxicants must be differentiated from that due to primary neurodegenerative diseases, head trauma, and respiratory illnesses such as COVID-19 (Hirsch and Zavala, 1999; Lechien et al., 2020; Peabody and Tinklenberg, 1985; Ratner et al., 2005). Occupational exposure to industrial solvents is associated with red-green color vision loss indicative of optic nerve damage (Mergler et al., 1988). Damage to optic nerves is also associated with methanol poisoning (Liberski et al., 2022). The effects of neurotoxicants on the optic nerve which must be differentiated from those due to a demyelinating disease such as MS can be documented with visual evoked potentials (VEPs) (Gong et al., 2003; Hageman et al., 1999; Pratt et al., 2000). Loss of vision in one eye is typically indicative of a non-neurotoxic etiology. Symptoms due to neurotoxic damage to the oculomotor nerves include bilateral diplopia and ptosis. Exposure to cholinesterase inhibitors (e.g., organophosphates) is associated with oculomotor nerve symptoms including bilateral diplopia and ptosis (Basrai et al., 2019). The iris sphincter and dilator muscles which control pupil size are innervated by cholinergic parasympathetic and adrenergic sympathetic fibers of the third cranial nerve originating in the Edinger-Westphal nucleus. Miosis is a common clinical manifestation of exposure to cholinesterase inhibitors (Maksimović et al., 2021). Unilateral symptoms of oculomotor nerve damage are typically suggestive of a non-neurotoxic etiology such as cranial nerve palsy due to stroke, trauma, or neoplasm. Trauma is the most common cause of adult onset trochlear nerve damage and therefore, this should be suspected in all patients presenting with diplopia due to trochlear palsy. Other etiological factors implicated in trochlear nerve damage include vascular disease and neoplasms. The breakdown of trichloroethylene to dichloroacetylene is associated with damage to the motor and sensory branches of the trigeminal nerve (Feldman et al., 1970). Trigeminal nerve injuries present as ocular pain, bilateral facial numbness and paresis (Buxton, 1967; Feldman et al., 1970). Other symptoms include eyelid ptosis, flattening of nasolabial folds, inability to furrow the brow, weakness in smiling and showing teeth and anesthesia of the intraoral cavity and tongue. The neuropathy underlying these symptoms appears to be due to damage to trigeminal nerve nuclei (Buxton, 1967). Recovery following cessation of exposure may be incomplete (Buxton, 1967; Feldman et al., 1970). The differential diagnosis includes trigeminal neuralgia which causes painful sensations on one side of the face. Symptoms of sixth cranial nerve dysfunction include horizonal gaze nystagmus. Damage to the abducens nerve is associated with Wernicke-Korsakoff syndrome due to thiamine deficiency, stroke, and demyelinating diseases such as Multiple Sclerosis (MS) which should all be included in the differential diagnosis. Facial nerve palsy has reported following accidental application of copious amounts of the neurotoxic topical anesthetic phenol into the ear (Santa Maria et al., 2016). Ototoxic effects of exposure to mixed solvents has been reported (Moshe et al., 2002). Most symptoms associated with ototoxic and vestibulotoxic chemicals are mediated by effects sensory hair cells of the inner ear (Campo et al., 2013). Vertigo, hearing loss and tinnitus are common clinical manifestations associated with exposures to ototoxic and vestibulotoxic chemicals such as styrene, the aminoglycoside antibiotics and the antineoplastic agent cisplatin (Sliwinska-Kowalska et al., 2020; Zhang et al., 2022). Neurotoxic damage to the ninth cranial nerve is associated with reduced gag and swallowing reflexes which are reportedly reduced in patients exposed to trichloroethylene (Buxton, 1967). The effects of neurotoxicants on the tenth cranial nerve included reduced gag and swallowing reflexes which are reportedly reduced in patients exposed to trichloroethylene (Buxton, 1967). The eleventh cranial nerve innervates the sternocleidomastoid and trapezius muscles. Symptoms of eleventh cranial nerve dysfunction include muscle weakness and torticollis. Symptoms of twelfth cranial nerve injury include tongue atrophy. When present symptoms of hypoglossal nerve dysfunction may be indicative of bulbar onset Amyotrophic Lateral Sclerosis (Goldstein and Edelman, 2018).

Clinical pearl

Pupillary abnormalities are often found in normal patients. The examiner must consider more common causes of exam abnormalities when performing a detailed examination.

5. Evaluation of motor system function for neurotoxicity

The neuromuscular examination evaluates the functional integrity of the motor system. The motor system includes the pyramidal and extrapyramidal systems. Both systems target lower motor neurons but the clinical manifestations associated with dysfunction of each system are unique. Symptoms of extrapyramidal system dysfunction include parkinsonism and dystonia while symptoms of pyramidal system dysfunction are characterized by weakness and paralysis. Upper and lower motor neurons are both vulnerable to the adverse effects of neurotoxicants. One of the most well-studied examples of a neurotoxicant affecting both upper and lower motor neurons is triorthocresyl phosphate (Vasilescu and Florescu, 1980). These patients presented with symptoms resembling those seen in amyotrophic lateral sclerosis. Functional recovery of lower motor neurons was documented based on nerve conduction studies but their symptoms persisted due to irreversible damage to upper motor neurons. The long peripheral nerve fibers that innervate muscles of the upper and lower extremities are vulnerable to disruption of axonal transport caused by electrophilic neurotoxicants such as ethylene oxide, carbon disulfide, *n*-hexane and acrylamide (Graham, 1999). Muscle atrophy due to denervation is seen in patients with neuropathy due to *n*-hexane (Feldman, 1999). Exposure to arsenic among smelter workers is associated with clinical and subclinical symptoms and signs of neuropathy which correlate with increased content of arsenic in urine, hair and nails (Feldman et al., 1979). When present, bilateral clinical evidence of denervation is often more consistent with peripheral nerve disease due to neurotoxicants than are unilateral signs and symptoms which may be due to a compression neuropathy. However, because compression neuropathies can be exacerbated by exposures to neurotoxicants it is very important to carefully document the exposure circumstances when an interaction is suspected. Deficits of motor control such as problems with coordination and initiation and cessation of movement are associated with damage to the extrapyramidal system including brainstem nuclei, the basal ganglia and the cerebellum. Typical clinical findings include tremor, bradykinesia, dysmetria, and ataxia. Damage to midbrain and brainstem structures also manifests as problems with coordinated control of conjugate eye movement, articulation of speech, and swallowing. Neurotoxicants implicated in disorders of motor control include carbon monoxide, mercury, manganese, and toluene (Cersosmimo and Koller, 2006; Criswell et al., 2012; Damasceno and de Capitani, 1994; Ikeda and Tsukagoshi, 1990; Lau et al., 2021; McCullough et al., 2001; Rutchik et al., 2021). Occupational exposure to ethylene oxide is associated with weakness in the distal extremities (Kuzuhara et al., 1983). Exposure to solvent mixtures has also been implicated in motor and sensory neuropathies (Jovanović et al., 2004; Moshe et al., 2002). When two or more neurotoxicants known to produce neuropathy such as *n*-hexane and methyl-*n*-butyl-ketone have been used by the patient, their combined effect, rather than that of either individually, should be given primary consideration (Graham, 1999). However, identifying the specific culpable agent is difficult because the associations between many solvents and the biologic effects of these are not well defined and the adverse effects of solvent-solvent interactions at both the toxicokinetic and toxicodynamic levels have not been fully elucidated (Rutchik and Wittman, 2004).

Movement disorders including tremor due to medications, substance abuse or withdrawal must also be differentiated from those due to occupational or environmental exposure to neurotoxicants (Bhidayasiri, 2005; Deik et al., 2012). Some patients with a history of exposure to neurotoxicants will present with functional movement disorders characterized by one or more symptoms of altered voluntary motor function which are incompatible with recognized neurological or medical conditions (Aybek and Perez, 2022). These patients may complain of balance problems which are much worse than those observed on the clinical examination. Unusual gait patterns, extreme slowness of movement and tremor that improves with distraction or shows entrainment with rhythmic tapping of another body part are indicative of a functional movement disorder (Aybek and Perez, 2022).

Clinical pearl

The skilled examiner will be able to assess a movement disorder with distraction and consider how a tremor, for example changes with use of the opposite limb or body part. Hoover sign is a well-known finding in patients with functional weakness. Increasingly patients are presenting to medical clinics with functional neurological syndromes.

6. Evaluation of sensory nervous system function for neurotoxicity

Patients with sensory nervous system dysfunction due to neurotoxicants often present with complaints of numbness and paresthesia (Jovanović and Jovanović, 2004; Pan et al., 2017). As with motor symptoms, the sensory disturbances due to neurotoxicants are often bilateral with a stocking and glove distribution (Feldman, 1999). Although many patients report an onset of symptoms in their lower extremities, workers who frequently immerse their hands in neurotoxic solvents may report an onset of symptoms in their upper extremities. These patients may show evidence of dermatitis as well (Brigham and Landrigan, 1985). Occupational exposure to ethylene oxide is associated with paresthesia and cutaneous sensory loss in the limbs (Kuzuhara et al., 1983). Solvent mixtures have been implicated in sensorimotor neuropathy (Jovanović et al., 2004; Moshe et al., 2002). Isolated sensory symptoms in a single upper or lower extremity should raise suspicions of a non-toxic etiology. Because compression neuropathies can be exacerbated by neurotoxicants when an interaction is suspected it is important to carefully document the exposure circumstances.

Clinical pearl

It is important to identify if a sensory finding is in a dermatomal pattern supportive of peripheral neuropathy, entrapment neuropathy, a central condition or one not consistent with either.

7. Evaluation of reflexes for neurotoxicity

Reflex testing is used to determine the intactness and functional continuity of afferent and efferent nerve pathways. The symmetry and intensity of the reflexes provide valuable diagnostic information. Bilateral reductions in knee and ankle reflex intensity are often seen with toxic neuropathies affecting the lower extremities. Trigeminal nerve reflexes which include the jaw jerk and corneal blink reflex, the latter of which is often reduced in patients with a history of exposure to trichloroethylene or dichloroacetylene (Buxton, 1967; Feldman et al., 1970; Feldman et al., 1992). Increased intensity of reflexes in the extremities suggests loss of inhibitory control of upper motor neurons due to a lesion within central nervous system (CNS). Asymmetric reflexes or a unilateral Babinski sign are suggestive of focal lesions and a non-neurotoxic etiology.

Clinical pearl

Patients who are being evaluated for headache, cognitive issues or cervical spine pain should be thoroughly evaluated with reflex assessment as myelopathy may be a surgical emergency.

8. Evaluation of cognitive functions for neurotoxicity

The OEM specialist needs to be familiar with the types of cognitive changes associated with exposure to neurotoxicants. The OEM specialist also needs to be able to effectively incorporate the results of neuropsychological tests of cognitive function into their differential diagnosis of toxic encephalopathy versus non-neurotoxic disease.

The Mini-Mental Status Examination (MMSE) was developed to assess for profound changes in cognitive function. The MMSE can therefore be used to differentiate dementia from the subtle behavioral effects often associated with neurotoxicant exposure (Ratner and Jabre, 2017). The severe acute changes in cognitive function due to CO poisoning can also be effectively assessed using the MMSE (Amitai et al., 1998; Kao and Nañagas, 2006; Ku et al., 2010; Messier and Myers, 1991). The patient's ability to comprehend and follow directions, perform coordinated motor functions, and perceive sensory stimuli should also be assessed during the intake examination. The MMSE is typically not sensitive enough to effectively evaluate the subtle changes in cognition associated with most cases of toxic encephalopathy of the type associated with chronic low level exposures to mixed solvents. Therefore, all patients exposed to neurotoxic chemicals who present with complaints of cognitive dysfunction should undergo formal neuropsychological testing in order to ascertain whether or not they are suffering from toxic encephalopathy.

Toxic encephalopathies are typically defined based on the duration of the exposure circumstances that gives rise to the cognitive deficits (Ratner et al., 2002; White and Feldman, 1987). *Acute toxic encephalopathy* is used to describe the persistent changes in cognitive function associated with a single serve acute exposure such as occurs with carbon monoxide poisoning. *Chronic toxic encephalopathy* describes the changes in cognition that emerge in association with months to years of exposure to neurotoxicants. Both types of encephalopathy can be associated with changes in mood and affect including depression and anxiety in addition to changes in cognitive function. Depression and anxiety also occur as secondary clinical manifestations following a diagnosis of neurotoxic disease. In such cases the depression and anxiety are situational due to loss of gainful employment or limitation on performance of activities of daily living.

The cognitive domains assessed with formal neuropsychological test batteries should include attention and executive function, memory, psychomotor function, visuospatial ability, language and verbal skills. Formal neuropsychological assessment has several advantages over the aforementioned bedside tests. First, these tests are better suited for detecting subtle changes in cognitive function which often persist following exposure to neurotoxicants. Second, estimate premorbid intellectual abilities based on educational history and work records can be related to the results of neuropsychological testing. Third, neuropsychological testing can detect functional deficits in specific cognitive domains such as speech and language function which may be less vulnerable to neurotoxicants but more vulnerable to localized brain lesions such as those associated with a stroke or neoplasm. Forth, serial neuropsychological assessments can be used to differentiate neurodegenerative disease from toxic encephalopathy based on evidence of functional recovery or progression after cessation of exposure (Ratner et al., 2002; White and Feldman, 1987). Finally, formal neuropsychological testing includes measures of effort which are used to determine the motivation of the patient to perform at their full potential (Ratner and Jabre, 2017; Ratner et al., 2002).

Dynamic cognitive processes such as attention and executive function, memory, psychomotor function and visuospatial ability are particularly vulnerable to the effects of neurotoxicants. *Executive function* is a higher order cognitive process such as social judgment and the ability of a patient to plan and carryout complex tasks and understand proverbs. Patients exposed to neurotoxicants often complain about *attention and concentration* deficits which interfere with their ability to focus on relevant stimuli (Ratner et al., 2020). Patients often describe this symptom as "brain fog." Impaired attention can also interfere with memory function when relevant information is not encoded and stored. Memory involves the encoding and consolidation of information. Neurotoxic chemicals can interfere with both aspects of memory function. Memory function can be divided into short-term and long-term memory. Encoding and consolidation of new memories (anterograde memory function) is more vulnerable to the effects of neurotoxicants than is recall of previously stored information (retrograde memory function). Acute exposure to neurotoxic chemicals and drugs can be associated with acquisition and consolidation deficits mediated by reversible disruptions of neurotransmission (Ratner and Farb, 2022). *Psychomotor* slowing and *visuospatial deficits* are also seen following exposure to certain neurotoxicants. Deficits in motor function induced by exposure to neurotoxic agents can be differentiated into those affecting movement per se and those affecting motor learning or psychomotor function. Slowed psychomotor processing speed due to toxic leukoencephalopathy and bradykinesia associated with parkinsonism can both interfere with the patient's performance on timed neurobehavioral tests (e.g., Digit Symbol). Motor aspects of writing may be affected in those patients with movement disorders such as tremor and bradykinesia. For example, tremor interferes with the patient's ability to draw an Archimedes spiral (Rauh et al., 2015). Language and verbal functioning are relatively resistant to the effects of neurotoxicant exposure. Although akinetic mutism has been reported in patients with carbon monoxide poisoning (Chen et al., 2016; Sakamoto et al., 1998) when present, performance deficits on tests of language and verbal function such as dysarthria and word finding problems (anomia) are often more likely to be consistent with cerebrovascular injury and AD respectively than with toxic encephalopathy (Feldman et al., 1999). (Note: neuropsychological testing is explained in greater detail in a dedicated chapter.)

Clinical pearl

While the goals of this type of evaluation are clinically assessing function, the clinician needs to understand that validity and consistency of testing is mandatory for results. Specific testing methods must be utilized. Differentiating anxiety, depression from cognitive difficulties is also very important for neuropsychological testing.

9. Neuroimaging findings associated with neurotoxicity

Neuroimaging studies are useful in the diagnosis of toxic encephalopathy and neurotoxicant induced parkinsonism. These tests also facilitate the process of ruling out other etiologies of the patient's symptoms such as stroke and neoplasm. Common findings due to neurotoxicant exposure include leukoencephalopathy, cerebral and cerebellar atrophy, and pallidal lesions

(Criswell et al., 2012; Feldman et al., 1999; Filley et al., 2004; Jin et al., 2017). The neuroimaging findings associated with exposure to neurotoxicants are also often bilateral and diffuse (Cersosmimo and Koller, 2006; Criswell et al., 2012; Feldman et al., 1999; Filley et al., 2004; Jin et al., 2017). Focal and unilateral findings suggest stroke, neoplasm, demyelinating or neurodegenerative processes such as AD which is commonly associated with disproportionate atrophy of the medial temporal lobes (Feldman et al., 1999). Periventricular white matter lesions in a subject with a poorly documented exposure or a history of a single low level acute exposure that did not result in a loss of consciousness should raise suspicions of MS (Filley and Kleinschmidt-DeMasters, 2001). (Note: Brain imaging is explained in greater detail in a dedicated chapter).

Clinical pearl

Neuroimaging test results may be sensitive but not specific to the exposure in question and may not be proven to be associated with brain dysfunction. Reviewing the literature on a specific imaging testing method may reveal that the prevalence in the general population has not been identified.

10. Neurophysiological diagnosis

The results of neurophysiological tests serve as objective measures of neurologic function which can also serve as de facto *biological markers of effect*. Electrophysiological changes in nerve conduction velocity, amplitude, and latency as well as changes in motor unit function (e.g., fibrillation potentials) are often measurable before the onset of overt signs and symptoms of neurotoxicity (He et al., 1989). These abnormalities which, often persist after cessation of exposure as well can be used to monitor recovery of function. Electromyographic studies of denervated muscle fibers due to neurotoxicant exposure must be differentiated from those due to ALS (Kuzuhara et al., 1983; Ratner et al., 2018). EMG can be used to document recovery from neurotoxicant induced neuropathy. *For example,* polyphasic potentials are seen on recording of a previously denervated muscle unit which has become reinnervated. Additional neurophysiological metrics sensitive to the central connections of the peripheral and cranial nerves include the: *F-Wave; H-Reflex;* and *blink reflex*. Neurophysiological tests of CNS functioning, include electroencephalography (EEG) and sensory evoked potentials (SEPs). Generalized and paroxysmal changes are common findings on

EEGs of neurotoxicant exposed workers (Sińczuk-Walczak et al., 2001). Leukoencephalopathy due to neurotoxicants such as carbon monoxide is associated with diffuse slowing characterized by increases in delta and theta activity (Xu et al., 2020). However, due to the lack of specificity of observed abnormalities the diagnostic value of quantitative EEG is limited (Keski-Säntti et al., 2008). As with all laboratory tests, the significance of the EEG findings depends on correlation with other clinical information such as renal and hepatic function. Neurotoxicant exposure-induced epileptiform activity must be differentiated from that associated with lesions due to a neoplasm, stroke, or traumatic brain injury. Marked asymmetry of the EEG pattern suggests lateralized neuropathology.

The EEG of patients with a history of neurotoxic exposure who present with psychogenic seizures will be normal during the psychogenic event (Aybek and Perez, 2022). Video EEG monitoring facilitates the diagnosis of psychogenic seizures (Gedzelman and LaRoche, 2014).

The SEPs consist of the visually evoked potentials, and the brainstem auditory evoked potentials; and somatosensory evoked potentials. *Visual evoked potentials* (VEPS) can be used to document the effects of exposure to neurotoxins on the optic nerve (Gong et al., 2003; Hageman et al., 1999; Pratt et al., 2000). *Brainstem auditory evoked potentials (BAEPs)* are useful for detecting sensorineural hearing loss due to ototoxic substances (Hormes et al., 1986; Otto and Fox, 1993). The *blink reflex* which interrogates the integrity of the 5th and 7th cranial nerves has been used to document dysfunction due to TCE exposure (Feldman et al., 1992).

Clinical pearl

Neurophysiological testing data must be reproducible, and methods must be transparent. Results must be consistent with abnormalities identified.

11. Biological markers of exposure

Biological marker data is time sensitive. Whenever a patient presents with a history of occupational or environmental exposures to neurotoxic chemicals the OEM specialist should always attempt to collect relevant urine, blood, and hair samples to document the exposures. Only conventional biological markers of exposure should be used for this purpose; biomarkers that have not approved for the diagnosis and treatment of neurological or neurotoxic disease should not be relied upon. Collection of biological markers is best achieved in conjunction with the use of a set of standard operating procedures designed to ensure that the sample is properly collected and stored (e.g., frozen). All biomarkers should be selected based on their sensitivity and specificity for documenting the specific exposures the patients report experiencing. Because structurally related chemicals share common metabolites it is important to determine the concentration of the parent molecule which is often a more specific biomarker of exposure (Chiu et al., 2007; Cichocki et al., 2016). The clinical relevance of biologic marker data should be interpreted cautiously particularly when the clinical manifestations are inconsistent with the exposure data (Ratner et al., 2020; Rutchik and Ratner, 2018). For example, a single positive test result for a neurotoxic chemical that is known to take months of chronic exposure to produce neurological symptoms should be interpreted with extreme caution. False positives can occur and therefore any questionable test results should be repeated (Rutchik and Ratner, 2018).

Appropriate sample timing in relationship to the exposure circumstance and the known half-life of the specific biological marker of interest must also be considered. Biological exposure indices (BEIs) have been established by the American Conference of Governmental Industrial Hygienists (ACGIH) for common toxic substances found in the workplace (ACGIH, 2021). Relevant specimens should be collected at the end of the shift or end of the workweek according to ACGIH recommendations (ACGIH, 2021). Unfortunately, it is not always possible to obtain biological markers of exposure due to the amount of time that has passed between cessation of exposure and examination of the patient. In such cases, exposure modeling preferably conducted by a certified industrial hygienist (CIH) can be used as a surrogate (Ratner et al., 2018; Reference Manual on Scientific Evidence, 2011).

Blood and urine levels of heavy metals can be useful for monitoring and documenting acute exposures. Because these levels decline quickly with cessation of exposure these biological markers of exposure are not well correlated with neurophysiological evidence of peripheral neuropathy (Thomson and Parry, 2006) or neuroimaging evidence of metal accumulation in the brain (Criswell et al., 2012, 2019). However, assessment for heavy metal poisoning based on non-validated post-chelation challenge urine heavy metal concentration is not a scientifically validated method and therefore according to the American College of Medical Toxicology this approach should not be used (AMCT, 2010; Ratner et al., 2020). Hair and nails samples are useful for documenting chronic and remote occupational and environmental exposures to metals and metalloids such as arsenic (Feldman et al., 1979).

12. Biologic markers of effect

In addition to the concentration of the parent molecule and its metabolites, neurotoxicant exposures are also associated with quantifiable effects on enzymes and tissues and other molecules which can be used as adjunct biomarkers of toxic effects. Biologic markers of effect are measurable biochemical, or physiological alterations within the nervous system that can be associated with an established or potential health impairment or disease. These include markers of early biologic effects, markers of altered structure and function changes more closely related to the development of disease, and markers of clinical disease and can be any qualitative or quantitative change from a baseline or expected level of appearance or performance that indicates impairment or damage resulting from a given chemical exposure (NRC, 1992). In the clinical setting, the results of neurophysiologic studies, neuropsychological tests, and neuroimaging studies serve as biologic markers of neurotoxicant effects which can be correlated chronologically with the exposure history and biological markers of exposure. Biological markers of effect often return toward baseline with cessation of exposure and thus, these have important prognostic value.

These biomarkers of effect are typically sensitive to toxicants and thus, are often the first indicators of changes in biological function which can be used predict the risk for other adverse effects of neurotoxicant exposure and monitor recovery of function with cessation of exposure. For example, lead (Pb) is known to disrupt heme biosynthesis leading to the formation of zinc protoporphyrin (ZPP) (Thomson and Parry, 2006). The blood lead level (BLL) most accurately reflects recent Pb exposures and thus, it cannot be used to reliably predict prior or cumulative exposures. Therefore, blood ZPP should be used in conjunction with the BLL for patients at risk for chronic occupational exposure to Pb. Because an iron deficiency or impaired iron utilization can also lead to ZPP formation it is important to exclude all other possible causes of this biomarker being elevated whenever it is used for this purpose. Along the same lines, the enzymatic activity of acetylcholinesterase and butyrylcholinesterase (a.k.a., pseudocholinesterase) are used as biological markers of effect in patients exposed to organophosphates and/or carbamate insecticides. Because inactivation of acetylcholinesterase can be lethal this biomarker of effect is considered to be a more specific prognostic marker (Masson and Lockridge, 2010; Newmark, 2019).

Peripheral nerve biopsies are also useful biological markers effect. Teased preparations and transverse sections reveal evidence of swollen nodes of Ranvier, demyelination and axonal loss. Ethylene oxide, *n*-hexane, methyl *n*-butyl ketone, and acrylamide neuropathies have all been documented with nerve biopsies (Chang et al., 1993; Davenport et al., 1976; Feldman, 1999; Kuzuhara et al., 1983; Ruff et al., 1981).

While some biomarkers of effect show relatively good specificity others are nonspecific. For example, neurofilament-light chain (NF-L) is a well-known clinical biomarker of many neurodegenerative diseases which is beginning to find use as a biomarker of nervous system injury due to neurotoxicants (Sano et al., 2021). Neurofilaments occur in three sizes neurofilament light (NfL), medium (NfM), and heavy (NfH) chains. Damage to nerves leads to the release of these proteins into the cerebrospinal fluid or blood where the concentrations of these proteins can be measured to monitor disease onset and progression (Karteri et al., 2022; Sandelius et al., 2018). However, while NfL chain is a highly sensitive and early biomarker of neuronal injury including that due to neurotoxicants, this biomarker also has low specificity and therefore, while it can provide additional information about disease progression and recovery within specific clinical contexts it should not be relied upon in isolation to formulate a clinical diagnose of neurotoxic disease (Karteri et al., 2022; Sano et al., 2021).

13. Biomarkers of susceptibility

The advent of commercial genetic testing has generated a new source of valuable data about individual susceptibility. Additional information gained from genetic testing aids the OEM specialist in understanding individual factors that can account for bioaccumulation of the parent molecule and/or its metabolites which in turn can influence the nature and severity of the symptoms experienced by the patient. The most well-known of these genetic risk factors is the genetic polymorphism for enzyme aldehyde dehydrogenase-2 which influences the detoxification of aldehydes. In the absence of genetic tests results the race of the patient and their sensitivity to ethanol (the later determined by asking if their face feels flush when they drink alcoholic beverages) can be used to predict whether they have this specific polymorphism. A positive history for alcohol abuse does not negate the concomitant effects of chronic exposure to organic solvents (Juntunen, 1982).

Other genetic polymorphisms implicated in gene-environment interactions that increase risk for adverse effects of neurotoxicants include the gene that codes for the enzyme glutathione-*S*-transferase (GST) which catalyzes the conjugation of reactive electrophilic alkylating agents with glutathione (Spencer and Chen, 2021; Lecomte et al., 2006). An adenine to guanine transition at codon 105 in exon 5 of the GSTP1 gene results in amino acid substitution from isoleucine to valine which decreases enzymatic activity (Lecomte et al., 2006).

The neurotoxicant trichloroethylene is metabolized by aldehyde dehydrogenase. Metabolism of ethanol depends on aldehyde dehydrogenase and thus, consumption of ethanol by workers and others exposed to trichloroethylene results in an increase in acetaldehyde and facial flushing (Pardys and Brotman, 1974). Patients who lack functional aldehyde dehydrogenase also experience facial flushing on exposure to this chlorinated hydrocarbon solvent (Watanabe, 2011).

Butyrylcholinesterase is a polymorphic nonspecific esterase that plays an important role in the metabolism of carbamates and organophosphates. Single nucleotide polymorphisms in the butyrylcholinesterase gene that contribute to enzymatic activity have been implicated in susceptibility to organophosphate poisoning (Dimov et al., 2012; Loewenstein-Lichtenstein et al., 1995; Howard et al., 2010). Paraoxonase-1 is another polymorphic enzyme implicated in susceptibility to organophosphate poisoning (Costa et al., 1990; Lee et al., 2003; Haley et al., 2022).

OEM specialists should also be aware of polygenic risk factors play a role in complex neurological disease such as parkinsons. When combined with an increased risk for occupational exposure to toxic chemicals polygenic risk factors can serve to increase individual susceptibility to the adverse effects of neurotoxicants (Cliff et al., 2017). For example, the combination of a positive family history for PD and decreased glutathione-*S*-transferase (e.g., activity due to genetic polymorphisms has been implicated in a younger age at onset of this PD among men exposed to herbicides (Wilk et al., 2006).

14. Interpretation of biologic marker data

In order for biological marker data to be useful in the clinical setting the appropriate sample must have been collected at the appropriate time in relationship to the exposure circumstances and the sample must have been handled and stored correctly. It is not possible for OEM specialists to effectively interpret data from biological specimens that were not properly handled. To avoid potential problems with biomarker analysis it is highly recommended that samples be processed by highly qualified laboratories with expertise in forensic toxicology such as NMS Labs in Horsham, Pennsylvania.

The *sensitivity* and *specificity* of biologic marker data must also be considered when interpreting results as these pertain to symptoms of neurotoxic disease. Sensitive biomarkers of exposure can be increased before adverse effects on nervous system function occur. If this limitation of the biomarker data is not effectively explained it can cause a patient to become unnecessarily concerned. Conversely, biological markers with high specificity such as inhibition of acetylcholinesterase activity correlate more closely with specific clinical manifestations (e.g., cholinergic crisis) and therefore, these can also have much greater prognostic value. With respect to markers of susceptibility when possible these should be related to the anticipated adverse effects such as facial flushing due to an aldehyde dehydrogenase polymorphism (Watanabe, 2011) or increased risk for peripheral neuropathy due a reduced ability to detoxify reactive metabolites (Spencer and Chen, 2021).

15. Time-exposure-symptoms-line

A *Time-Exposure-Symptoms-Line (TESL)* should be constructed from the exposure history and symptom onset data provided by the patient (Feldman et al., 1999; Ratner et al., 2020) (Fig. 2). The TESL is a visual tool that helps explain the chronological relationship between the exposure history and the patient's symptoms. The OEM specialist should document the beginning and end dates for the patient's exposure to all known neurotoxicants even if these exposures were not specifically associated with the onset of any symptoms. The TESL should include dates of sampling for biomarkers of exposure as well as the dates of environmental monitoring for air and/or water concentrations of all known neurotoxic chemicals. If the patient has coworkers or family members who also experienced symptoms the dates when their symptoms emerged should also be included in the TESL.

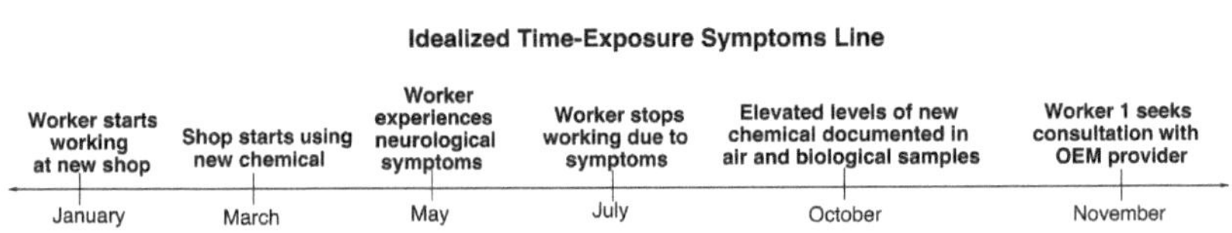

Fig. 2 Time-exposure symptoms line.

16. Conclusions

The clinical manifestations associated with exposure to neurotoxic chemicals must be effectively differentiated from those due to non-neurotoxic neurological disorders. This goal can be achieved with a careful intake interview and analysis of the chronological relationship between the reported exposure circumstances and the emergence of symptoms. The clinical findings should be interpreted in the context of developmental, academic, occupational, personal and familial medical and psychiatric histories. The use of a diagnostic algorithm and TESL will aid the OEM specialist in organizing data in a visual format which facilitates diagnosis and can be used to explain the observed relationships between the exposure and symptoms to the patient, other medical specialists, and even to legal counsel if necessary for a worker compensation claim. Follow-up examinations and serial neuropsychological testing are useful for differentiating neurotoxic disease from progressive neurodegenerative processes, documenting recovery and to directing treatment and vocational retraining strategies.

References

American College of Medical Toxicology, 2010. American College of Medical Toxicology position statement on post-chelator challenge urinary metal testing. J. Med. Toxicol. 6 (1), 74–75. https://doi.org/10.1007/s13181-010-0039-0.

American Conference of Governmental Industrial Hygienists (ACGIH), 2021. TLVs and BEIs. Threshold Limit Values for Chemical Substances and Physical Agents and Biological Exposure Indices. ACGIH Worldwide, Cincinnati, OH.

Amitai, Y., Zlotogorski, Z., Golan-Katzav, V., Wexler, A., Gross, D., 1998. Neuropsychological impairment from acute low-level exposure to carbon monoxide. Arch. Neurol. 55 (6), 845–848. https://doi.org/10.1001/archneur.55.6.845. PMID: 9626776.

Aybek, S., Perez, D.L., 2022. Diagnosis and management of functional neurological disorder. BMJ 376, o64. https://doi.org/10.1136/bmj.o64.

Basrai, Z., Koh, C., Celedon, M., Warren, J., 2019. Clinical effects from household insecticide: pyrethroid or organophosphate toxicity? BMJ Care Rep., e230966.

Bhidayasiri, R., 2005. Differential diagnosis of common tremor syndromes Postgrad. Med. J., 756–762.

Brigham, C.R., Landrigan, P.J., 1985. Safety and health in boatbuilding and repair. Am. J. Ind. Med. 8 (3), 169–182. https://doi.org/10.1002/ajim.4700080303.

Buxton, H., 1967. Polyneuritis cranialis associated with industrial trichloroethylene poisoning. J. Neurol. Neurosurg. Psychiatry 30 (6), 511–518. https://doi.org/10.1136/jnnp.30.6.511.

Campo, P., Morata, T.C., Hong, O., 2013. Chemical exposure and hearing loss. Dis. Mon. 59 (4), 119–138. https://doi.org/10.1016/j.disamonth.2013.01.003.

Cersosmimo, M.G., Koller, W.C., 2006. The diagnosis of manganese-induced parkinsonism. Neurotoxicology, 340–346.

Chang, C.M., Yu, C.W., Fong, K.Y., Leung, S.Y., Tsin, T.W., Yu, Y.L., Cheung, T.F., Chan, S.Y., 1993. N-hexane neuropathy in offset printers. J. Neurol. Neurosurg. Psychiatry 56 (5), 538–542. https://doi.org/10.1136/jnnp.56.5.538.

Chen, S.Y., Lin, C.C., Lin, Y.T., Lo, C.P., Wang, C.H., Fan, Y.M., 2016. Reversible changes of brain perfusion SPECT for carbon monoxide poisoning-induced severe akinetic mutism. Clin. Nucl. Med. 41 (5), e221-7. https://doi.org/10.1097/RLU.0000000000001121.

Chiu, W.A., Micallef, S., Monster, A.C., Bois, F.Y., 2007. Toxicokinetics of inhaled trichloroethylene and tetrachloroethylene in humans at 1 ppm: empirical results and comparisons with previous studies. Toxicol. Sci. 95 (1), 23–36. https://doi.org/10.1093/toxsci/kfl129.

Cichocki, J.A., Guyton, K.Z., Guha, N., Chiu, W.A., Rusyn, I., Lash, L.H., 2016. Target organ metabolism, toxicity, and mechanisms of trichloroethylene and perchloroethylene: key similarities, differences, and data gaps. J. Pharmacol. Exp. Ther. 359 (1), 110–123. https://doi.org/10.1124/jpet.116.232629.

Cliff, J., Jorgensen, A.L., Lord, R., Azam, F., Cossar, L., Carr, D.F., Pirmohamed, M., 2017. The molecular genetics of chemotherapy-induced peripheral neuropathy: a systematic review and meta-analysis. Crit. Rev. Oncol. Hematol. 120, 127–140. https://doi.org/10.1016/j.critrevonc.2017.09.009.

Costa, L.G., McDonald, B.E., Murphy, S.D., Omenn, G.S., Richter, R.J., Motulsky, A.G., Furlong, C.E., 1990. Serum paraoxonase and its influence on paraoxon and chlorpyrifos-oxon toxicity in rats. Toxicol. Appl. Pharmacol. 103 (1), 66–76. https://doi.org/10.1016/0041-008x(90)90263-t.

Criswell, S.R., Perlmutter, J.S., Huang, J.L., Golchin, N., Flores, H.P., Hobson, A., Aschner, M., Erikson, K.M., Checkoway, H., Racette, B.A., 2012. Basal ganglia intensity indices and diffusion weighted imaging in manganese-exposed welders. Occup. Environ. Med. 69 (6), 437–443. https://doi.org/10.1136/oemed-2011-100119.

Criswell, S.R., Nielsen, S.S., Warden, M.N., Flores, H.P., Lenox-Krug, J., Racette, S., Sheppard, L., Checkoway, H., Racette, B.A., 2019. MRI signal intensity and parkinsonism in manganese-exposed workers. J. Occup. Environ. Med. 61 (8), 641–645. https://doi.org/10.1097/JOM.0000000000001634.

Damasceno, B.P., de Capitani, E.M., 1994. Cerebellar atrophy related to chronic exposure to toluene. Case Rep. Arq. Neuropsiquiatr. 52 (1), 90–92. https://doi.org/10.1590/s0004-282x1994000100017.

Davenport, J.G., Farrell, D.F., Sumi, M., 1976. "Giant axonal neuropathy" caused by industrial chemicals: neurofilamentous axonal masses in man. Neurology 26 (10), 919–923. https://doi.org/10.1212/wnl.26.10.919.

Deik, A., Saunders-Pullman, R., Luciano, M.S., 2012. Substance of abuse and movement disorders: complex interactions and comorbidities. Curr. Drug Abuse Rev. 5 (3), 243–253. https://doi.org/10.2174/1874473711205030243.

Dimov, D., Kanev, K., Dimova, I., 2012. Correlation between butyrylcholinesterase variants and sensitivity to soman toxicity. Acta Biochim. Pol. 59 (2), 313–316.

Doty, R.L., 2015. Neurotoxic exposure and impairment of the chemical senses of taste and smell. Handb. Clin. Neurol. 131, 299–324. https://doi.org/10.1016/B978-0-444-62627-1.00016-0.

Feldman, R.G., 1999. Approach to Diagnosis. Occupational and Environmental Neurotoxicology. Lippincott-Raven, New York, pp. 13–29 (Chapter 3).

Feldman, R.G., Pransky, G.S., 1988. Neurologic considerations in worker fitness evaluation. In: Comprehensive Approaches to Worker Fitness and Risk Evaluations, Occupational Medicine: State of the Art Review. Vol. 3. Hanley and Belfus Medical Publishers, Philadelphia, pp. 299–308.

Feldman, R.G., Mayer, R.M., Taub, A., 1970. Evidence for peripheral neurotoxic effect of trichloroethylene. Neurology 20 (6), 599–606. https://doi.org/10.1212/wnl.20.6.599.

Feldman, R.G., Niles, C.A., Kelly-Hayes, M., Sax, D.S., Dixon, W.J., Thompson, D.J., Landau, E., 1979. Peripheral neuropathy in arsenic smelter workers. Neurology 29 (7), 939–944. https://doi.org/10.1212/wnl.29.7.939.

Feldman, R.G., Niles, C., Proctor, S.P., Jabre, J., 1992. Blink reflex measurement of effects of trichloroethylene exposure on the trigeminal nerve. Muscle Nerve 15 (4), 490–495. https://doi.org/10.1002/mus.880150413.

Feldman, R.G., Ratner, M.H., Ptak, T., 1999. Chronic toxic encephalopathy in a painter exposed to mixed solvents. Environ. Health Perspect. 107 (5), 417–422. https://doi.org/10.1289/ehp.99107417.

Filley, C.M., Kleinschmidt-DeMasters, B.K., 2001. Toxic leukoencephalopathy. N. Engl. J. Med. 345 (6), 425–432. https://doi.org/10.1056/NEJM200108093450606.

Filley, C.M., Halliday, W., Kleinschmidt-DeMasters, B.K., 2004. The effects of toluene on the central nervous system. J. Neuropathol. Exp. Neurol. 63 (1), 1–12. https://doi.org/10.1093/jnen/63.1.1.

Gamache, P.L., Haj Salem, I., Roux-Dubois, N., Le Bouthillier, J., Gan-Or, Z., Dupré, N., 2019. Exposure to pesticides and welding hastens the age-at-onset of Parkinson's disease. Can. J. Neurol. Sci. 46 (6), 711–716. https://doi.org/10.1017/cjn.2019.248.

Gedzelman, E.R., LaRoche, S.M., 2014. Long-term video EEG monitoring for diagnosis of psychogenic nonepileptic seizures. Neuropsychiatr. Dis. Treat. 10, 1979–1986. https://doi.org/10.2147/NDT.S49531.

Goldstein, E.D., Edelman, B.H., 2018. Misdiagnosis: hypoglossal palsy mimicking bulbar onset amyotrophic lateral sclerosis. eNeurologicalSci 14, 6–8. https://doi.org/10.1016/j.ensci.2018.11.028. PMID: 30555942; PMCID: PMC6275204.

Gong, Y., Kishi, R., Kasai, S., Katakura, Y., Fujiwara, K., Umemura, T., Kondo, T., Sato, T., Sata, F., Tsukishima, E., Tozaki, S., Kawai, T., Miyama, Y., 2003. Visual dysfunction in workers exposed to a mixture of organic solvents. Neurotoxicology 24 (4–5), 703–710. https://doi.org/10.1016/S0161-813X(03)00034-2.

Graham, D.G., 1999. Neurotoxicants and the cytoskeleton. Curr. Opin. Neurol. 12 (6), 733–737. https://doi.org/10.1097/00019052-199912000-00013.

Hageman, G., van der Hoek, J., van Hout, M., van der Laan, G., Steur, E.J., de Bruin, W., Herholz, K., 1999. Parkinsonism, pyramidal signs, polyneuropathy, and cognitive decline after long-term occupational solvent exposure. J. Neurol. 246 (3), 198–206. https://doi.org/10.1007/s004150050334.

Haley, R.W., Kramer, G., Xiao, J., Dever, J.A., Teiber, J.F., 2022. Evaluation of a gene-environment interaction of *PON1* and low-level nerve agent exposure with Gulf War illness: a prevalence case-control study drawn from the U.S. Military Health Survey's national population sample. Environ. Health Perspect. 130 (5), 57001. https://doi.org/10.1289/EHP9009.

He, F.S., Zhang, S.L., Wang, H.L., Li, G., Zhang, Z.M., Li, F.L., Dong, X.M., Hu, F.R., 1989. Neurological and electroneuromyographic assessment of the adverse effects of acrylamide on occupationally exposed workers. Scand. J. Work Environ. Health 15 (2), 125–129. https://doi.org/10.5271/sjweh.1878.

Hightower, J.M., Moore, D., 2003. Mercury levels in high-end consumers of fish. Environ. Health Perspect. 111 (4), 604–608. https://doi.org/10.1289/ehp.5837.

Hirsch, A.R., Zavala, G., 1999. Long-term effects on the olfactory system of exposure to hydrogen sulphide. Occup. Environ. Med. 56 (4), 284–287. https://doi.org/10.1136/oem.56.4.284.

Hormes, J.T., Filley, C.M., Rosenberg, N.L., 1986. Neurologic sequelae of chronic solvent vapor abuse. Neurology 36 (5), 698–702. https://doi.org/10.1212/wnl.36.5.698.

Howard, T.D., Hsu, F.C., Grzywacz, J.G., Chen, H., Quandt, S.A., Vallejos, Q.M., et al., 2010. Evaluation of candidate genes for cholinesterase activity in farmworkers exposed to organophosphorus pesticides: association of single nucleotide polymorphisms in BCHE. Environ. Health Perspect. 118 (10), 1395–1399. https://doi.org/10.1289/ehp.0901.

Ikeda, M., Tsukagoshi, H., 1990. Encephalopathy due to toluene sniffing. Report of a case with magnetic resonance imaging. Eur. Neurol. 30 (6), 347–349. https://doi.org/10.1159/000117371.

Jin, J.H., Lee, E.S., Choi, J.Y., Kim, J.S., 2017. Isolated cerebellar atrophy due to rodenticide (β-fluoroethyl acetate) intoxication. J. Neurol. Sci. 373, 208–209. https://doi.org/10.1016/j.jns.2016.12.068.

Jovanović, J., Jovanović, M., 2004. Neurotoksicni efetki organskih rastvaraca kod radnika industrije boja i lakova [neurotoxic effects of organic solvents among workers in paint and lacquer manufacturing industry]. Med. Pregl. 57 (1–2), 22–25. https://doi.org/10.2298/mpns0402022j. Serbian.

Jovanović, J.M., Jovanović, M.M., Spasić, M.J., Lukić, S.R., 2004. Peripheral nerve conduction study in workers exposed to a mixture of organic solvents in paint and lacquer industry. Croat. Med. J. 45 (6), 769–774.

Juntunen, J., 1982. Alcoholism in occupational neurology: diagnostic difficulties with special reference to the neurological syndromes caused by exposure to organic solvents. Acta Neurol. Scand. Suppl. 92, 89–108.

Kao, L.W., Nañagas, K.A., 2006 Mar. Toxicity associated with carbon monoxide. Clin. Lab. Med. 26 (1), 99–125. https://doi.org/10.1016/j.cll.2006.01.005.

Karteri, S., Bruna, J., Argyriou, A.A., Mariotto, S., Velasco, R., Alemany, M., Kalofonou, F., Alberti, P., Dinoto, A., Velissaris, D., Stradella, A., Cavaletti, G., Ferrari, S., Kalofonos, H.P., 2022. Prospectively assessing serum neurofilament light chain levels as a biomarker of paclitaxel-induced peripheral neurotoxicity in breast cancer patients. J. Peripher. Nerv. Syst. https://doi.org/10.1111/jns.12493.

Keski-Säntti, P., Kovala, T., Holm, A., Hyvärinen, H.K., Sainio, M., 2008. Quantitative EEG in occupational chronic solvent encephalopathy. Hum. Exp. Toxicol. 27 (4), 315–320. https://doi.org/10.1177/0960327107082231.

Ku, H.L., Yang, K.C., Lee, Y.C., Lee, M.B., Chou, Y.H., 2010. Predictors of carbon monoxide poisoning-induced delayed neuropsychological sequelae. Gen. Hosp. Psychiatry 32 (3), 310–314. https://doi.org/10.1016/j.genhosppsych.2009.11.005.

Kuzuhara, S., Kanazawa, I., Nakanishi, T., Egashira, T., 1983. Ethylene oxide polyneuropathy. Neurology 33 (3), 377–380. https://doi.org/10.1212/wnl.33.3.377.

Lau, Y.H., Mawardi, A.S., Zain, N.R., Viswanathan, S., 2021. Toluene-induced leukodystrophy from glue sniffing. Pract. Neurol. 21 (5), 439–441. https://doi.org/10.1136/practneurol-2021-002942.

Lechien, J.R., Chiesa-Estomba, C.M., De Siati, D.R., Horoi, M., Le Bon, S.D., Rodriguez, A., et al., 2020. Olfactory and gustatory dysfunctions as a clinical presentation of mild-to-moderate forms of the coronavirus disease (COVID-19): a multicenter European study. Eur. Arch. Otorhinolaryngol. 277 (8), 2251–2261. https://doi.org/10.1007/s00405-020-05965-1.

Lecomte, T., Landi, B., Beaune, P., Laurent-Puig, P., Loriot, M.A., 2006. Glutathione S-transferase P1 polymorphism (Ile105Val) predicts cumulative neuropathy in patients receiving oxaliplatin-based chemotherapy. Clin. Cancer Res. 12 (10), 3050–3056. https://doi.org/10.1158/1078-0432.CCR-05-2076.

Lee, B.W., London, L., Paulauskis, J., Myers, J., Christiani, D.C., 2003. Association between human paraoxonase gene polymorphism and chronic symptoms in pesticide-exposed workers. J. Occup. Environ. Med. 45 (2), 118–122. https://doi.org/10.1097/01.jom.0000052953.59271.e1.

Lewis, L.D., Smith, B.W., Mamourian, A.C., 1997. Delayed sequelae after acute overdoses or poisonings: cranial neuropathy related to ethyleneglycol ingestion. Clin. Pharmacol. Ther. 61 (6), 692–699. https://doi.org/10.1016/S0009-9236(97)90105-3.

Liberski, S., Kaluzny, B.J., Kocięcki, J., 2022. Methanol-induced optic neuropathy: a still-present problem. Arch. Toxicol. 96 (2), 431–451. https://doi.org/10.1007/s00204-021-03202-0.

Loewenstein-Lichtenstein, Y., Schwarz, M., Glick, D., Nørgaard-Pedersen, B., Zakut, H., Soreq, H., 1995. Genetic predisposition to adverse consequences of anti-cholinesterases in 'atypical' BCHE carriers. Nat. Med. 1 (10), 1082–1085. https://doi.org/10.1038/nm1095-1082.

Maksimović, Ž.M., Jović-Stošić, J., Vučinić, S., Perković-Vukčević, N., Vuković-Ercegović, G., Škrbić, R., Stojiljković, M.P., 2021. Acute organophosphate and carbamate pesticide poisonings - a five-year survey from the National Poison Control Center of Serbia. Drug Chem. Toxicol. 7, 1–9. https://doi.org/10.1080/01480545.2021.2012481.

Masson, P., Lockridge, O., 2010. Butyrylcholinesterase for protection from organophosphorus poisons: catalytic complexities and hysteretic behavior. Arch. Biochem. Biophys. 494 (2), 107–120. https://doi.org/10.1016/j.abb.2009.12.005.

McCullough, J.E., Dick, R., Rutchik, J., 2001. Chronic mercury exposure examined with a computer-based tremor system. J. Occup. Environ. Med. 43 (3), 295–300. https://doi.org/10.1097/00043764-200103000-00022.

Mergler, D., Bélanger, S., De Grosbois, S., Vachon, N., 1988. Chromal focus of acquired chromatic discrimination loss and solvent exposure among printshop workers. Toxicology 49 (2–3), 341–348. https://doi.org/10.1016/0300-483x(88)90017-0.

Messier, L.D., Myers, R.A., 1991. A neuropsychological screening battery for emergency assessment of carbon-monoxide-poisoned patients. J. Clin. Psychol. 47 (5), 675–684. https://doi.org/10.1002/1097-4679(199109)47:5<675::aid-jclp2270470508>3.0.co;2-h.

Moshe, S., Bitchatchi, E., Goshen, J., Attias, J., 2002. Neuropathy in an artist exposed to organic solvents in paints: a case study. Arch. Environ. Health 57 (2), 127–129. https://doi.org/10.1080/00039890209602927.

National Research Council (NRC), 1992. Environmental Neurotoxicology. National Academy Press, Washington DC.

Newmark, J., 2019. Therapy for acute nerve agent poisoning: an update. Neurol Clin Pract. 9 (4), 337–342. https://doi.org/10.1212/CPJ.0000000000000641.

Otto, D.A., Fox, D.A., 1993. Auditory and visual dysfunction following lead exposure. Neurotoxicology 14 (2–3), 191–207. Summer-Fall.

Pan, J.H., Peng, C.Y., Lo, C.T., Dai, C.Y., Wang, C.L., 2017. n-Hexane intoxication in a Chinese medicine pharmaceutical plant: a case report. J. Med. Case Rep. 11 (1), 120. https://doi.org/10.1186/s13256-017-1280-9.

Pardys, S., Brotman, M., 1974. Letter: trichloroethylene and alcohol: a straight flush. JAMA 229 (5), 521–522.

Peabody, C.A., Tinklenberg, J.R., 1985. Olfactory deficits and primary degenerative dementia. Am. J. Psychiatry 142 (4), 524–525. https://doi.org/10.1176/ajp.142.4.524c.

Pratt, H., Karim, N., Bleich, N., Mittelman, N., 2000. Short latency visual evoked potentials in occupational exposure to organic solvents. Neurophysiol. Clin. 30 (5), 306–312. https://doi.org/10.1016/s0987-7053(00)00230-6.

Ratner, M.H., 2016. A critical review of the interactions between genetics and neurotoxicant exposure in age at onset of neurodegenerative disease. Curr. Top. Toxicol. 12, 1–10.

Ratner, M.H., Farb, D.F., 2022. Probing the neural circuitry targets of neurotoxicants *in vivo* through high density silicon probe brain implants. Front. Toxicol. 4. https://doi.org/10.3389/ftox.2022.836427.

Ratner, M.H., Feldman, R.G., 2005. Environmental toxins and Parkinson's disease. In: Pfeiffer, R.F., Ebadi, M. (Eds.), Parkinson's Disease. CRC Press, Boca Raton, pp. 51–62.

Ratner, M.H., Fitzgerald, E., 2017. Understanding of the role of manganese in parkinsonism and Parkinson disease. Neurology 88 (4), 338–339. https://doi.org/10.1212/WNL.0000000000003543.

Ratner, M.H., Jabre, J.F., 2017. Neurobehavioral Toxicology. Reference Module in Neuroscience and Biobehavioral Psychology. Elsevier, B.V., pp. 1–11.

Ratner, M.H., Feldman, R.G., White, R.F., 2002. Neurobehavioral toxicology. In: Ramachandran, V.S. (Ed.), Encyclopedia of the Human Brain. Vol. 3. Elsevier Science, New York, pp. 423–439.

Ratner, M.H., Nelson, S., Feldman, R.G., 2005. Smell and Taste Disorders. In: Levy, B.S., Wagner, G.R., Rest, K.M., Weeks, J.L. (Eds.), Preventing Occupational Disease and Injury. American Public Health Association Publication, pp. 454–457.

Ratner, M.H., Farb, D.H., Ozer, J., Feldman, R.G., Durso, R., 2014. Younger age at onset of sporadic Parkinson's disease among subjects occupationally exposed to metals and pesticides. Interdiscip Toxicol. 7 (3), 123–133. https://doi.org/10.2478/intox-2014-0017.

Ratner, M.H., Jabre, J.F., Ewing, W.M., Abou-Donia, M., Oliver, L.C., 2018. Amyotrophic lateral sclerosis-a case report and mechanistic review of the association with toluene and other volatile organic compounds. Am. J. Ind. Med. 61 (3), 251–260. https://doi.org/10.1002/ajim.22791.

Ratner, M.H., Ewing, W.M., Rutchik, J.S., 2020. Neurological effects of chronic occupational exposure to alcohol mists and vapors in a machinist. Toxicology Communications 4 (1), 43–48. https://doi.org/10.1080/24734306.2020.1768341.

Rauh, V.A., Garcia, W.E., Whyatt, R.M., Horton, M.K., Barr, D.B., Louis, E.D., 2015. Prenatal exposure to the organophosphate pesticide chlorpyrifos and childhood tremor. Neurotoxicology 51, 80–86. https://doi.org/10.1016/j.neuro.2015.09.004.

Reference Manual on Scientific Evidence: Third Edition. Washington, DC: The National Academies Press. 2011. doi: 10.17226/13163.

Ruff, R.L., Petito, C.K., Acheson, L.S., 1981. Neuropathy associated with chronic low level exposure to n-hexane. Clin. Toxicol. 18 (5), 515–519. https://doi.org/10.3109/15563658108990277.

Rutchik, J., Ratner, M.H., 2018. Is it possible for late-onset schizophrenia to masquerade as manganese psychosis? J. Occup. Environ. Med. 60 (4), e207–e209. https://doi.org/10.1097/JOM.0000000000001296.

Rutchik, J., Ratner, M.H., 2019. Should age at onset of Parkinsonism be the end point of interest in investigations of the link between exosomal α-synuclein and manganese exposure in welders? J. Occup. Environ. Med. 61 (12), 530–531. https://doi.org/10.1097/JOM.0000000000001720.

Rutchik, J.S., Wittman, R.I., 2004. Neurologic issues with solvents. Clin. Occup. Environ. Med. 4 (4), 621–656. https://doi.org/10.1016/j.coem.2004.03.010. v-vi.

Rutchik, J., Bowler, R.M., Ratner, M.H., 2021. A rare case of Holmes tremor in a worker with occupational carbon monoxide poisoning. Am. J. Ind. Med. 64 (5), 435–449. https://doi.org/10.1002/ajim.23235.

Sakamoto, K., Murata, T., Omori, M., Kimura, H., Nishio, M., Murata, I., Koshino, Y., Itoh, S., Ishii, Y., Isaki, K., 1998. Clinical studies on three cases of the interval form of carbon monoxide poisoning: serial proton magnetic resonance spectroscopy as a prognostic predictor. Psychiatry Res. 83 (3), 179–192. https://doi.org/10.1016/s0925-4927(98)00038-9.

Sandelius, Å., Zetterberg, H., Blennow, K., Adiutori, R., Malaspina, A., Laura, M., Reilly, M.M., Rossor, A.M., 2018. Plasma neurofilament light chain concentration in the inherited peripheral neuropathies. Neurology 90 (6), e518–e524. https://doi.org/10.1212/WNL.0000000000004932.

Sano, T., Masuda, Y., Yasuno, H., Shinozawa, T., Watanabe, T., Kakehi, M., 2021. Blood neurofilament light chain as a potential biomarker for central and peripheral nervous toxicity in rats. Toxicol. Sci. 185 (1), 10–18. https://doi.org/10.1093/toxsci/kfab122.

Santa Maria, P.L., Corrales, C.E., Sevy, A.B., Jackler, R.K., 2016. Iatrogenic phenol injury causing facial paralysis with tympanic membrane and ossicular necrosis. Otol. Neurotol. 37 (4), 385–387. https://doi.org/10.1097/MAO.0000000000000979.

Schottenfeld, R.S., Cullen, M.R., 1985. Occupation-induced posttraumatic stress disorders. Am. J. Psychiatry 142 (2), 198–202. https://doi.org/10.1176/ajp.142.2.198.

Shin, H.W., Chung, S.J., 2012. Drug-induced parkinsonism. J. Clin. Neurol. 8 (1), 15–21. https://doi.org/10.3988/jcn.2012.8.1.15.

Sińczuk-Walczak, H., Jakubowski, M., Matczak, W., 2001. Neurological and neurophysiological examinations of workers occupationally exposed to manganese. Int. J. Occup. Med. Environ. Health 14 (4), 329–337.

Sliwinska-Kowalska, M., Fuente, A., Zamyslowska-Szmytke, E., 2020. Cochlear dysfunction is associated with styrene exposure in humans. PLoS One. 15 (1), e0227978. https://doi.org/10.1371/journal.pone.0227978.

Spencer, P.S., Chen, X., 2021. Role of protein adduction in toxic neuropathies of exogenous and endogenous origin. Toxics 9 (5), 98. https://doi.org/10.3390/toxics9050098.

Thomson, R.M., Parry, G.J., 2006. Neuropathies associated with excessive exposure to lead. Muscle Nerve 33 (6), 732–741. https://doi.org/10.1002/mus.20510.

Tvedt, B., Skyberg, K., Aaserud, O., Hobbesland, A., Mathiesen, T., 1991. Brain damage caused by hydrogen sulfide: a follow-up study of six patients. Am. J. Ind. Med. 20 (1), 91–101. https://doi.org/10.1002/ajim.4700200109.

Vasilescu, C., Florescu, A., 1980. Clinical and electrophysiological study of neuropathy after organophosphorus compounds poisoning. Arch. Toxicol. 43 (4), 305–315. https://doi.org/10.1007/BF00366186.

Watanabe, H., 2011. Hypersensitivity syndrome due to trichloroethylene exposure: a severe generalized skin reaction resembling drug-induced hypersensitivity syndrome. J. Dermatol. 38 (3), 229–235. https://doi.org/10.1111/j.1346-8138.2010.01155.x.

White, R.F., Feldman, R.G., 1987. Neuropsychological assessment of toxic encephalopathy. Am. J. Ind. Med. 11 (4), 395–398. https://doi.org/10.1002/ajim.4700110402.

Wilk, J.B., Tobin, J.E., Suchowersky, O., Shill, H.A., Klein, C., Wooten, G.F., Lew, M.F., Mark, M.H., Guttman, M., Watts, R.L., Singer, C., Growdon, J.H., Latourelle, J.C., Saint-Hilaire, M.H., DeStefano, A.L., Prakash, R., Williamson, S., Berg, C.J., Sun, M., Goldwurm, S., Pezzoli, G., Racette, B.A., Perlmutter, J.S., Parsian, A., Baker, K.B., Giroux, M.L., Litvan, I., Pramstaller, P.P., Nicholson, G., Burn, D.J., Chinnery, P.F., Vieregge, P., Slevin, J.T., Cambi, F., MacDonald, M.E., Gusella, J.F., Myers, R.H., Golbe, L.I., 2006. Herbicide exposure modifies GSTP1 haplotype association to Parkinson onset age: the GenePD study. Neurology 67 (12), 2206–2210. https://doi.org/10.1212/01.wnl.0000249149.22407.d1.

Xu, S.Y., Li, C.X., Li, L.Y., Song, Y., Sui, Y., 2020. Wallerian degeneration of bilateral cerebral peduncles after acute carbon monoxide poisoning. BMC Neurol. 20 (1), 96. https://doi.org/10.1186/s12883-020-01677-5.

Zhang, X., Trendowski, M.R., Wilkinson, E., Shahbazi, M., Dinh, P.C., Shuey, M.M., Center, R.G., Feldman, D.R., Hamilton, R.J., Vaughn, D.J., Fung, C., Kollmannsberger, C., Huddart, R., Martin, N.E., Sanchez, V.A., Frisina, R.D., Einhorn, L.H., Cox, N.J., Travis, L.B., Dolan, M.E., 2022. Pharmacogenomics of cisplatin-induced neurotoxicities: hearing loss, tinnitus, and peripheral sensory neuropathy. Cancer Med. https://doi.org/10.1002/cam4.4644. Mar 23.

CHAPTER FOUR

Aerotoxic syndrome: A new occupational disease caused by contaminated cabin air?

G. Hageman[a,*], S.J. Mackenzie Ross[b], J. Nihom[a], and G. van der Laan[c]
[a]Medical Spectrum Twente, Hospital Enschede, The Netherlands
[b]Research Department of Clinical, Educational and Health Psychology, University College London, London, United Kingdom
[c]Foundation Learning and Developing Occupational Health, Leusden, The Netherlands
*Corresponding author: e-mail address: g.hageman6@kpnplanet.nl

Contents

Advances in Neurotoxicology, Volume 7
ISSN 2468-7480
https://doi.org/10.1016/bs.ant.2022.04.001

1. Introduction

The term "aerotoxic syndrome" was proposed in 1999 by Balouet and Winder to describe a constellation of symptoms reported by pilots and cabin crew following exposure to hydraulic fluids, engine oil and pyrolysis products during flight (Balouet and Winder, 1999). Cabin air on most commercial aircraft is supplied from the engines or auxiliary power unit. Air is drawn from outside and then circulated around the engine where it is heated and pressurized to a breathable level before being "bled off" and pumped into the aircraft (see Fig. 1).

Cabin air can be contaminated by a large number of toxic substances such as ingestion of exhaust from other aircraft, engine oil and hydraulic fluid leaks, combusted and pyrolised materials (e.g., carbon monoxide) and application of de-icing fluids and insecticides. Contamination can occur in two ways: (1) chronic repeated low-level exposure due to jet aircraft design which allows continuous leaking of lubrication oil past wet seals, occurring

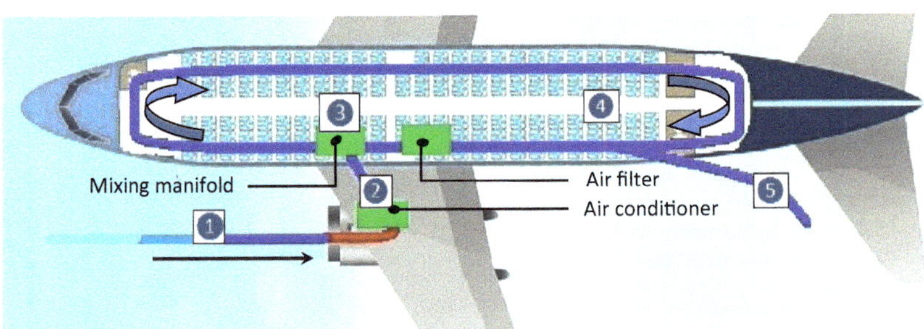

1. Air enters the compressor stage of the aircraft's jet engine, where it becomes very hot as it is pressurized.

2. The hot, compressed air then passes to air conditioning units where it is cooled.

3. The outside air is passed to the mixing manifold, where it is mixed 50/50 with recirculated cabin air that has been cleaned with high-efficiency filters.

4. Air from the mixing manifold is supplied to the cabin on a continuous basis from the overhead outlets.

5. As outside air enters the aircraft, an equal amount of air is discharged. The air in the entire cabin is replaced every two to three minutes.

Fig. 1 Bleed air.

without a detectable fume event, (2) acute high-level exposure associated with odors and sometimes a visible haze, due to changes of engine power or air supply configuration. Less frequently due to general system failure such as seal bearing leakage, worn seals, overfilling of oil and of hydraulic reservoirs (Howard et al., 2017; Michaelis et al., 2021).

Over the last two decades several case studies and health surveys have been published describing adverse health effects reported by aircrew and passengers, thought to be related to exposure to contaminated air. Symptoms frequently reported by aircrew following a fume event include eye-, nose-, throat- and airway-irritation, nausea, headaches, dizziness, cognitive impairment and fatigue. Occasionally aircrew, including pilots, have been completely incapacitated by fumes and were forced to perform emergency landings. Once exposure stops often resolution of symptoms is reported. In case of repeated exposure, it takes longer for aircrew to recover after each episode. Once a threshold is passed in number of repeated episodes, symptoms may persist and become chronic (Mackenzie Ross, 2008).

Although the aviation industry recognizes that some people experience acute symptoms following fume events—with transient discomfort—debate continues as to whether exposure to contaminated air can cause long-term health effects (UK Civil Aviation Authority Guidance for passengers and healthcare professionals, 2015). Air quality monitoring studies, undertaken over the last decade, have examined a small number of chemicals and gasses and concluded levels were too low to cause adverse effects (Crump et al., 2011; de Ree et al., 2014; Wolkoff et al., 2016). However, these on-board studies have a number of methodological weaknesses, the most important being that most studies did not capture a fume event and only a limited number of chemicals were studied, without due consideration of the synergistic effects of chemical mixtures. Also, these studies did not explore the possibility that cumulative low-level exposure may be as harmful as a single episode of high level exposure. Finally individual differences in rate or even ability to metabolize certain compounds, were not taken into consideration. This means the health risks may be higher than assessed hitherto.

Nonetheless many researchers continue to argue that the health risks of exposure to contaminated cabin air are negligible and they suggest chronic ill health is more likely to be caused by factors, unrelated to toxic gasses such as psychological factors (including mass hysteria, nocebo effects, psychosomatic conditions). In addition, it is argued that specific work-related factors have to be taken into consideration, such as shift work, jet lag, relative hypoxia, exposure to cosmic radiation and pressure changes (Cable, 2003; COT report, 2007; van Drongelen et al., 2015; Wolkoff et al., 2016). The symptoms reported by aircrew following a contaminated air event are diverse, occurring in many other medical conditions and common even in the general population. Similar symptoms have been reported by aircrew who were not exposed to contaminated air (McNeely et al., 2018).

Furthermore, many researchers object to the term 'aerotoxic' because it implies causation when this is yet to be established beyond any degree of certainty. Diagnostic tests like brain imaging and investigations of peripheral nerve conduction and the autonomic nervous system often fail to reveal abnormalities in many individuals (de Graaf et al., 2014). Considerable scientific uncertaintly still surrounds this issue and given the limited number of high-quality research studies undertaken over the last two decades, we consider further research is warranted. In this chapter we review the possibility that:

1) Aerotoxic syndrome is the result of long-term, low-level exposure to a mixture of toxic compounds, including various organophosphates, solvents and carbonmonoxide.
2) These toxic compounds are attached to ultrafine particles in bleed air, crossing oil seals, and entering cabin air, and are absorbed by the human body via inhalation.
3) Symptoms develop in pilots and cabin crew, especially in those with a reduced capacity to metabolize toxic compounds as a result of genetic polymorphisms.

2. Incidence of fume events

Fume events are defined as any incident associated with smells, smoke or mist inside an airplane with impairment to the health of occupants (Anderson, 2021). Events reported range from hazy smoke entering the cabin to the smell of dirty socks or burned rubber. This definition may include a fog of ultrafine (<100 nm) droplets in bleed air, which can occur under standard operating conditions (Amiri et al., 2017; Michaelis et al., 2021). Fume events can occur at any time during operation of the aircraft, although most likely (80%) during take-offs and landings. In 2013, the International Air Transport Association published on its analysis of safety reports from more than 150 airlines, covering 31 million flights during the period 2008–2012 (IATA, 2013). Of these, 3444 reports were categorized as "smoke, fumes and odor", which is equivalent to about one event per 10.000 flights. Engine-related fumes were the most commonly reported source. Smell events do not necessarily indicate the presence of toxic contaminants, often the source of a smell event cannot be identified (Anderson, 2021; Rosenberger, 2018; Schuchardt et al., 2019). The incidence of fume events has been difficult to determine because cabin air is not routinely monitored for the presence of chemical contaminants. In addition, underreporting of symptoms is common among aircrew. Nevertheless, the aviation industry accepts that fume events occur on commercial aircraft; some aircraft types (BAe 146, Airbus 320 and Boeing 757) report more events than others. In a study of 33 aircraft models in the period 2007–2012, the highest fume event frequency for one aircraft model (Boeing 757) was 7.8 per 10.000 flights (Shehadi et al., 2016). The reported frequency of fume events varies widely, see Table 1. Regulatory authorities estimate fume events happen on 0.2–0.5% of flights. If only those events are counted cabin crew judged worthwhile to report, one event occurs in every 460 flights (Cannon, 2016).

Table 1 Frequency of reported fume events and reported complaints.

Author, year	Period	Aircraft type	Frequency	Remarks
National Research council, 2002	One year, 2002	Boeing 737, BAe 146	0,09 (B737)-3,88 (BAe 146) per 1000 flights	
Michaelis, 2003	Questionnaire October 2001: 106 respondents; 93 of 106 after fume event	B757 (102 of 106) B737 A320	1674 fume events; most reported complaints: ENT irritation, headache, dizziness, impaired concentration, fatigue, nausea, vomiting, diarrhoea	Only 61 of 1674 incidents reported
Singh, 2004	1998–2003	Australian airforce	0,5 incidents per 1000 flight hours	
Murawski, 2008	January 2006-June 2007	US commercial aircraft	0,86 events per day	470 incident reports; 51 % caused flight delay or cancellation

Continued

Table 1 Frequency of reported fume events and reported complaints.—cont'd

Author, year	Period	Aircraft type	Frequency	Remarks
IATA, 2013	2008–2012	150 airlines, 31 million flights	1 event per 1000 flights	3444 incident reports, smoke, fumes and odor
German Federal Bureau of Aircraft Accidents Investigation, 2014	2006–2013	Transport aircraft, Airbus 319–330, Boeing 737–757, Antonov	663 fume events, (460 smell-, 188 smoke) in 6 years	In 15 cases health impairment
Heutelbeck et al., 2016	2015	Not reported	11 flight crew, admitted < 5 days after fume event: headache, fatigue, respiratory distress, cogn. impairment, salivation, gastro-intestinal symptoms	Low NTE activity (3,1-6,3 nmol, normal values 3-24) in 10 of 11 flightcrew
Shehadi, 2016	2007–2012	33 aircraft models Airbus, Boeing, Douglas, Mc Donnell-Douglas	Mean: 2,1 incidents per 10.000 flights; Boeing 757: 7,8 per 10.000 flights	Events widely distributed across all commercial aircraft models
Klerlein and Loizeau, 2019	Questionnaire after fume event, March 2017-March 2018. Of 610 aircrew involved, 376 responded, 354 were included	81 flights Air France	185 of 354: no complaints 169 of 354 (48 %): ENT irritation, headache, dyspnea, lightheadedness	Only 53 of 169: medical investigation by health care provider
Anderson, 2021	2002–2011	US airlines	1–2 events per 10.000 flights, max 8 per 10.000 flights	12.417 fumes-smoke incident reports

Aircrew complain of ill health in around 30% of fume event reports (Michaelis, 2016). In a one-year survey of Air France from March 2017 to 2018, 81 flights with an onboard fume event were noted, involving 610 aircrew; of these 376 (61.6%) answered an online questionnaire. The mean delay between fume event and response was 11 days. Symptoms reported included eye, nose and throat irritation, headache, lightheadedness and dyspnea. Only 14% of aircrew attended the airport medical service or occupational health service (Klerlein and Loizeau, 2019). More severe symptoms were reported in another survey of fume related health problems, involving 106 pilots who were members of the British Airline Pilots Association (Michaelis, 2003). These symptoms were fatigue, decrease in performance, confusion, impaired concentration and short term memory, muscle weakness and diarrhea. In this study the duration of fume events lasted seconds to hours. Similar symptoms were reported in a German study of 11 flight crew members, investigated within 5 days after a fume event. The main symptoms were fatigue, headache, impaired concentration and memory, gastrointestinal symptoms, paresthesias, chestpain or dyspnea, vertigo, disorientation, sweating, sensation of thirst and dry mouth but also increased salivation (Heutelbeck et al., 2016). Another German study reported fume events in the cabin of transport aircraft. A total of 648 fume events occurred between 2006 and 2013, 460 smell and 188 smoke events, with health impairment in 15 cases (German Federal Bureau of Aircraft Accident investigation, 2014).

3. Bleed air contaminants

Hydraulic fuels and engine oils contain a large number of potentially toxic chemicals. A recent cabin air quality study, carried out on 177 flights involving 61 bleed air aircraft and eight bleed air free aircraft B787, quantified approximately 100 compounds (Schuchardt et al., 2019). Over the last decade various air monitoring studies have been conducted to identify these chemical compounds released into the cabin following an engine oil leak (see Tables 2 and 3). Measurements include various organophosphate compounds (tricresylphosphate (TCP), triphenylphosphate (TPhP), dibutylphenylphosphate (DBPP) and tri-n-butyl phosphate) and volatile organic compounds (toluene, benzene, formaldehyde, acetic acid, acetone, ethanol) (Solbu et al., 2011; Winder and Balouet, 2002). Cabin air may be polluted by pyrolytic degradation products of engine oil, or due to thermal decomposition of hydraulic fluids. Other causes of contamination arise from

Table 2 In-flight measurements conducted since 2011.

Author, year	Target compound	Number of flights/ samples	Aircraft type	Findings	Remarks
Crump et al., 2011	VOC, CO, TCP, ToCP, TnBP, toluene, xylene	100 flights	B757 cargo, B757, A319, A320/1, BAe 146,	In >95% of cabin air samples no total TCP or ToCP detected. Highest ToCP level of 22.8 $\mu g/m^3$ during climb of a B757. Mean total TCP 0.14 $\mu g/m^3$. Maximum 28.5 $\mu g/m^3$. Highest TnBP level 21.8 $\mu g/m^3$. Low concentrations of toluene, xylene	(1) The suitability of the sampling and analysis method applied for TCP determination were questioned. (2) No fume events recorded
Denola et al., 2011	TCP and other organo-phosphates	46 flights, 78 samples	Three types from Australian Defense force: fighter trainer, cargo transporter and fighter bomber, 46 different aircraft	In 11 of 78 samples TCP levels detected from 0.12–4.99 $\mu g/m^3$. In 2 samples high TCP levels of 21.7 and 51.3 $\mu g/m^3$. Other peaks were TiBP, TnBP and TPhP.	During the flight with TCP level 51.3 $\mu g/m^3$ smoke in the cabin was reported.
Solbu et al., 2011	Organo-phosphates	95 Samples from 47 flights	40 Aircraft (jet engine and propellor airplanes, helicopters)	TCP detected in 4% of air samples, only in propellor airplanes and helicopters; levels ranging from <75 ng/m^3 to 290 ng/m^3. TnBP in all jet engine and propellor airplanes and in 58% of helicopters: 24–4100 ng/m^3. DBPP in the jet and propellor airplanes: <75 to 310 ng/m^3. TPhP in one propellor airplane: 110 ng/m^3	TCP detected in 39% of the wipe samples and in all 6 HEPA filters from the jet engine airplanes, 1.1–4.3 ng/g per hour. No ToCP detected in air and wipe samples.

Spengler et al., 2012	TCP, VOC, CO, ozone, carbon-dioxide	83 Commercial flights	6 Aircraft models (2 Airbus, 4 Boeing, from 3 US airlines	TCP was detected in 1 of 71 samples. Acetaldehyde was found in 81% of 70 samples, acetone in 79%, acrolein in 71%, formaldehyde in 49%, propionaldehyde in 17%	ToCP not detected
Houtzager et al., 2013	TCP	80 Air sampes	4 B737–700 s, 3 B737–800 s, 3B737–900 s, and from 2 B737–700 s and 800 s while running the APU.	In 37 of 80 samples TCP; 0.5 to 155 ng/m^3, average 6.9 ng/m^3. Highest TCP levels during climb and descent.	TCP levels in wipes 0.01 to 0.06 ng/cm^2
Guan et al., 2014	VOC	107 Flights, 76 China domestic and 31 inter-national flights	A320, A321, B737, A330, B747, B777	Benzene, toluene, xylene, ketones, aldehydes (acetone), esters, alcohols, alkanes and alkenes were detected	Concentrations varied in the three different flight phases, peaking before take-off or during cruise
Wang et al. (2014a,b)	VOC	84 Air samples, in 5 flight-phases, in 14 flights	B737–800, China domestic flights	19 VOCs were detected. Highest levels for D-limonene, decanal, nonanal, toluene, xylene, acetic acid, benzene and 6-methyl-5-hepten-2-one	Only one third of VOC emission attributed to fuels, oil, combustion, 29% to (drink)services, 10% to perfumes, cosmetics, cleaning agents

Continued

Table 2 In-flight measurements conducted since 2011.—cont'd

Author, year	Target compound	Number of flights/ samples	Aircraft type	Findings	Remarks
Rosenberger et al., 2016	Aldehydes	353 Samples, 108 flights	26 Aircraft, 11 A380, 15 A321	Formaldehyde 0.4–44 $\mu g/m^3$ (median 4.9) and acetaldehyde 0.3–90 $\mu g/m^3$ (median 4.6) were detected	15 Smell events were reported, measurements in these phases did not differ from phases without smell events
Schuchardt et al., 2019	Aldehydes,VOC, CO, carbon-dioxide, organophosphates, including TCP	840 Samples, in 5 flight phases, in 177 flights	A320, A321, A340, A380, B747, B767, B787	Mean total VOC concentration: 122–370 $\mu g/m^3$; low formaldehyde and acetaldehyde. Organophosphates: 0.74–5.36 $\mu g/m^3$. Highest: TnBP, and in A321 TCiPP, low TCP	Increased levels of propylene glycol in winters due to de-icing. TCP also detected in B787. (non-bleed-air)

TnBP = tri-n-butyl phosphate, TiBP = tro-isobutylphosphate, TPhP = triphenylphosphate, DBPP = dibutylphenylphosphate, TCiPP = trichloroisopropylphosphate, VOC = volatile organic compounds, CO = carbonmonoxide.

Table 3 Summary of exposure data from in-flight measurement studies conducted since 2011 (in μg/m³).

Contaminent of concern	Mean/range	Maximum concentration value	References
Organophosphates	0.74–5.36	7.45	Crump et al. (2011); Rosenberger et al. (2016); Rosenberger (2018); Denola et al. (2011); Schuchardt et al. (2019); Solbu et al. (2011)
Tricresylphosphate (TCP)	0.14	51.3	
Tri-ortho-cresylphosphate (ToCP)	0.07	22.8	
Tri-n-butylphosphate (TnBP)		21.8	
Diphenylphosphate (DPP)		0.3	
Trichloro-isopropylphosphate (TCiPP)	0.429		
Total Volatile Organic Compounds (T-VOCs)	122–370	7230	Rosenberger (2018)
Aldehydes	21.6 (13–35)	429	Rosenberger (2018); Rosenberger et al. (2016); Chen et al. (2021)
Formaldehyde	0.4–44; median 4.9	134	
Acetaldehyde	0.3–90; median 4.6	234	
Solvents			Wang et al. (2014a,b); Guan et al. (2014); Chen et al. (2021)
Benzene	14.78	77.9	
Ethylbenzene	7.04	45.1	
Toluene	29.84	209.3	
Xylene	4.55	62.9	
Acetone	14	384	
Ozone	0–117	302	Rosenberger et al. (2016); Bekö et al. (2015)
Flame retardants			
Polybrominated phenylesters	0.04–20 ng/m³	2100 ng/m³	Allen et al. (2013)
Carbonmonoxide	0–3 ppm	> 5 ppm	Rosenberger, 2018; Rosenberger et al., 2016
Insecticides: Permethrin, d-Phenothrin	120–300 μg pyrethrins/person	830 μg pyrethrins/person	Pang et al. (2020),
Ultrafine particles	417–100.000 particles/cm³	500.000 particles/cm³	Li et al. (2014); Zhai et al. (2014); Michaelis et al. (2021)

kerosene fumes ingested during taxiing or (waiting for) take-off, and the application of de-icing products or insecticides in the cabin. Cold air drawn from outside is heated to aircraft engine temperatures of more than 500 °C This may alter the composition of the original oil and hydraulic fluids and create other toxic semivolatile compounds such as alkylated TCP, trimethylolpropane phosphate (TMPP) and tri-butyl phosphate (Chen et al., 2021; de Boer et al., 2015). Products of incomplete combustion, such as carbon monoxide are also present in engine oil fumes and potentially neurotoxic. The synergistic effects of these compounds is as yet unknown.

All published air quality studies report detection of small amounts of contaminants in cabin air under normal operating conditions. Yet to date none revealed a fume event of sufficient severity to trigger formal reporting procedures (Harrison and Mackenzie Ross, 2016). Interestingly, mean concentrations of volatile organic compounds and aldehydes are reported to be lower in the B787 aircraft than in the bleed air aircraft.

4. T(*O*)CP and other organophosphates

With proven neurotoxicity TCP is the most comprehensively studied chemical substance in air quality studies. TCP/ToCP was used as an adulterant in Jamaica ginger, a medicinal alcohol extract used for stomach problems, commonly abused as an illicit source of alcohol. In the 1930's in the United States it led to "ginger jake paralysis" in 20.000–50.000 people. Other major toxic outbreaks involving more than 10.000 victims were described in Morocco and China (Wang et al., 1995).

In the aviation industry, this compound is used as an anti-wear, extreme pressure additive in lubricants and hydraulic fluids. In gasoline it is used as a flame retardant and lead scavenger. TCP is a potentially neurotoxic organophosphate (OP) compound which inhibits human butyrylcholinesterase (BChE) as well as acetylcholinesterase (AChE), thus disrupting neuronal function. Although TCP is an organophosphate, it has no application as a pesticide, though the neurotoxic mechanisms are considered to be similar (Costa, 2006). TCP is a mixture of various cresyl isomers, i.e. ortho-, meta- and para-TCP. Tri-ortho-cresylphosphate (ToCP) is more toxic than the meta- and para-forms. Isomers containing mono-ortho-cresyl phosphate are considered to be most toxic, followed by the di-ortho and tri-ortho compounds.

The inhibition of red cell AChE and of plasma BChE by the metabolite of ToCP, 2-(2-cresyl)-4H-1,3,2-benzodioxaphosphorin-2-oxide, (or cresyl saligenic phosphate, CBDP) is probably the most important underlying mechanism for ToCP neurotoxicity. ToCP and especially CBDP are known to affect Neuropathy Target Esterase (NTE) and cause a delayed neuropathy called OPIDN. Non-cholinergic toxicity has been proven in animal experiments with very low concentrations of ToCP. One possible mechanism being impairment of glutamate signaling (Hausherr et al., 2014). The recommended safe exposure limit is considered to be breathing 0.1 mg/m^3 of ToCP for 30 min by a 70 kg individual (Schopfer et al., 2010). The maximum total-TCP concentrations in flight deck air reported in various studies are lower, 50–100 ng/m^3, with exceptions as high as 290 ng/m^3 (Crump et al., 2011; de Boer et al., 2015).

The Dutch Organization for applied scientific research TNO measured concentrations of TCP isomers inside the cockpit during 20 flights of nine different Boeing 737-type aircraft. In addition, wipe samples were taken from the glare shield before and after the flight (de Ree et al., 2014). In 10 out of 20 flights TCP isomers could be detected, with a maximum of 155 ng/m^3 during climb. The wipe samples demonstrated small amounts of TCP, at levels below 0.1 ng/cm^2. ToCP levels were below the limit of detection. In an UK-study cabin air samples were measured in 100 flights of several types of aircraft: Boeing 757 cargo and passenger, BAe146, Airbus A319 and A320. TCP levels could be detected in the range of 0–28.5 μg/m^3, ToCP levels in the range of 1–23 μg/m^3 (Crump et al., 2011). In a few studies data are presented of TCP concentrations during a probable fume event (Rosenberger et al., 2016; Schuchardt et al., 2019). However these maximum TCP concentrations are lower, compared to the Crump data: TCP concentrations of 1.67 μg/m^3were measured, especially in the takeoff phase (Schuchardt et al., 2019).

T(*o*)CP is not detectable in human blood due to rapid metabolism, but there is an assay for TCP based on the active-site serine of butyrylcholinesterase reacting with the active metabolite of TCP, cresyl saligenic phosphate (Liyasova et al., 2011). Similarly new methods have been developed for the quantitation of ortho-cresyl phosphate adducts to BChE in human serum. The metabolite CBDP undergoes hydrolysis to form orthocresylphosphoserine, providing a biomarker of CBDP exposure (Johnson et al., 2015). To quantify the exposure to OP's, including T(*o*)CP, specific metabolites in urine can be measured. In a total of 332 urine samples taken from pilots and cabincrew, who reported fume events during their previous flight, ToCP was under the limit of detection (Schindler et al., 2013). Only one sample contained a low-level of TCP-metabolites.

Metabolite levels of tributylphosphate, tris-(2-chloroethyl) phosphate and triphenylphosphate were found to be significantly higher. This was confirmed in a recent air quality study, in which tri-n-butylphosphate, which originates from the hydraulic system, was the most prominent OP-contaminant (Schuchardt et al., 2019). OP-mean concentrations were in the range of 0.74–5.36 μg/m^3.

The above mentioned concentrations are inconsistent with T(*o*)CP being the sole neurotoxic agent of concern and suggest other OP compounds and other chemical substances or gasses may be to blame.

5. Mechanisms of long-term, low-dose organophosphate neurotoxicity

Three mechanism associated with central and peripheral neurotoxicity of OP's can be differentiated based on symptomatology and time of occurrence.

1) **AChE inhibition** may be involved if OP's account for ill health reported by aircrew following a contaminated air event. This results in accumulation of acetylcholine and overstimulation of the nicotinic and muscarinic ACh receptors leading to cholinergic effects: salivation, miosis, abdominal cramps and in case of high exposure reduced consciousness and flaccid paralysis. OP's deactivate cholinesterases by attaching an alkylphosphate group to the hydroxyl group of a serine residue at the enzyme's active site. Recovery from such inhibition generally takes 10–14 days (Davies et al., 2000).
2) **Non-cholinergic mechanisms.** However OP's can damage the nervous system as a result of non-cholinergic mechanisms. This most often occurs after repetitive exposure but can even result from a single intoxication. For example inhibition of NTE can cause OP-induced delayed neurotoxicity (OPIDN). The pathological substrate is a Wallerian type (or dying back) axonal degeneration accompanied by secondary demyelinisation. The most distal portion of the longest nerve tracts are affected in both the central and peripheral nervous system (Abou-Donia, 2003). The clinical picture consists of mild sensory disturbances, ataxia, muscle fatigue and twitching. Improvement is inconsistent and may require months or years. The earliest cases of tri-o-cresyl phosphate induced OPIDN have been documented in 1899, attributed to the use of creosote oil for treatment of tuberculosis (Abou-Donia, 1981).

3) **"OP-induced chronic neurotoxicity" (OPICN).** OP's are also known to cause neuropsychiatric symptoms in OPICN. It may result from excessive cholinergic activity, or non-cholinergic mechanisms as OPs are known to affect other neurotransmitters, including those involved in mood regulation. OP's also affect other physiological processes such as axonal transport and mitochondrial function and may initiate a neuro-inflammatory response by release of cytokines from activated microglia and astrocytes, with apoptotic neuronal cell death (Banks and Lein, 2012; El Rahman et al., 2018). This leads to slow neurodegeneration in various brain areas—cortex, cerebellum, (hypo)thalamus, amygdala—resulting in chronic neuropsychiatric, neurologic and behavioral problems (Abou-Donia, 2003). Clinically, OPICN is manifested by fatigue, apathy, confusion, memory deficits, insomnia, and personality changes. Mood disorders may pose a diagnostic challenge, ranging from anxiety and mild depression to suicidal attempt in severe cases (El Rahman et al., 2018). However most cases present with subtle long-term neuropsychological symptoms following a history of cumulative low level exposure without previous signs of acute toxicity (Brown and Brix, 1998; Mackenzie Ross et al., 2007, 2010). Indeed studies by Terry et al., found OP's can cause adverse health effects at exposure levels below those required to cause cholinesterase inhibition (Terry, 2012).

6. Solvents

There are several other potential contaminants in jet oil which may play an aetiological role in aerotoxic syndrome. In-flight measurements of volatile organic compounds have been reported regularly, see Table 1 and 2. In a review of 47 papers, with measurements taken from 2251 flights the most frequently measured compounds were solvents and aldehydes: toluene, benzene, ethylbenzene and formaldehyde (Chen et al., 2021). Average concentrations of benzene at 5.9 $\mu g/m^3$ were higher than the permissible levels of air quality standards. In 107 randomly selected commercial flights, on average 59 different compounds in each flight were detected, with a total of 346 compounds (Guan et al., 2014). The compounds found in highest abundace on all flights were benzene series (e.g. toluene) and acetone. Other volatile compounds such as tetrachloroethene arise from the dry cleaning of fabrics in the cabin, and 1.4 dichlorobenzene evaporates as component of disinfectant for lavatories (Guan et al., 2014). Finally, de-icing fluids—consisting of ethylene- and propylglycols—may contaminate incoming engine-air (Solbu et al., 2011).

Exposure to organic solvents occurs in a wide range of circumstances. In industrialized countries, the prevalence of occupational exposure to solvents is around 8% (Dick, 2006). A small subsample of workers who are exposed to organic solvents for a long time, e.g., daily exposure for 5 years or more (housepainters, car sprayers, shoe manufacturers), have been found to develop Chronic Solvent induced Encephalopathy (CSE). CSE is characterized by symptoms of cognitive impairment (reduced information processing speed, memory and concentration difficulties, fatigue, irritability, and mood changes (van Valen et al., 2012, 2018). In a next paragraph we elaborate on "painter's disease", compared to aerotoxic syndrome.

7. OZONE

Following Federal Aviation Regulations cabin ozone levels should not exceed 250 parts per billion by volume (ppbv) when the aircraft is above 32,000 ft. A time-weighted average ozone concentration should not exceed 100 ppbv for any 3-h period when the aircraft is above 27,000 ft. (Spengler et al., 2012). Although many planes are equipped with "converters", that promote the decomposition of ozone in the cabin air, high levels of ozone are occasionally reported, reaching 256–275 ppbv (Bekö et al., 2015; Bhangar et al., 2008; Spengler et al., 2012). Ozone concentrations are significantly higher for flights on higher latitude routes, versus flights that follow lower latitudes (Spengler et al., 2012). This is due to ozone production from solar ultraviolet radiation that is highest in the tropics, and on the other hand the result of the large-scale air circulation in the stratosphere that transports tropical ozone towards the poles. Exposure to high ozone concentrations is associated with adverse respiratory effects, especially in infants or adults with cardiopulmonary conditions (Bhangar et al., 2008). Most commonly reported symptoms are dry mouth, lips and eye symptoms, fatigue, headache, breathing discomfort and asthma-exacerbation (Bekö et al., 2015).

8. Flame retardants

Brominated flame retardants are added in seats, carpets, plastics and electronics of aircraft to prevent or slow down the outbreak of fire. High concentrations of polybrominated diphenylesters have been reported in dust samples, collected from airplanes (Allen et al., 2013; Strid et al., 2014).

9. Carbon monoxide

Another potential contaminant in cabin air is carbonmonoxide (Solbu et al., 2011). Pilots exposed to fume events, with reductions in acetyl- or butyl cholinesterases and NTE activity (Heutelbeck et al., 2016; Michaelis et al., 2017), also had carboxyhemoglobin levels above the normal range (Michaelis et al., 2017). Carbonmonoxide exposure is well known in aircraft fueling vehicles, or due to exhaust fumes from the combustion engines, and was reported by pilots of small airplanes (Cessna, Piper) in civil aviation. The most common physical symptoms of chronic low-level carbon-monoxide toxicity are headache, dizziness, breathing difficulties, nausea, and flu-like symptoms. These symptoms are non-specific and often dismissed as having a viral cause (Clarke et al., 2012). Carbon monoxide is a neurotoxin that can cause immediate or delayed cognitive impairment (Sykes and Walker, 2016).

10. Insecticides

The recent severe outbreaks of Corona and Zika virus disease showed the important role of airtravel in the rapid spread of newly emerging infections, resulting in pandemics (Eldin et al., 2020; Mangili and Gendreau, 2005). Of major concern are vectors (mosquitos) of malaria, dengue, chikungunya, Zika and yellow fever. Airlines are obligated to spray pesticides inside the aircraft cabin to kill insects that may be on board. Disinsection is undertaken as a public health measure to address the potential threat posed by insects to the health of humans, animals, plants and agriculture and to comply with quarantaine regulations of various countries. The insecticides used are Permethrin and d-Phenothrin, belonging to the synthetic pyrethroids type 1. Disinsection can be applied at several moments of the flight: (1) spraying of the entire internal surfaces of the passenger cabin and cargo at least once in 2 months. This is done shortly before passengers embark the aircraft, using Permethrin 0.2 and 0.5 g/m^2 on floors (excluded are food preparation areas), (2) pre-flight spraying, using Permethrin 2% aerosol, 35 $g/100\,m^3$ (aircrew on board), (3) before the flight takes off, using d-Phenothrin 2% aerosol, 35 $g/100\,m^3$ (passengers and aircrew on board), (4) while the aircraft starts its descent to its destination (passengers and aircrew on board). Of the countries that require disinsection, about half require the insecticides to be applied while passengers are on board, the other half permits the use of an aerosolized spray before passengers are boarded (Sutton et al., 2007). Concerns have been expressed about the efficacy of disinsection for preventing the international spread of vectors via air travel, apart from the possible toxicity for flight crew. However, a systematic review showed that disinsection with the recommended permethrin or d-Phenothrin concentrations was highly effective against a broad range of arthropods (Pang et al., 2020). Only three papers reported symptoms in passengers or crew possibly associated with insecticide spraying in cabins. Two papers concern occupational related symptoms of 12 and 33 aircrew members, respectively (Kilburn, 2003; Sutton et al., 2007). The other paper describes a case-report of a single passenger, who developed anaphylaxis symptoms immediately after application of d-Phenothrin (van den Driessche et al., 2010). Most common symptoms in the case series of 12 flight attendants were respiratory problems, headache, irritability, tingling of hands or feet. In the other case series, 33 self selected flight attendants, who were regularly exposed to synthetic pyrethroids during flight, were compared with 202 non-exposed persons. Using questionnaires and a test battery an association was observed between exposure to pyrethroid insecticides and changes in mood and balance.

The application in aircraft of Permethrin and d-Phenothrin is not considered to pose undue risk to human (World Health Organization, 2019). The evidence linking exposure to insecticide spraying with negative health impact is limited. However, flight attendants working on international flights connected to Australia (where disinsection is mandatory) had significant higher urinary permethrin metabolite levels than those on domestic flights. In pre-flight spraying, exposure by inhalation was the main route (almost 90%) versus oral and dermal exposure. It is stated that disinsection leads to exposure at levels comparable to agricultural pesticide applicators (Wei et al., 2012, 2013).

To reduce on-board transmission of SARS-CoV-2 aircraft manufacturers recommend the use of 70% isopropylalcohol as a disinfectant for surfaces in the cockpit, cabin (tray tables, armrests, seat covers, lavatories) and cargo holds, as well as alcohol-based hand sanitizer dispensers for frequent hand disinfection, near the boarding door.

11. Nanoparticles

Air quality monitoring studies near the Amsterdam Schiphol airport and the Frankfurt International airport have demonstrated that airports are major sources of ultrafine particles (UFPs), due to aircraft exhaust

(Keuken et al., 2015; Ungeheuer et al., 2021). Particulate samples collected near a runway of an international airport in Japan showed that organic compounds in the ambient nanoparticles were dominated by nearly intact forms of jet engine oil (Fushimi et al., 2019). Using liquid chromatography for compound separation, pentaerythritol- and trimethylolpropane esters were detected—base stocks of lubrication oil—as well as jet oil additives, including TCP and trimethylolpropane phosphate (Ungeheuer et al., 2021).

These particles are also present in cabin air. High concentrations of UFPs are to be expected because of the high temperatures up to 1700 °C attained in aircraft jet engines leading to pyrolysis of engine oil (Amiri et al., 2017). Oil contamination in the compressor will result in an aerosol, i.e., a fog of ultrafine (<100 nm) droplets in bleed air, under most operating conditions (Amiri et al., 2017). Oil leaks permit UFPs to cross oil seals used in turbine engines (Howard et al., 2018). Inflight cabin events were associated with ingress of external particles. Major events were defined as 15-min averages of >10.000 particles/cc; minor events as 15-min averages between 500 and 10.000 particles/cc (Spengler et al., 2012). Aerosol composition data showed non-exhaust, non-airport sources, especially contaminated oil.

There are several studies reporting UFP measures in cabin air (Li et al., 2014; Michaelis et al., 2021). In a recent study UFPs were counted on four short-haul flights on Airbus A320 and A319 (Michaelis et al., 2021). It was explored whether particles number was associated with the age of the aircraft engines, ranging from 7 months to 14 years, and with engine or APU power changes in air supply. UFP concentrations ranged from as low as 35 to as high as 97.800 particles/cm^3. Maximum concentrations in the four flights ranged from 31.300–97.800 particles/cm^3. These peak concentrations were reached after a full-power take-off, or when APU bleed air was selected on. Two of these peak concentrations were linked by aircrew to a noticeable oil smell. Following maximum concentrations frequently rapid decreases of UFPs were recorded, as a result of the high air exchange rate in aircraft. The lowest level of UFPs was identified during steady state engine operation, during the cruise phase of flight. Although seal leakage is expected to increase with sealing wear, particle numbers were not associated with the age of the engine of these aircraft. It was concluded that counting UFPs may have potential as a rapid means to identify bleed air contamination (Michaelis et al., 2021). In a systematic review of five studies sampling 148 flights, the number of ultrafine particles showed a wide variation, counting 417–100.000/cm^3 (Hayes et al., 2021). Peaks of particle concentrations were found in climbing, descending and cruising phases in several flights (Spengler et al., 2012). At elevated power levels, median particle size is in the range of 50–70 nm, incidentally particles larger than 2000 nm are identified (Zhai et al., 2014). So the bulk of particles form at less than 100 nm. These fine particles have a low deposition rate. At very low contamination rates, ultrafine particles can be generated even in the size of 10 nm (Amiri et al., 2017).

The above findings confirm the presence of UFP aerosols in cabin air. More studies presenting chemical analysis of these particles are urgently needed. UFPs bind with organophosphates and other toxic compounds. After inhalation these particles are preferentially deposited in the alveolar part of the lungs. They cross the air-blood barrier by pinocytosis and are able to transport the chemicals across the blood-brain barrier (Howard et al., 2018). In addition the olfactory pathway may provide a direct access of particles to the brain. Continual presence with superimposed peaks of particles in cabin air may be responsible for chronic respiratory and neurological problems. Cumulative toxicity is a probable explanation, considering a working life time of aircrew up to 20.000 flying hours (He et al., 2021; Howard et al., 2018). Hydraulic fluid fumes may be more toxic per unit mass than engine oil fumes, due to a higher concentration of organophosphates (2800 $\mu g/m^3$) and smaller particle size (He et al., 2021). So, the presence of an aerosol of ultrafine particles in bleed air may be an additional causal mechanism, with cumulative (toxic) exposure leading to aerotoxic syndrome (Howard et al., 2018).

12. **Symptoms of aerotoxic syndrome** (see Tables 4A, 4B, and 4C)

We previously described a case series of three patients (two pilots, one flight attendant) with probable aerotoxic syndrome (Hageman et al., 2020a,b).

The most frequently reported symptoms were headache, balance problems, fatigue, gastro-intestinal complaints and cognitive impairment. These symptoms appeared to onset during the flying career. A temporal relationship was noted between time spent on board aircraft and symptom onset or exacerbation, and symptoms reduced or resolved during holidays and days off. All three patients reported smell events during their flying career, but none reported distinct fume events. In one of these patients reduced levels of butyrylcholinesterase were measured after a flight indicating possible exposure to organophosphate compounds. All three aircrew were found to have elevated neuronal and glial auto-antibodies, compared to sera of

Table 4A Health surveys and questionnaires on symptoms of pilots and cabin crew, 2000–2020.

Nr	Author, year	Study population	Aircraft type	Remarks
1	Winder and Balouet (2000)	10 cases: pilots, first officers, pursers, flight attendants, involved in a fume event	3 models of airplanes of 5 airlines	Both immediate and long-term/ residual symptoms
2	Winder and Balouet (2001)	7 cases, pilots and flight attendants, exposed to cabin contamination	3 models of airplanes	
3	Cox and Michaelis (2002)	21 cases, 19 pilots, 2 flight attendants	BAe 146	Increase of symptoms while on duty: 84%; improvement on holidays or days off: 90%
4	Winder et al. (2002)	50 cases, 16 pilots, 34 cabin crew	BAe 146 and Airbus 320 aircraft	Onset of symptoms immediately while flying (96%), in 74% persisting complaints for 6 months
5	Michaelis (2003)	106 cases, all pilots, 93 of 106 involved in a fume event	Boeing 757	
6	Harper (2005)	60 cases, 39 pilots, 19 flight attendants, 2 not specified, after exposure to contaminated cabin air	BAe 146, Boeing 757 and Airbus	
7	Somers (2005)	39 cases, flight crew	BAe 146	Temporal relationship between onset of symptoms and exposure history
8	Mackenzie Ross (2008)	27 cases, all pilots	BAe 146 and Boeing 757	
9	Abou-Donia et al. (2013)	34 cases, pilots and flight attendants, after exposure to bleed air contamination; 4000–16.500 flying hours	n.a.	Symptoms 2–4 weeks after exposure
10	Michaelis et al. (2017)	219 cases, 77: no health problems, 142: symptoms	BAe 146	88%: aware of exposure to contaminated air, 7% visible smoke or mist; 32% median to chronic symptoms
11	Hageman et al. (2020a,b)	3 cases, 2 pilots, 1 flight attendant	Boeing 737,747 and Airbus A 330	

n.a. = not available.

Table 4B Most often reported symptoms.

Symptoms	**1[a]**	**2**	**3**	**4**	**5**	**6**	**7**	**8**	**9**	**10**	**11**
Irritation of eyes, nose, throat	x	x	x	x	x	x	x	x			x
Salivation	x			x							
Nausea, vomiting	x	x	x	x	x		x	x	x	x	
Flu-like symptoms								x			
Headache	x	x	x	x	x	x	x	x	x	x	x
Fatigue	x	x	x		x	x		x	x	x	x
Lethargy							x				
Disorientation	x	x									
Dizziness	x	x	x	x	x	x	x	x		x	
Cognitive impairment			x		x	x	x	x		x	x
Memory impairment	x	x	x	x					x	x	
Confusion		x			x					x	
Balance/coordination loss	x	x	x	x	x		x		x		x
Tremor	x										
Irritability		x									
Blurred vision	x	x		x	x					x	
Breathing difficulties	x	x	x	x		x	x	x	x	x	x
Chemical hypersensitivity		x		x			x				
Chest pain	x										
Palpitations	x			x					x		x
GI-complaints, cramps			x	x		x	x	x			x
Diarrhea	x			x	x		x				
Loss of sensation, tingling		x	x		x		x		x		

[a]Study number see Table 3 A.

healthy controls, consistent with central nervous system injury. All three had genetic polymorphisms of paraoxonase (PON-1), and two of cytochrome P450, leading to reduced ability to metabolize toxic compounds (Hageman et al., 2020a,b).

The symptoms reported have much in common with those reported in other case-series and surveys of aircrew, thought to have been exposed to

Table 4C Acute and chronic symptoms of pilots and cabin crew (Michaelis et al., 2017; Winder and Balouet, 2001).

Acute signs and symptoms	Chronic symptoms
Irritation of eyes, nose, throat	Irritation of eyes, nose, throat
Neurology: Headache, dizziness, confusion, balance problems, blurred vision, nystagmus, tremor, impaired memory and concentration, paresthesias	Headache, dizziness, slowed mental processing, impaired memory and concentration, difficulty multi-tasking, balance problems, paresthesias, numbness
Gastrointestinal: Nausea, vomiting, diarrhea	diarrhea
Respiratory: Breathing difficulties, shortness of breath, tightness in chest	Breathing difficulties, cough, tightness in chest
Cardiological: Increased heart rate, palpitations	Increased heart rate, palpitations
General: fatigue	Fatigue, multiple chemical hypersensitivity, muscle pain

contaminated air. The earliest report in the literature of effects of contaminated air on board aircraft already dates back to 1977 (Montgomery et al., 1977). The authors described a military navigator who was incapacitated due to neurological and gastro-intestinal distress, after synthetic lubricating oil fumes had entered the air supply. Routine medical investigations did not reveal any abnormalities, no treatment was instigated, and he recovered within 24 h.

12.1 Questionnaire surveys

Prior to the 1980s, contaminated cabin air was masked by the impact of tobacco smoke. It was not until the enactment of smoking bans on aircraft in the early 1990s that attention turned to other potential contaminants. Since then, a number of case-series and surveys of aircrew have appeared in the literature. In a survey of 50 Australian aircrew, participants were asked to identify health symptoms experienced while flying, and factors which may have caused them. Most respondents reported that their symptoms occurred after exposure to fumes in the cabin (Winder et al., 2002). Some recovered once they vacated the plane, but symptoms persisted for 1–6 months in 82%; and 16% required hospitalization. The most frequently reported symptoms included eye and skin irritation, breathing difficulties, nausea and vomiting, headache, dizziness, balance problems, fatigue and cognitive impairment. The authors coined the term 'aerotoxic syndrome' for the association of symptoms with exposure to hydraulic fluids and to engine oil fumes mixing with cabin air.

Cox and Michaelis (2002) undertook a small-scale survey of Australian aircrew who reported an increase in health complaints (most notably headaches, respiratory difficulties, eye and skin irritation, fatigue and cognitive impairment) since flying the BAe-146 aircraft type. Almost half of the respondents mentioned an association with exposure to engine oil fumes. In a comparable British survey of 106 Boeing-757 pilots, a deterioration in health during their flying career was reported linked to a similar constellation of symptoms (Michaelis, 2003). Remarkably, 88% reported at least one fume event during their flying career and the total number of incidents reported amounted to 1674. Yet only 61 events (17% of the total) were officially reported to regulatory authorities. A large health survey among 4011 flight attendants, conducted by the Occupational Health Research Consortium in Aviation (OHRCA) in 2014, showed that almost 50% of flight attendants reported one or more symptoms, in which respiratory and neurological symptoms accounted for 23% and 17%, respectively (OHRCA, 2014). More recently Michaelis et al. published the findings from a survey of 219 British BAe-146 pilots, of whom 34% reported frequent exposure to contaminated air and 63% reported "adverse effects, ranging from acute to long-term symptoms". In the same paper a forensic analysis discloses 15 cabin air quality incidents in which aircrew reported symptoms leading to impairment or even complete in-flight incapacitation (Michaelis et al., 2017).

12.2 Case series

Over the last two decades a number of case-series have been published describing the health status of aircrew with a history of exposure to contaminated air on commercial aircraft. All describe a constellation of physical health complaints comparable to those reported in the surveys by Winder in 2002 or Michaelis in 2003 (Burdon and Glanville, 2005; Harper, 2005; Mackenzie Ross, 2008; Mackenzie Ross et al., 2011; Somers, 2005).

13. Evidence of central nervous system injury

13.1 Neuropsychological studies

Several neuropsychological studies have been undertaken in different countries across the globe over the last two decades, exploring the nature and severity of cognitive complaints reported by many aircrews in questionnaire surveys, see Table 5.

Table 5 Neuropsychological findings in aircrew with probable aerotoxic syndrome.

Author, year	Description of cases	Study design	Findings	Remarks
Coxon (2002)	2 pilots, 6 flight attendants, all female, age 24–56 years, (mean 36.1). Flying career: 2–12 years. Complaints after oil emissions from the BAe 146	Battery of neuro-psychological tests	Impaired: reaction time, information processing speed, fine motor skills, grip strength and auditory verbal and working memory.	No statistical analysis, no control group
Heuser et al. (2005)	26 flight attendants with complaints after "neurotoxic injury", 3 male, 23 female. (of whom) 8 had impaired smell	Occupational clinic, neuropsychological tests ($n = 26$), and PET scan ($n = 12$)	Memory/learning disability in 18/26, 2 with reading disorder, 4 had depression. All 12 PET scans abnormal: decreased frontal brain function (consistent with toxic encephalopathy).	4 abnormal PET without neuropsychological abnormalities
Mackenzie Ross et al. (2006)	1 pilot, "middle aged", with chronic complaints following contaminated air events: skin, gastrointestinal, fatigue, light headed, coordination↓, cognitive performance↓	Case report, clinical, laboratory and neurophysiological data	Normal scores, but submaximal on information processing speed, working memory span, verbal learning and mental flexibility. No mood disorder, somatisation or conversion disorder.	Also mild sensory neuropathy, autonomous dysfunction and increased autoantibodies against brain specific proteins
Mackenzie Ross (2008)	18 pilots, 16 male, 2 female; mean age 48.4 years (36–62), symptoms after fume events on a Boeing 757 or BAe 146 aircraft. Working: 9. Sick leave: 4. Retired: 5 Mean lifetime flying hours: 11.642, range 3000–25,000 Mean lifetime flying years: 22, range 5,5–40	Cohort study, clinical interview and detailed neuropsychological assessment	Mean full scale IQ: 119 Tests of psychomotor speed, attention, executive function and working and verbal memory below expected level. Significant correlation between exposure indices (total number of flying years) and scores on tests of verbal memory and executive function	Self selected cohort, crude indices of exposure. None of the pilots failed the malingering test

Continued

Table 5 Neuropsychological findings in aircrew with probable aerotoxic syndrome.—cont'd

Author, year	Description of cases	Study design	Findings	Remarks
Mackenzie Ross et al. (2011)	29 UK pilots, 15: BAe 146 or B 757, 14: B 737; one or more fume events in their career; mean age 50,4 years, mean years of flying 20 years	Case series, neuropsychological investigation	Reduced performance on tests of attention, psychomotor speed and visual sequencing	A random sample of 29 of 70 pilots, who returned a questionnaire, was asked to undergo neuropsychological assessment
Coxon (2014)	3 pilots: 1: 20 years flying, Fokker 27,50,70 and 100, BAe 146; fatigue, deterioration of memory 2: 30 years flying, Boeing 747, 777, headache, "brain fog" 3: 30 years flying, no further data	Case series, neuro-psychological investigation	Impaired: processing speed (in 2 pilots),executive function (3 pilots), new learning skills (3 pilots). Poor: reaction time to stimuli (2 pilots), visual memory (1 pilot)	No controls
Reneman et al. (2016)	12 aircrew (pilots average flying hours: 8130), 10 male, 2 female. Average age: 44,4 years (29–55). 11 matched controls (racing car drivers)	Questionnaire, MRI (DTI-MRI, fMRI) and detailed psychometric neuropsychological tests	More self reported cognitive complaints and depressive symptoms in aircrew. Significant group differences on reaction speed measures and interference. Neurocognitive impairment in only 3/12 aircrew.	DTI-MRI: small regions of affected white matter microstructure.

PET: positron emission tomography.
DTI-MRI: diffusion tensor imaging- magnetic resonance imaging.
fMRI: functional magnetic resonance imaging.

In 2002 Coxon published findings from a case series of eight Australian aircrew who reported mental confusion, concentration difficulties, and memory problems following alleged exposure to oil emissions on BAe 146 aircraft. All were female, age range 24–56 years and they had worked on this aircraft type for between 2 and 12 years. Psychometric assessment revealed evidence of reduced choice and sequential reaction time, psychomotor speed, grip strength and auditory verbal memory, compared to normative data, in all but one; reduced auditory working, memory span, concentration and recall of visual information in 4, and 5 aircrew showed impairment on a sequencing task (Picture Arrangement). Although this was a case series in which no control group was available for comparison and no statistical analyses were undertaken, Coxon concluded that there were sufficient grounds to warrant further investigation. She recommended a wider-scale study of BAe 146 aircrew was undertaken.

In 2005 Heuser et al. reported findings from an evaluation of 26 North American flight attendants with complaints after probable exposure to cabin fumes on one or multiple occasions. In most cases fume events resulted in flight interruption, emergency room visits or sick leave. All underwent a neurological examination and neuropsychological assessment and 12 underwent PET brain imaging. The interval between fume event and PET scan was not reported. Unfortunately, the authors did not specify the psychometric tests used or test scores but reported that all flight attendants exhibited neuropsychological deficits (particularly memory impairment) and all were diagnosed with toxic encephalopathy. Neurological examination was abnormal in 9, and mildly abnormal in 6 flight attendants showed a bilateral postural tremor, had impaired balance, coordination, smell and taste. PET imaging showed imbalance of function between cortical and subcortical areas and frontal and occipital areas. The authors recommend further research and that a protocol be created, outlining the evaluations flight personnel should undergo, following exposure to cabin fumes.

In 2006 Mackenzie Ross published a single case study involving a UK airline pilot who developed chronic ill health over an 11-year period of flying, during which he often smelt oily fumes. Routine medical investigations failed to identify the cause of his symptoms. The pilot subsequently suffered a marked deterioration in cognitive function after experiencing a contaminated air event of sufficient severity to provoke light headedness, eye and throat irritation in both the pilot and co-pilot. They filed an Air Safety Report on landing. The pilot resumed flying within a few days but developed insomnia, fatigue, coordination problems, joint weakness, excessive sweating and poor memory and concentration and he was referred to medical specialists. Neurological examination found evidence of coordination problems, positive Romberg sign, mild sensory neuropathy, and autonomic nervous system dysfunction. Serum autoantibody tests showed increased autoantibodies against nervous system proteins. Neuropsychological assessment did not find evidence of global intellectual decline; however, performance on tests of information processing speed, auditory working memory span, auditory verbal memory and mental flexibility was below expected levels compared to premorbid estimates. The pilot passed a test of malingering, was not suffering from mood or anxiety disorder and had no history of disease, injury or substance abuse which might otherwise account for his symptoms. His neuropsychological profile was similar to that reported by Coxon in Australian aircrew. The relevant aviation authority withdrew the pilot's license on psychiatric grounds despite the fact he was not suffering from a psychological disorder and had never been examined by a psychiatrist.

In 2008 Mackenzie Ross published findings from a case series of 27 UK pilots who underwent neuropsychological and adult mental health assessment undertaken by 12 different examiners, blind to the exposure status of the pilots. Seven were referred by medical specialists and 20 referred themselves. All described distinctive oily chemical smells, the intensity of which would vary depending on phase of flight. Ten pilots had not formally reported these incidents because they assumed the smell was part of the normal working environment and not something to be concerned about. Two pilots reported being threatened by colleagues when they suggested to report fume events, and seventeen reported fume events at some point in their career. Pilots reported alarming failures at work such as being unable to retain or confusing numerical information from Air Traffic Control, forgetting whether they had lowered the undercarriage, completing tasks in the wrong sequence, concentration and decision-making difficulties. Mackenzie Ross excluded nine pilots with a medical or psychiatric history which might otherwise explain these complaints. In the remaining 18 pilots, language, perceptual skills, and general intellectual ability were relatively well preserved, but performance on tests of auditory verbal memory, processing speed, auditory working memory, and executive functioning was below expected levels in 14/11/10/9 pilots, respectively. This was consistent with their subjective complaints of attentional problems and mental slowing in everyday life. Nine were still flying, 4 were on sick leave and 5 had retired on ill health grounds. The neuropsychological profile seen in pilots was different from that seen in a control sample and significant correlations were noted between exposure indices (e.g., total number of years or hours flying) and scores on tests of verbal memory and executive function. The deficits identified in this study could not be attributed to malingering or mood disorder. Firm conclusions regarding a causal link with contaminated air could not be determined because only crude indices of exposure were available (i.e., self-report) and the sample was small and self-selected. The author concluded further research was warranted which should include consideration of other aspects of flying which could have an adverse effect on health and cognitive function.

In 2011 Mackenzie Ross et al. reported findings from a neuropsychological study of 29 randomly selected UK pilots. The study was completed in two phases. In phase one 1500 pilots on a union database were contacted and asked to complete questionnaires on work history, exposure history and physical health. In the second phase, 29 pilots were randomy selected. Twenty-two pilots reported experiencing at least one fume event at some point in their career, with most reporting multiple exposures. This sub-sample was asked to undergo a neuropsychological assessment. Performance on psychometric tests was compared to national norms based on over 2000 adults. Pilots overall IQ was higher than the UK average, which is not surprising given their occupation. However, the pilots neuropsychological profile deviated from that seen in the standardization sample and mirrored that seen in other neurotoxic conditions and in the previously reported case series of 27 pilots. In particular they showed reduced performance on tests of attention, psychomotor speed and visual sequencing. The authors concluded this has implications for flight safety and that further research is warranted.

In 2014 Coxon described three case studies involving pilots repeatedly exposed to fume events over very long careers (17–30 years) who reported a variety of physical symptoms (fatigue, skin rashes, eye irritation, headaches, sinusitis) and a gradual decline in cognitive function. Neuropsychological assessment found evidence of impaired processing speed, executive functioning and new learning/memory. All three pilots eventually ceased flying due to ill health. They had complained of ill health and cognitive problems many times beforehand, but their symptoms were usually dismissed as 'temporary problems'. These cases highlight how neurotoxic effects can take time to fully evolve. Coxon concludes it is of utmost importance that every worker exposed to neurotoxic chemicals, who complains of cognitive impairment, is thoroughly investigated and has follow up evaluations at regular intervals. Slowed processing speed and executive dysfunction would impair a pilot's ability to handle aircraft in an emergency situation.

In 2016 Reneman et al. published findings from a case series of 12 aircrew, 10 male, 2 female, who complained of cognitive impairment, who underwent neuropsychological assessment and MRI brain imaging including Diffuse Tensor Imaging, Spectroscopy and Functional MRI. Results were compared with those obtained from 11 healthy volunteers, matched for gender age and IQ. Only 3 of the 12 aircrew showed lower scores on only 2 (reaction time and susceptibility to interference) out of 20 psychometric tests. On MRI white matter macrostructure did not differ between the groups, but small differences were noted in microstructure in specific brain regions in the corpus callosum. Higher cerebral blood flow was noted in left occipital cortex; and hypoactivation in the right prefrontal cortex when performing an executive task. Extent of cognitive impairment correlated with subjective reports and white matter integrity; but none of these variables correlated with flight hours. No significant differences in brain metabolites or volume were observed between the groups. Given the obvious aviation safety implications, the authors conclude further research is warranted in which alternative explanations should be considered such as irregular work schedules and dysregulation of biological clocks.

Neuropsychological studies undertaken to date have documented cognitive deficits in aircrew (primarily involving reduced reactions times/processing speed, memory impairment and executive dysfunction), associated with repeated exposure to engine oil emissions in aircraft. Every study has methodological limitations in terms of sample size, lack of comparison groups and objective indices of exposure, making it difficult to establish causation. Most studies are cross-sectional. Nevertheless, these initial findings highlight the need for further research because any form of central nervous system dysfunction is dangerous in a safety critical environment and constitutes a flight safety risk. Unfortunately, industry, regulatory authorities and government departments appear reluctant to fund or commission a large-scale epidemiological study to determine whether exposure to oil and hydraulic fluid causes ill health and cognitive impairment. This despite numerous advices of researchers and even government scientific advisory committees, across the globe (COT report, 2007).

13.2 Brain imaging

DTI-MRI of 12 aircrew with cognitive complaints, revealed minor abnormalities of the white matter microstructure (Reneman et al., 2016).

These were not as prominent as imaging abnormalities recognized in other chronic toxic encephalopathies, like painters' disease (Visser et al., 2008). Using MRI-spectroscopy no significant differences between aircrew and controls were observed, in brain (choline-)metabolites nor in brain volume (Reneman et al., 2016). In another toxicology study, 12 flight attendants were analyzed for of ill health and cognitive complaints following a history of exposure to contaminated air. Positron Emission Tomography (PET) showed abnormal results in all subjects, with decreased activity in the frontal regions and increased function in occipital areas and of the limbic system (Heuser et al., 2005). Such a pattern is often seen in other neurotoxic conditions, ascribed to an imbalance in function between cortical (decrease) and subcortical (increase) areas, and imbalance of frontal (decrease) versus occipital (increase) functions.

13.3 Autoantibodies to nervous system proteins

The blood-brain barrier (BBB) protects the brain from auto-immune reactions and generally prevents developing B-cells from being exposed to unique brain antigens. In case of degraded integrity of the BBB an immune response may be provoked. Neuronal and glial proteins, which are normally present in the central nervous system, may leak into the bloodstream. Once in the systemic circulation, these antigens activate a variety of cells, culminating in B-cell activation, rendering autoantibodies production to the targeted proteins (Abou-Donia and Brahmajothi, 2020). In a wide variety of neurological diseases and syndromes serum autoantibodies have been implicated as biomarkers of specific neuronal degeneration in the central nervous system (CNS) (Diamond et al., 2009). For example, autoantibodies have been identified as reliable biomarkers for neuronal damage in dementia and traumatic brain injury (Kim et al., 2018). The last few years the accent has been on a large spectrum of paraneoplastic and other antineuronal auto-antibodies with associated diseases (Lancaster and Dalmau, 2012). The concentration of some individual autoantibodies may be disease-specific (Levin et al., 2010). The presence of serum autoantibodies may have more diagnostic utility than the brain derived protein biomarkers themselves. They are particularly useful in the chronic stage because they remain in the body for a relatively long period.

Higher levels of circulating autoantibodies compared to matched controls were found in flight crew who complained of ill health following exposure to contaminated air (Abou-Donia et al., 2013; Abou-Donia and Brahmajothi, 2020). From our own studies a pilot was tested after a set of flights spending 45 h in the cockpit during 10 days. Significant increases in autoantibodies were noted after exposure to polluted air and correlated with the onset of symptoms. After cessation of flying for a year, this pilot's clinical condition improved and levels of serum autoantibodies normalized. Thus, a temporal relationship between exposure to air emissions, clinical condition, and levels of serum autoantibodies to nervous system specific proteins was found in this case. This suggests that autoantibody testing may be a valuable marker of nervous system injury in aerotoxic syndrome. In our cases MBP, MAP-2 and GFAP were most abnormal (Abou-Donia et al., 2013; Hageman et al., 2020a,b).

14. Involvement of the peripheral nervous system

In surveys of health symptoms in pilots and cabin crew with probable aerotoxic syndrome sensory complaints, such as paresthesia, tingling and numbness are reported in 20–70% of cases (Cox and Michaelis, 2002; Michaelis, 2003; Michaelis et al., 2017; Somers, 2005; Winder et al., 2002). These typical sensory symptoms contrast with the predominantly motor symptoms of the delayed type OP-polyneuropathy. This supports the existence of a distinct toxic sensory polyneuropathy in patients with suspected aerotoxic syndrome, probably not related to OP toxicity. In most of these patients with a corresponding pattern of complaints, nerve conduction studies were unremarkable (de Graaf et al., 2014; Heutelbeck, 2019). In a few (neurographic) studies however a mild sensory polyneuropathy was found (Mackenzie Ross et al., 2006; Michaelis et al., 2017; Winder et al., 2002).

Solvents in cabin air may be responsible for this sensory type polyneuropathy. Solvents, such as ethylene oxide and n-hexane produce a peripheral neuropathy with characteristic pathological features. These agents are known to cause cross linking of neurofilaments, focal axonal swelling, slow axonal transport, segmental demyelination and secondary axonal swelling. Most reports on solvents-induced polyneuropathy concern n-hexane, used as industrial solvent and degreasing agent (Pan et al., 2017; Puri et al., 2007).

Over the last 20 years peripheral nerve impairment has been investigated in a few studies of patients with probable aerotoxic syndrome. In one study various complaints such as restless legs, muscular jerking and tingling sensations were correlated with fume events (Heutelbeck, 2019). In some cases, balance problems, muscle fasciculations and reduced sensation were noted. Paresthesia of the hands were reported in about 30% of flight crew members (Abou-Donia et al., 2013; Somers, 2005). One frequently used test is vibration sensitivity, which evaluates peripheral somatosensory function. In chronic low-dose OP-exposure a decreased sensitivity is described in farmers and sheep dippers (Pilkington et al., 2001). However gross neurological status (balance, motor and sensory function, reflexes) is normal in most suspected cases of aerotoxic syndrome (de Graaf et al., 2014).

Characteristic symptoms of *small fiber neuropathy*, a subtype of polyneuropathy of thin myelinated and unmyelinated nerve fibers, are unfrequently reported by pilots and cabin crew. These symptoms are painful burning paresthesia, hypersensitivity to touch and temperature-changes and autonomic complaints (Sommer et al., 2018). There is one study on the prevalence of small fiber neuropathy in patients with probable aerotoxic syndrome. In nearly all patients in this studygroup skin biopsy showed that the intra-epidermal nerve fiber density was significantly decreased. Therefore this sensory, typically painful disease may well be responsible for both sensory and autonomic complaints in pilots and cabin crew (Heutelbeck, 2019).

A skin biopsy with quantification of intraepidermal nerve fiber density is the most reliable tool to confirm the diagnosis of small fiber neuropathy (Devigili et al., 2019; Lauria et al., 2010). Both solvents and ToCP exposure may lead to a reduced density of epidermal nerve fibers (Sommer et al., 2018).

15. Respiratory symptoms

Respiratory symptoms are common among aircrew and have been reported in various studies (Burdon, 2012; Burdon and Glanville, 2005; Cox and Michaelis, 2002; Mackenzie Ross et al., 2006; Michaelis et al., 2017; Winder et al., 2002; Winder and Balouet, 2001). These symptoms include nasal or throat irritation, sinus problems, tightness of chest, cough, wheezing, influenza-like symptoms during flight, and dyspnea. In some cases, a hypersensitivity for chemical substances is evoked; for instance a cough triggered by odors or irritants (Burdon and Glanville, 2005; Roig et al., 2021). In case series spirometry is often normal or may show mild upper airway obstruction and a positive bronchodilator test. In a retrospective case series in BAe-146 flight crew, pulmonary diffusing capacity was abnormal in six of 14 pilots and flight-attendants (Burdon and Glanville, 2005). Decreased vital capacity and forced expiratory volume (FEV1) were found in cabin crew after a fume event (Heutelbeck, 2017). In the latter study most of the about 350 patients showed normal results on routine spirometry, but more subtle abnormalities were detected using additional pulmonary testing. Small airway obstruction, a reduced alveolar-arterial gradient, and mild abnormalities of diffusion capacity and oxygenation during exercise (ergo-spirometry) are described (Heutelbeck, 2017). In most cases with respiratory symptoms after a fume event a bronchial hyperreactivity, irritant-induced asthma or reactive airways dysfunction syndrome (RADS) is diagnosed. These symptoms subside within a few months in most cases, although in some cases complaints may persist for years (Roig et al., 2021).

16. Cardiological symptoms

Cardiological symptoms and signs mentioned in aircrew are increased heart rate, palpitations and dysrhythmias (Winder and Balouet, 2000). There are no systematic studies on cardiotoxicity in aircrew after fume events. A cross-sectional study did not show an increased prevalence of cardiac events outcomes in flight attendants (McNeely et al., 2018). However, in a post-mortem study of a 43-year old airline pilot, a T-lymfocyte infiltration appeared in heart muscle tissue, consistent with lymfocytic myocarditis (Abou-Donia et al., 2014). Remarkably a similar infiltration was found in peripheral nerves, concurrent with axonal degeneration and demyelination of the brain and spinal cord.

In animal models a variation of cardiovascular complications has been reported: electrocardiografic abnormalities, myocardial infarction, cardiac failure with hemodynamic instability, and histopathological changes (apoptosis, degeneration) (Georgiadis et al., 2018). Experimental animal studies suggest that chronic low level OP exposure upregulates inflammatory mediators, leading to neurotoxicity as well as cardiotoxicity in the form of "catecholamine mediated myocarditis" (Banks and Lein, 2012). This underlying mechanism is confirmed in human studies. An examination of hearts from 13 patients who died as a result of organophosphate poisoning revealed myocarditis, pericarditis and interstitial inflammation (Anand et al., 2009).

17. Gastro-intestinal symptoms

Nausea has been reported very commonly by aircrew during or shortly after fume events. Other gastrointestinal complaints include salivation, vomiting, diarrhea, abdominal cramps and pain (Cox and Michaelis, 2002; Mackenzie Ross et al., 2006; Michaelis et al., 2017; Winder et al., 2002; Winder and Balouet, 2001). These symptoms may be regarded as muscarinic, and are also seen in chronic occupational OP exposure, particularly in farmers and sheep dippers (Lee et al., 2003). Sheep farmers with chronic ill health present with a syndrome known as dipper's flu, with headache and gastrointestinal symptoms (Cherry et al., 2011).

18. A genetic susceptibility to organophosphates

A genetic susceptibility to OP's may explain why not all aircrew are equally affected. Over the last decade studies on chronic, low-level OP neurotoxicity revealed inter-individual differences in the ability to metabolize OP's and other organic substances as a result of genetic polymorphisms (Kaur et al., 2017; Khattab et al., 2016; Sunay et al., 2015). This may explain the difference in individual reports of ill health following exposure to contaminated air. A genetic susceptibility to OP's has been reported too in other occupational groups, with focus on the Paraoxonase (PON)-1 enzyme, involved in the detoxification of OP's (Albers et al., 2010; Cherry et al., 2011). As such, people differ in terms of PON-1 activity.

In addition, genetically based diverging levels of cytochrome-P450 activity account for differences in hepatic metabolic capacity up to 50–100 fold (Polimanti et al., 2012). Individuals expressing a very high P450 activity and a very low PON-1 activity may have a summative 4000 fold increase in threshold for toxicity.

18.1 PON-1

Human paraoxonase (PON)-1 is an essential enzyme that is synthesized mostly in the liver and secreted into the blood. PON-1 plays a major role in metabolism and deactivation of OP's. The PON-1 enzyme is characterized as an organophosphate hydrolase and it has a structural resemblance to ToCP. The PON-1 status (paraoxonase levels across PON-1 genotypes) is a predictor for OP-sensitivity, indicating if persons are more susceptible to the adverse effects of organophosphates (Furlong et al., 2006). The importance of plasma PON-1 activity level in protecting against OP exposure has been clearly demonstrated in animal experiments (Li et al., 1993; Shih et al., 1998). PON contains three isoforms (PON-1, PON-2 and PON-3) and is encoded by a cluster of genes. The gene is located on chromosome locus 7q21.3, in a region near the locus of the gene for AChE (Sunay et al., 2015). The plasma paraoxonase activity in human populations exhibit a polymorphic distribution, and individuals with high, intermediate or low paraoxonase activity can be identified (Costa et al., 2005). So far more than

400 single-nucleotide polymorphisms have been detected in human PON-1 gene. However, most of the studies focus on two amino acid polymorphisms in PON-1. One of these is the alteration of glutamine to arginine at the codon 192 (PON 192Q). The other is a single-nucleotide transversion from methionine to leucine at codon 55. Distribution of PON 192Q and 55 M/L polymorphism frequencies vary greatly across populations. The 192Q and 55 M alleles are more frequent and have low Paraoxonase-activities (Josse et al., 2001). These two polymorphisms were present in all three patients with probable aerotoxic syndrome, in a recent case series (Hageman et al., 2020a,b). Serum paraoxonase status may be a major factor in determining resistance to organophosphates in medical protocols.

18.2 CYP-450

Cytochrome-P450 enzymes mediates the hepatic metabolization of OP's. In the human genome so far 57 genes encoding for CYP-450 enzymes have been identified. Which of these 57 cytochrome P450 enzymes is involved in the metabolism of contaminants in engine oil is largely unknown (Furlong, 2011), however various studies have made some progress in characterization of specific CYP-450 enzymes (Khattab et al., 2016; Reinen et al., 2015). The genes can be classified into different families, according to the amino acid similarities of the encoding proteins. In particular, CYP-450 genes in families 1–3 encode for 22 different isoforms, mainly involved in the metabolism of drugs and toxins (Polimanti et al., 2012). More specifically CYP-1A and 1B, CYP-2D and 2E1, and CYP-3A are involved in neurotoxin breakdown (Ferguson and Tyndale, 2011). There is evidence that variations in the subtype CYP-2D6 play an important role in susceptibility to chronic OP intoxication, but so do CYP-1A2, CYP-2C19, and CYP-3A4 (Khattab et al., 2016, Mutch and Williams, 2006). In the metabolism of ToCP CYP 450-1A2 and 3A4 were found to be responsible (Reinen et al., 2015) Variation in alleles may lead to four phenotypes: poor, intermediate, rapid and ultra-rapid metabolizers. In a recent case study two of the three patients had CYP-polymorphisms, probably consistent with overexpression of enzymes mediating the conversion of OP's to its reactive metabolites (Hageman et al., 2020a,b).

19. Establishing a link with exposure

In 2000 a committee of the National Research Council (NRC) of the United States convened to study sources of cabin air contamination in aircraft and advise on control systems. Their report confirmed the occurrence of contamination and the committee proposed a series of recommendations, including improved air quality measurement (National Research council committee on air quality in passenger cabins of commercial aircraft, board on environmental studies and toxicology, 2002). Inspired by this report the US congress instructed the Federal Aviation Administration to fund research for the improvement of cabin air quality. On carrying out this project researchers could not come to a mutual agreement with airline companies to apply air sampling devices. Most aircraft still have not been equipped with air monitoring devices. Large scale implementation of this equipment is essential to increase the likelihood of capturing a fume event. Besides objective and reliable measures such as biomarkers of both exposure and nervous system injury would be useful. These measures are indispensable to link chemical exposure to health outcomes. Researchers need to explore the possible effects of both cumulative low-level exposure to contaminants, as well as the effect of fume events. Comparable to other occupational groups, particularly those involving work with pesticides, both acute poisoning events and cumulative low-level exposure may be harmful (e.g., Farahat et al., 2003; Kamel and Hoppin, 2004; Mackenzie Ross et al., 2013; Pilkington et al., 2001; Zhang et al., 2016) (Fig. 2).

It should be considered that more than one chemical contaminant is involved in the etiology of ill health. To date the primary focus in this area has been on TCP; this OP comprises about 3% of engine oil (Michaelis, 2011). Concern is raised in view of the fact that OP's – used as chemical weapons or in pesticides—are known to be neurotoxic (Mackenzie Ross et al., 2013; White et al., 2016). Only a small number of air quality monitoring studies have been undertaken over the last decade. The conclusion is that levels of contaminants detected in cabin air fall within safe exposure standards and are *unlikely* to be responsible for alleged aerotoxic syndrome (de Ree et al., 2014). Furthermore, Schindler et al. (2013) analyzed levels of T(*o*)CP metabolites in the urine of 332 aircrew who had reported fume or smell events on their last flight; ToCP was not detected, and TCP in only a single sample. TCP was unlikely to account for their health complaints. The continued focus to date on TCP has led to the neglect of other contaminants; many other potentially toxic substances are contained in jet engine oil.

Beyond variant OP's (triphenylphosphate (TPP), dibutylphenylphosphate (DBPP), chloorisopropyl phosphate and tri-n-butyl phosphate (Solbu et al., 2011; Winder and Balouet, 2002), one should consider solvents like toluene, xylenes, and formaldehyde. Chronic solvent-induced encephalopathy (CSE) is a syndrome similar to aerotoxic syndrome in many aspects (see Table 6). CSE is recognized as an occupational disease by the International Labour

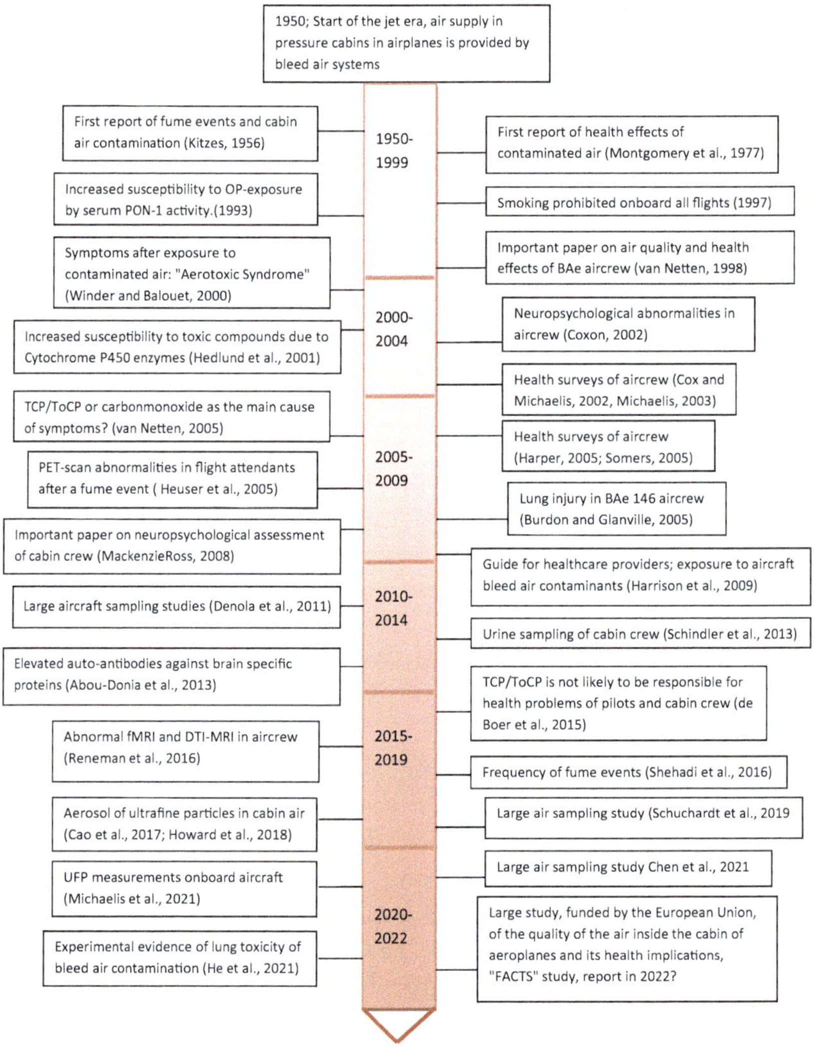

Fig. 2 Timeline, studies of cabin air contamination and health effects.

Organization (ILO, 2010), based on a landmark paper in the Lancet on the effects of solvents on the nervous system (White and Proctor, 1997) and the diagnosis is supported by later neuro-imaging findings in patients suspected of "painter's disease" (Alkan et al., 2004; Tang et al., 2011; Visser et al., 2008).

Furthermore, pilots exposed to fume events have been found to have raised carboxyhemoglobin levels, reflecting exposure to carbon monoxide

Table 6 Recognition as an occupational disease of aerotoxic syndrome, compared to solvent induced chronic toxic encephalopathy.

	Aerotoxic syndrome	Solvent-induced chronic toxic encephalopathy	References
Exposure to	Organophosphates, solvents, CO	Solvents	Harrison and MackenzieRoss (2016); Meyer-Baron et al. (2008)
Population involved	Pilots, cabin crew	Painters, carsprayers, furniture manufacturers, printers	Michaelis (2016); van Valen et al. (2018)
Duration of exposure	No data, >5000 flying hours?	>5 years of daily occupational exposure	van Valen et al. (2012)
Incidents of high exposure	Fume events	Working with solvents in an enclosed space	Shehadi et al. (2016); Lee et al. (2019)
Complaints, symptoms	Headache, fatigue, memory and concentration problems, nasal or throat irritation, sinusproblems, tightness of chest, gastrointestinal and muscle complaints, palpitations	Headache, fatigue, impaired memory and concentration, irritability, mood changes.	Michaelis (2011), Hageman et al. (2020a,b), van Valen et al. (2018).
Diagnostic criteria	Criteria of probable aerotoxic syndrome described in 2020	WHO criteria 1985, European expert group: consensus paper on neuropsychological criteria, 2012	Hageman et al. (2020a,b); van der Hoek et al. (2000); van Valen et al. (2012).
Clinical neurological examination	Occasionally balance problems, muscle fasciculations and reduced sensation.	In some cases optic atrophy (methanol), decreased smell, impaired balance, or signs of sensory neuropathy; in rare cases parkinsonism	Heutelbeck (2019); Dick (2006)

Continued

Table 6 Recognition as an occupational disease of aerotoxic syndrome, compared to solvent induced chronic toxic encephalopathy.—cont'd

	Aerotoxic syndrome	Solvent-induced chronic toxic encephalopathy	References
Neuropsychological testing	Reduced performance on tests of working memory, processing speed / reaction time and mental flexibility.	Impaired speed of information processing, memory and learning; correlation between exposure duration and cognitive complaints	Coxon (2002); Heuser et al. (2005); Mackenzie Ross (2008, Mackenzie Ross et al., 2006); van Hout et al. (2006)
Neurophysiological abnormalities	EMG: Nerve conduction unremarkable in most cases. In a few studies a mild sensory polyneuropathy. Skin biopsy: intra-epidermal nerve fiber density significantly decreased.	Increased latencies and reduced amplitudes of visual evoked responses. EMG: reduction in nerve conduction velocity, prolongation of distal latencies, mildly positive sharp waves and polyphasic potential in n-hexane exposure.	Winder et al. (2002); Mackenzie Ross et al. (2006); Michaelis et al. (2017); Heutelbeck (2019); Verberk et al. (2004); Pan et al. (2017)
MRI	DTI-MRI: slight abnormalities of the white matter microstructure (reduced FA values) in the CC, corona radiata, superior longitudinal fasciculus and thalamic radiation. fMRI: reduced brain activation (precuneus and prefrontal cortex) on a executive function task. MR spectroscopy: no significant metabolite changes	MRI: decreased volume of the genu of the corpus callosum. DTI-MRI: fractional anisotropy reduced in thalamus, caudate nucleus, correlating with attention and psychomotor speed abnormalities. Functional MRI (fMRI) during a working memory test: lower activation of anterior cingulate, prefrontal and parietal cortices MR spectroscopy: metabolite changes of basal ganglia, thalamus and parietal white matter	Reneman et al. (2016); Visser et al. (2008); Tang et al. (2011; Haut et al. (2006), Alkan et al. (2004)

SPECT/PET	PET: decreased activity in the frontal regions and increased function of occipital areas and the limbic system, especially the extended amygdala region	IBZM-SPECT: reduced striatal dopamine binding	Heuser et al. (2005); Visser et al. (2008)
Auto-antibodies against brain specific proteins	Increased auto-antibodies against central nervous system proteins	No data	Abou-Donia et al. (2013), Abou-Donia and Brahmajothi (2020)
Genetic susceptibility	High CYP P450 and low PON-1 activity	CYP2E1 polymorphism and glutathion S-transferase M1–0 genotype	Furlong (2011); Kezic et al. (2006)
Recognition as an occupational disease	No recognition in aviation medicine	Fully recognized	Wolkoff et al. (2016); Cannon (2016); ILO, (2010); White and Proctor (1997)

EMG: electromyography, DTI-MRI: diffusion tensor magnetic resonance imaging, FA: fractional anisotropy, PET: Positron Emission Tomography.
IBZM-SPECT: single photon emission tomography with a highly specific dopamine D2 ligand, CYP 450: cytochrome P450, PON-1: Human paraoxonase-1.

(Michaelis et al., 2017). The most common symptoms of chronic low-level carbon-monoxide toxicity are similar to complaints reported by aircrew; headache, dizziness and nausea, however there may be a broader spectrum of symptoms, complaints, either temporary or persistent (Clarke et al., 2012; Solbu et al., 2011; Sykes and Walker, 2016).

Hydraulic and de-icing fluids can contaminate incoming engine-air (Solbu et al., 2011). Hydraulic fluids are made up of tributyl phosphates and triphenyl phosphates, while de-icing fluids consist of ethylene- and propylglycols. Finally, high concentrations of polybrominated diphenylesters, a component of flame retardants have been reported in dust samples, collected in aircraft (Allen et al., 2013).

Therefore, air quality studies must take into consideration a wide spectrum of potential toxic chemical compounds, which might combine at high temperatures to form even more toxic compounds. Researchers also need to consider that individual susceptibility leads to different expressions of illness. Increased risk is influenced by age, previous health status or genetic variability to detoxify chemicals.

20. A probable diagnosis of aerotoxic syndrome?

In this chapter we have reviewed research on various physical and neurological complaints in aircrew who report deterioration in health throughout their flying career. Causation from cabin air contamination has yet to be established. In 1965 Hill published a paper to identify necessary steps to pass from an observed association between an environmental feature and a health problem to causation (Hill, 1965). He listed the following research criteria to establish causation between exposure and disease:

1) a temporal relationship must exist between exposure and the onset of symptoms,
2) a dose-response relationship should be apparent
3) removal from exposure should modify the effect,
4) sound epidemiological studies must exist which demonstrate a strong association between exposure and disease outcome,
5) consistent findings should be documented in different populations, in different study designs, and at different times,
6) the cause-effect relationship must be biologically plausible, consistent with laboratory findings, and relatively specific,
7) in agreement with current knowledge and
8) evidence of analogous disorders, caused by similar agents should be obtained.

These criteria are discussed in detail for aerotoxic syndrome by Howard et al. (2017); Ramsden (2012). The first paper was focussed on TCP, the latter on a mixture of toxic compounds. Aerotoxic syndrome was never reported prior to the introduction of engine bleed air pressurization systems, though it was detected soon afterwards (Howard et al., 2017). In addition to temporality, consistency of the pattern of symptoms exists and epidemiological studies and animal experimental data supports a causal link. However, aerotoxic syndrome does not meet all of these criteria (Verbeek, 2012). As mentioned previously, the symptoms reported by aircrew are non-specific and common in the general population (Wolkoff, 2013). As far as we are aware, no studies have been undertaken to date, which explore the sensitivity and specificity of the symptoms which are thought to reflect aerotoxic syndrome.

The role of specific other work-related factors in aviation which might account for ill health (including cognitive impairment) have yet to be considered, such as shift work, jet lag, low ambient pressure, sleep deprivation, temperature, low humidity, noise, vibration, work/life stress and separation from family (Bor et al., 2017a,b; Lindgren et al., 2007; Mackenzie Ross, 2017; McGuire et al., 2013). What is needed is an epidemiological study comparing exposed and unexposed populations of aircrew to tease apart the aetiological role of general work factors vs toxic exposure in causing ill health.

21. Proposal of diagnostic criteria for aerotoxic syndrome

Given the continuing scientific uncertainty of the impact on health of exposure to contaminated air on commercial aircraft, further research is warranted. There is an urgent need for a clearly defined, internationally accepted medical protocol for the evaluation of health complaints among aircrew. Internationally agreed diagnostic criteria must be developed, to facilitate disease recognition and allow prospective, prognostic and treatment studies to be undertaken. We recommend multicentre and multidisciplinairy research programs for this purpose.

We proposed the following criteria for the diagnosis of probable aerotoxic syndrome (Hageman et al., 2020a,b):

1) Aircrew report symptoms, such as headache, fatigue, loss of balance, gastro-intestinal complaints, palpitations and cognitive complaints;
2) Symptoms should show a temporal relationship with flying hours and onset shortly after a fume event or directly after flying, and improve after cessation of flying.

3) The same symptoms should occur repeatedly after flying and not under other circumstances (Ramsden, 2012).
4) Objective evidence of exposure should be sought such as flight log data (flying hours), air incident reports, engineering records, on-board air monitoring or swipe measurements of chemical contaminants. Biomarkers of exposure should be taken such as reductions in acetyl- or butylcholinesterases activity or increases in carboxyhaemaglobin levels in blood serum. Other supportive findings are present such as a high P450 activity or low PON-1 activity.
5) Objective evidence of nervous system injury is available following medical tests, neuropsychological assessment, brain imaging or serum neuronal and glial autoantibody-tests (Abou-Donia et al., 2013).
6) Other causes of ill health should be excluded by medical history, physical and neurological examination, laboratory investigation and brain imaging. Neuropsychological assessment should exclude a depressive disorder, somatisation disorder, and malingering.

We propose the following medical evaluation protocol, see Fig. 3:

In cases of probable aerotoxic syndrome a stepwise diagnostic procedure should be performed in which the first step is a comprehensive medical history taking, including a thorough occupational history (flying hours, aircraft types, fume events), and a physical neurological examination (Heutelbeck, 2019; Harrison et al., 2009; Burdon and Glanville, 2005; Michaelis et al., 2017; Mackenzie Ross et al., 2006; Liyasova et al., 2011; Schindler et al., 2014; Hageman et al., 2020a,b).

The second step includes a bloodtest and chest X-ray.

In a probable case, the third step, depends on the presented symptoms, but may include respiratory function tests, electrocardiography (ECG), echocardiography, neuropsychological assessment, autonomic nervous system tests (with a tilt table test), electromyo- and -neurography, and brain MRI.

Finally genetic testing for PON-1 polymorphisms and overexpression of Cytochrome P450 enzymes may demonstrate genetic susceptibility.

22. Discussion-future study directions

Although the studies cited above document ill health among aircrew, causation continues to be debated. A number of weaknesses limit the conclusions that can be drawn, such as sample bias, inadequate measurements and potential response bias, non-specificity of symptoms and lack

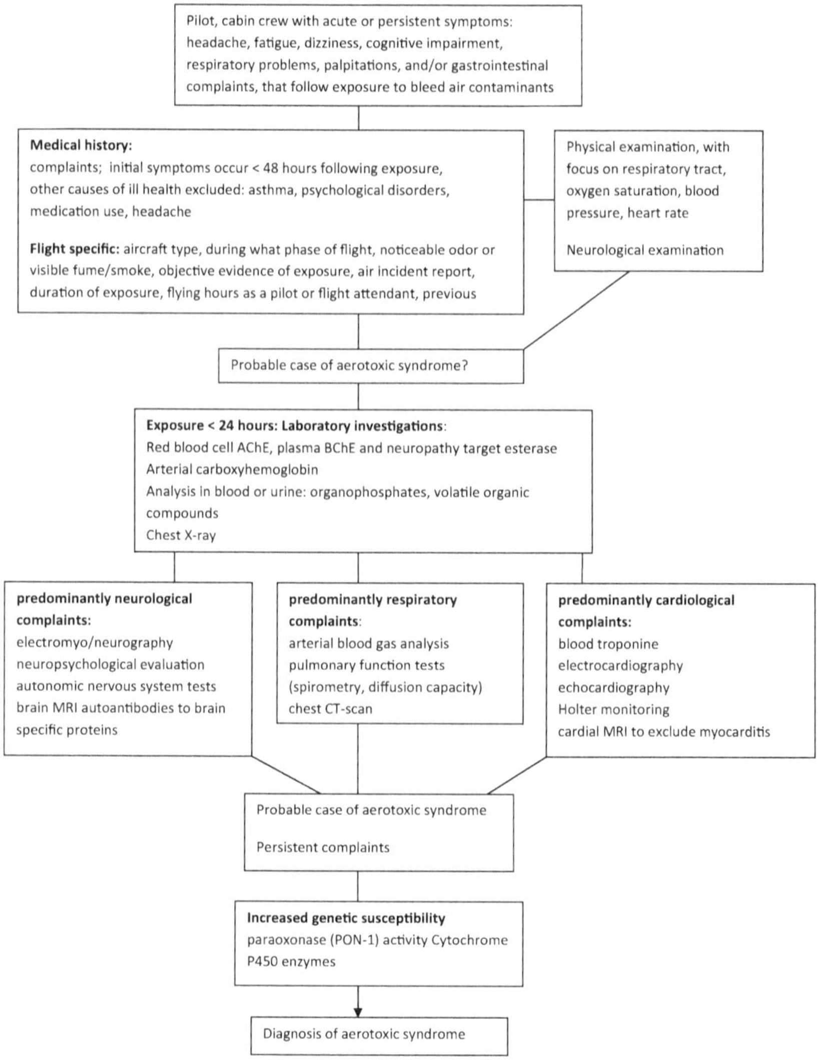

Fig. 3 Flow-chart, assessment of cases with probable aerotoxic syndrome.

of objective evidence of exposure. To date, all previous studies involve small, self-selected samples of aircrew. Researchers have undertaken questionnaire surveys of aircrew using unvalidated measures and response rates have been low (5–50%). Selection bias is involved as aircrew present themselves to physicians with symptoms they attribute to exposure to contaminated air. As a result, it remains unclear how representative these populations are of aircrew in general. The symptoms reported are diverse (involving several organ systems) and non-specific and are highly prevalent in other clinical populations and even in the general population. Few studies include any comparison or control groups to enable researchers to determine the specificity of symptoms and association with exposure. Objective evidence of exposure is frequently lacking as cabin air is not routinely monitored for the presence of chemical contaminants. On the other hand, underreporting of health complaints is common among aircrew.

Signs and symptoms in aircrew are probably related to repeated low dose exposure to a mixture of toxic compounds. Chronic pre-exposure to repeated low dose OP's may lead to increased vulnerability to subsequent higher (fume-event) dose of organophosphate exposure (Howard et al., 2017). This may explain why a high proportion of aircrew frequently attends a hospital after a fume event, while passengers do not appear to be affected. It has been pointed out that the pattern of symptoms in aerotoxic syndrome cannot be explained by inhibition of AChE or BChE alone (Costa, 2018). When effects are seen at organophosphate doses that cause minimal or no inhibition of esterases, possible neurotoxic effects of long-term, low-dose exposure should be studied. The assumption that symptoms of aerotoxic syndrome are based on OP-induced delayed neuropathy is a failure of hazard characterization (Howard, 2020). OPIDN is not seen in aircrew. It is more likely that symptoms of aircrew are the result of OP induced chronic neurotoxicity (OPICN) (Abou-Donia, 2003). OPICN induced by low-level inhalation of organophosphates present in jet engine lubricating oils and the hydraulic fluids of aircraft could explain the long-term neurologic deficits, and the neuropsychiatric and behavioral problems (Abou-Donia, 2003). In addition to OP's, exposure to other toxic compounds, such as solvents and carbonmonoxide in cabin air, may generate synergistic effects.

What is needed is a prospective epidemiological survey of a large and representative randomly enrolled sample of aircrew, in which symptoms are described and qualified in exposed and unexposed groups and compared to rates in other clinical and general populations. Such an exposure-health outcome study will require airline participation and facilitation. Clinically sensitive and meaningful measures with circumscript sensitivity and specificity should be used to establish whether the profile of symptoms seen in aircrew who have been exposed to contaminated air, differs from that observed in other cohorts. To the best of our knowledge only one paper has been published to date which has attempted to do so. Mackenzie Ross et al. (2011) compared the profile of neuropsychological deficits observed in aircrew with a history of exposure to contaminated air, with that of farm workers exposed to organophosphate pesticides, and plotted this against the cognitive profiles of healthy population. They found a specific profile of abnormal cognitive performance in pilots mirroring that of exposed farmers. Although these results hint at an association between exposure to neurotoxic chemicals and cognitive impairment, they do not provide incontrovertible evidence of causation in the case of ill aircrew. Scientific progress demands objective and reliable measures of exposure. Consequently the "aircraft cabin of the future" should have (1) improved engine seal designs to minimize oilleakage, (2) continuous air quality monitoring, with in-situ real time and delayed analysis, (3) bleed air sensors, distributed across the cabin, and (4) new cabin air filter technology to improve cabin air quality (Kos et al., 2018; OHRCA, 2014; Rosenberger, 2018). The new generation airplanes, like the Boeing 787 Dreamliner has a bleedless air distribution system. The air in the Dreamliner is funneled into electrical compressors near the wing, rather than through the engines. Airline-companies should invest in the development of bleedless aircraft as part of safety in aviation.

References

Abou-Donia, M.B., 1981. Organophosphorous ester-induced delayed neurotoxicity. Annu. Rev. Pharmacol. Toxicol. 21, 511–548.

Abou-Donia, M.B., 2003. Organophosphorus ester-induced chronic neurotoxicity. Arch. Environ. Health 58 (8), 484–497.

Abou-Donia, M.B., Abou-Donia, M.M., ElMasry, E.M., Monro, J.A., Mulder, M.F.A., 2013. Autoantibodies to nervous system-specific proteins are elevated in sera of flight crew members: biomarkers for nervous system injury. J. Toxicol. Environ. Health 76 (Part A), 363–380.

Abou-Donia, M.B., van de Goot, F.R.W., Mulder, M.F.A., 2014. Autoantibody markers of neural degeneration are associated with post-mortem histopathological alterations of a neurologically injured pilot. J. Biol. Phys. Chem. 14, 34–53.

Abou-Donia, M.B., Brahmajothi, M.V., 2020. Novel approach for detecting the neurological or behavioral impact of physiological episodes (PEs) in military aircraft crews. Mil. Med. 185, 383–389.

Albers, J.W., Garabrant, D.H., Berent, S., Richardson, R.J., 2010. Paraoxonase status and plasma butyrylcholinesterase activity in chlorpyrifos manufacturing workers. J. Exp. Sci. Environ. Epidemiol. 20, 79–89.

Alkan, A., Kutlu, R., Hallac, T., Sigirci, A., Emul, M., Pala, N., Altinok, T., Aslan, M., Sarac, K., Ozcan, C., 2004. Occupational prolonged organic solvent exposure in shoemakers: brain MR spectroscopy findings. Magn. Reson. Imaging 22, 707–713.

Allen, J.G., Stapleton, H.M., Vallarino, J., McNeely, E., McClean, M.D., Harrad, S.J., Rauert, C.B., Spengler, J.D., 2013. Exposure to flame retardant chemicals on commercial airplanes. Environ. Health 12, 17.

Amiri, S.N., Jones, B., Mohan, K.R., Weisel, C.P., Mann, G., Roth, J., 2017. Study of aldehydes, carbonmonoxide and particulate contaminants generated in bleed air simulator. J. Aircr. 54, 1364–1374.

Anand, S., Singh, S., Saikia, U.N., Bhalla, A., Sharma, Y.P., Singh, D., 2009. Cardiac abnormalities in acute organophosphate poisoning. Clin. Toxicol. 47, 230–235.
Anderson, J., 2021. Sources of onboard fumes and smoke reported by U.S. airlines. Aerospace 8, 122.
Banks, C.N., Lein, P.J., 2012. A review of experimental evidence linking neurotoxic organophosphorus compounds and inflammation. Neurotoxicology 33 (3), 575–584.
Balouet, J.-C., Winder, C., 1999. Aerotoxic syndrome in air crew as a result of exposure to airborne contaminants in aircraft. In: Paper presented at the American Society of testing and materials symposium on air quality and comfort in airliner cabins, New Orleans, 27–28 October 1999.
Bekö, G., Allen, J.G., Weschler, C.J., Vallarino, J., Spengler, J.D., 2015. Impact of cabin ozone concentrations on passenger reported symptoms in commercial aircraft. PLoS One 10 (5), e0128454.
Bhangar, S., Cowlin, S.C., Singer, B.C., Sextro, R.G., Nazaroff, W.W., 2008. Ozone levels in passenger cabins of commercial aircraft on north American and transoceanic routes. Environ. Sci. Technol. 42, 3938–3943.
Bor, R., Droog, A., Albuquerque, C., Dickens, P., Eriksen, C., Harris, P., Oakes, M., Mackenzie Ross, S., Farndon, H., 2017b. Aviation and Aerospace Psychology: Pilot Mental Health and Wellbeing. Position Statement 2017. The British Psychological Society.
Bor, R., Eriksen, C., Oakes, M., Scragg, P., 2017a. Pilot Mental Health Assessment and Support. A practitioner's Guide. Taylor & Francis group, Routledge.
Brown, M.A., Brix, K.A., 1998. Review of health consequences from high-, intermediate- and low-level exposure to organophosphorus nerve agents. J. Appl. Toxicol. 18, 393–408.
Burdon, J., Glanville, A., 2005. A lung injury following hydrocarbon inhalation in BAe 146 aircrew. In: Proceedings BALPA Air safety and cabin air quality international aero industry conference, London.
Burdon, J., 2012. Lung injury following hydrocarbon inhalation in aircrew. J. Biol. Phys. Chem. 12, 98–102.
Cable, G.G., 2003. In-flight hypoxia incidents in military aircraft: causes and implications for training. Aviat. Space Environ. Med. 74, 169–172.
Cannon, F., 2016. Aircraft cabin air contamination and aerotoxic syndrome-a review of the evidence. Nanotechnol. Percept. 12, 1–27.
Cao, Q., Xu, Q., Liu, W., Lin, C.-H., Wei, D., Baughcum, S., Norris, S., Chen, Q., 2017. In-flight monitoring of particle deposition in the environmental control systems of commercial airliners in China. Atmos. Environ. 154, 118–128.
Chen, R., Fang, L., Liu, J., Herbig, B., Norrefeldt, V., Mayer, F., Fox, R., Wargocki, P., 2021. Cabin air quality on non-smoking commercial flights: a review of published data on airborne pollutants. Indoor Air 31, 926–957.
Cherry, N., Mackness, M., Mackness, B., Dippnall, M., Povey, A., 2011. "dippers" flu and its relationship to PON 1 polymorphisms. Occup. Environ. Med. 68, 211–217.
Clarke, S., Keshishian, C., Murray, V., Kafatos, G., Ruggles, R., Coultrip, E., Oetterli, S., Earle, D., Ward, P., Bush, S., Porter, C., 2012. For the carbon monoxide in emergency departments (COED) working group. Screening for carbonmonoxide exposure in selected patient groups attending rural and urban emergency departments in England: a prospective observational study. BMJ Open 2, e000877.
Costa, L.G., Vitalone, A., Cole, T.B., Furlong, C.E., 2005. Modulation of paraoxonase (PON 1) activity. Biochem. Pharmacol. 69, 541–550.
Costa, L.G., 2006. Current issues in organophosphate toxicology. Clin. Chem. Acta. 366, 1–13.

Costa, L.G., 2018. Organophosphorus compounds at 80: some old and new issues. Toxicol. Sci. 162 (1), 24–35.
COT report, 2007. In: Statement of the Review of the Cabin Air Environment, Ill Health in Aircraft Crews, and the Possible Relationship between Smoke/Fume Events in Aircraft. UK Committee on the Toxicity of Chemicals in Food. Consumer Products, and the Environment, London, England.
Cox, L., Michaelis, S., 2002. A survey of health symptoms in BAe 146 aircrew. J. Occup. Health Safety-Aust. NZ 18 (4), 305–312.
Coxon, L.W., 2002. Neuropsychological assessment of a group of BAe 146 aircraft crew members exposed to jet engine oil emmissions. J. Occup. Health Safety-Aust. NZ 18 (4), 313–319.
Coxon, L.W., 2014. Delayed cognitive impairment and pilot incapacitation following contaminated air inhalation. J. Biol. Phys. Chem. 14, 107–110.
Crump, D., Harrison, P., Walton, C., 2011. Aircraft Cabin Air Sampling Report. Institute of Environment and Health, Cranfield University. https://dspace.lib.cranfield.ac.uk/handle/1826/5305.
Davies, R., Ahmed, G., Freer, T., 2000. Chronic exposure to organophosphates: background and clinical picture. Adv. Psychiatr. Treat. 6, 187–192.
Denola, G., Hanhela, P.J., Mazurek, W., 2011. Determination of tricresyl phosphate air contamination in aircraft. Ann. Occup. Hyg. 55 (7), 710–722.
de Boer, J., Antelo, A., van der Veen, I., Brandsma, S., Lammertse, N., 2015. Tricresyl phosphate and the aerotoxic syndrome of flight crew members-current gaps in knowledge. Chemosphere 119, S58–S61.
de Graaf, L.J., Hageman, G., Gouders, B.C.M., Mulder, M.F.A., 2014. The Aerotoxic syndrome: fact or fiction (in Dutch). Ned. Tijdschr. Geneeskd. 158 (20), 889–894.
de Ree, H., van den Berg, M., Brand, T., Mulder, G.J., Simons, R., Veldhuijzen vanZanten, B., Westerink, R.H.S., 2014. Health risk assessment of exposure to tricresyl phosphates (TCPs) in aircraft: a commentary. Neurotoxicology 45, 209–215.
Devigili, G., Rinaldo, S., Lombardi, R., Cazzato, D., Marchi, M., Salvi, E., Eleopra, R., Lauria, G., 2019. Diagnostic criteria for small fibre neuropathy in clinical practice and research. Brain 142, 3728–3736.
Diamond, B., Huerta, P.T., Mina-Osorio, P., Kowal, C., Volpe, B.T., 2009. Losing your nerves? Maybe it's the antibodies. Nat. Rev. Immunol. 9, 449–456.
Dick, F.D., 2006. Solvent neurotoxicity. Occup. Environ. Med. 63, 221–226.
Eldin, C., Lagier, J.-C., Mailhe, M., Gautret, P., 2020. Possible aircraft transmission of Covid-19 in-flight from the central African republic to France. Travel Med. Infect. Dis. 35, 101643.
El Rahman, H.A.A., Salama, M., El-Hak, S.A.G., El-Harouny, M.A., ElKafrawy, P., Abou-Donia, M.B., 2018. A panel of autoantibodies against neural proteins as peripheral biomarker for pesticide-induced neurotoxicity. Neurotox. Res. 33, 316–336.
Farahat, T.M., Abdelrasoul, G.M., Amr, M.M., Shebl, M.M., Farahat, F.M., Anger, W.K., 2003. Neurobehavioral effects among workers occupationally exposed to organophosphorus pesticides. Occup. Environ. Med. 60, 279–286.
Furlong, C.E., Holland, N., Richter, R.J., Bradman, A., Ho, A., Eskenazi, B., 2006. PON 1 status of farmworker mothers and children as a predictor of organophosphate sensitivity. Pharmacogenet. Genomics 16, 183–190.
Ferguson, C.S., Tyndale, R.F., 2011. Cytochrome P450 enzymes in the brain: emerging evidence of biological significance. Trends Pharmacol. Sci. 32 (12), 708–714.
Furlong, C.E., 2011. Exposure to triaryl phosphates: metabolism and biomarkers of exposure. J. Biol. Phys. Chem. 11, 165–171.
Fushimi, A., Saitoh, K., Fujitani, Y., Takegawa, N., 2019. Identification of jet lubrication oil as a major component of aircraft exhaust nanoparticles. Atmos. Chem. Phys. 19, 6389–6399.

Georgiadis, N., Tsarouhas, K., Tsitsimpikou, C., Vardavas, A., Rezaee, R., Germanakis, I., Tsatsakis, A., Stagos, D., Kouretas, D., 2018. Pesticides and cardiotoxicity. Where do we stand? Toxicol. Appl. Pharmacol. 353, 1–14.

German Federal Bureau of Aircraft Accident investigation, 2014. Study of Reported Occurrences in Conjunction with Cabin Air Quality in Transport Aircraft. (Report no BFU 803.1-14).

Guan, J., Gao, K., Wang, C., Yang, X., Lin, C.-H., Lu, C., Gao, P., 2014. Measurements of volatile organic compounds in aircraft cabins. Part 1: methodology and detected VOC species in 107 commercial flights. Build. Environ. 72, 154–161.

Hageman, G., Pal, T.M., Nihom, J., MackenzieRoss, S.J., van den Berg, M., 2020a. Three patients with probable aerotoxic syndrome. Clin. Toxicol. 58 (2), 139–142.

Hageman, G., Pal, T., Nihom, J., MackenzieRoss, S., van den Berg, M., 2020b. Short communication: Aerotoxic syndrome, discussion of possible diagnostic criteria. Clin. Toxicol. 58 (5), 414–416.

Harper, A., 2005. A survey of health effects in aircrew exposed to airborne contaminants. J. Occup. Health. Safety-Aust.-NZ 21, 433–439.

Harrison, R., Murawski, J., McNeely, E., Guerriero, J., Milton, D., 2009. Exposure to aircraft bleed air contaminants among airline workers, a guide for health care providers. Document April,.

Harrison, V., Mackenzie Ross, S.J., 2016. An emerging concern: toxic fumes in airplane cabins. Cortex 74, 297–302.

Harrison, V., MackenzieRoss, S., 2016. Anxiety and depression following cumulative low-level exposure to organophosphate pesticides. Environ. Res. 151, 528–536.

Hausherr, V., van Thriel, C., Krug, A., Leist, M., Schobel, N., 2014. Impairment of glutamate signaling in mouse central nervous system neurons in vitro by tri-ortho-cresyl phosphate at noncytotoxic concentrations. Toxicol. Sci. 142, 274–284.

Haut, M.W., Kuwabara, H., Ducatman, A.M., Hatfield, G., Parsons, M.W., Scott, A., Parsons, E., Morrow, L.A., 2006. Corpus callosum volume in railroad workers with chronic exposure to solvents. J. Occup. Environ. Med. 48, 615–624.

Hayes, K., Megson, D., Doyle, A., O'Sullivan, G., 2021. Occupational risk of organophosphates and other chemical and radiative exposure in the aircraft cabin: a systematic review. Sci. Total Environ. 796, 148742.

He, R.-W., Houtzager, M.M.G., Jongeneel, W.P., Westerink, R.H.S., Cassee, F.R., 2021. In vitro hazard characterization of simulated aircraft cabin bleed-air contamination in lung models using an air-liquid interface (ALI) exposure system. Environ. Int. 156, 106718.

Hedlund, E., Gustafsson, J.A., Warner, M., 2001. Cytochrome P 450 in the brain: a review. Curr. Drug Metab. 2 (3), 245–263.

Heuser, G., Aguilera, O., Heuser, S., Gordon, J.D., 2005. Clinical evaluation of flight attendants after exposure to fumes in cabin air. In: BALPA Conference Proceedings, pp. 105–110.

Heutelbeck, A.R.R., Bornemann, C., Lange, M., Seeckts, A., Müller, M.M., 2016. Acetylcholinesterase and neuropathy target esterase activities in 11 cases of symptomatic flight crew members after fume events. J. Toxicol. Environ. Health, Part A 79, 1050–1056.

Heutelbeck, A.R.R., 2017. Health disorders and biomonitoring results in aircraft crew members after "fume events". In: Abstract, International aircraft cabin air conference, London, 19–20 September.

Heutelbeck, A.R.R., 2019. Progress report: diagnostics of health disorders and bio monitoring in aircraft crew members after "fume events". Preliminary results after analyzing patients files.International aircraft cabin air conference 2017. J. Health Pollut. 9 (24), S1–S142.

Hill, A.B., 1965. The environment and disease; association or causation. Proc. R. Soc. Med. 58, 295–300.

Howard, C.V., Michaelis, S., Watterson, A., 2017. The aetiology of "aerotoxic syndrome"- a toxico-pathological viewpoint. Open Acc. J. Toxicol. 1 (5), 1–3.
Howard, C.V., Johnson, D.W., Morton, J., Michaelis, S., Supplee, D., Burdon, J., 2018. Is a cumulative exposure to a background aerosol of nanoparticles part of the causal mechanism of aerotoxic syndrome. J. Nanomed. Nanosci. 1, JNAN-139.
Houtzager, M.M.G., Havermans, J.G.B.A., Bos, J.G.H., 2013. Investigation of Presence and Concentration of Tricresylphosphates in Cockpits of KLM Boeing 737 Aircraft During Normal Operational Conditions. TNO Report TNO2013, R11976.
Howard, C.V., 2020. Inappropiate use of risk assessment in addressing health hazards posed by civil aircraft cabin air. Open Acc. J. Toxicol. 4 (2), 65–71.
IATA, 2013. Global safety information center analysis: smoke and fumes (smells in the cabin and on the flight deck). In: Safety Trend Evaluation, Analysis, and Data Exchange System. International Air Transport Association, Montreal, Canada.
International Labour Organization, 2010. Identification and Recognition of Occupational Diseases: Criteria for Incorporating Diseases in the ILO List of Occupational Diseases. Occupational Safety and Health Series, vol 74 International Labour Office, Geneva.
Johnson, D., Carter, M.D., Crow, B.S., Isenberg, S.L., Graham, L.A., Erol, H.A., Watson, C.M., Pantazides, B.G., van der Schans, M.J., Langenberg, J.P., Noort, D., Blake, T.A., Thomas, J.D., Johnson, R.C., 2015. Quantitation of ortho-cresyl phosphate adducts to butyrylcholinesterase in human serum by immunomagnetic-UHPLC-MS/MS. J. Mass Spectrom. 50, 683–692.
Josse, D., Lockridge, O., Xie, W., Bartels, C.F., Schopfer, L.M., Masson, P., 2001. The active site of human paraoxonase (PON1). J. Appl. Toxicol. 21 (suppl 1), S7–S11.
Kamel, F., Hoppin, J.A., 2004. Association of pesticide exposure with neurologic dysfunction and disease. Environ. Health Perspect. 112, 950–958.
Kaur, G., Jain, A.K., Singh, S., 2017. CYP/PON genetic variations as determinant of organophosphate pesticides toxicity. J. Genet. 96 (1), 187–201.
Keuken, M.P., Moerman, M., Zandveld, P., Henzing, J.S., Hoek, G., 2015. Total and size-resolved particle number and black carbon concentrations in urban areas near Schiphol airport (the Netherlands). Atmos. Environ. 104, 132–142.
Kezic, S., Calkoen, F., Wenker, M.A.M., Jacobs, J.J.L., Verberk, M.M., 2006. Genetic polymorphism of metabolic enzymes modifies the risk of chronic solvent-induced encephalopathy. Toxicol. Ind. Health 22, 281–289.
Khattab, A.M.T., Zayed, A.A., Ahmed, A.I., AbdelAal, A.G., Mekdad, A.A., 2016. The role of PON 1 and CYP2D6 genes in susceptibility to organophosphorus chronic intoxication in Egyptian patients. Neurotoxicology 53, 102–107.
Kilburn, K.H., 2003. Effects of onboard insecticide use on airline flight attendants. Arch. Environ. Health Int. 58, 284–291.
Kim, H.J., Tsao, J.W., Stanfill, A.G., 2018. The current state of biomarkers of mild traumatic brain injury. JCI. Insight. https://doi.org/10.1172/jci.insight.97105.
Kitzes, G. 1956. Cabin air contamination problems in jet aircraft. Aviat. Med. 2, 53–58.
Klerlein, M., Loizeau, M., 2019. Onboard fume events: short term health consequences in aircrew. Aerosp. Med. Hum. Perform. 90 (3), 204.
Kos, J., Greene, G., Jentink, H.W., Hodgkinson, J., Lourenço, C., Oliveira, M., Reis, R.J., Weeks, M., 2018. On-Board Air Quality-Final Report on the Effect of New Materials. European Commission, Future Sky Safety. Document D7.14.
Lancaster, E., Dalmau, J., 2012. Neuronal autoantigens-pathogenesis, associated disorders and antibody testing. Nat. Rev. Neurol. 8, 380–390.
Lauria, G., Hsieh, S.T., Johansson, O., Kennedy, W.R., Leger, J.M., Mellgren, S.I., Nolano, M., Merkies, I.S.J., Polydefkis, M., Smith, A.G., Sommer, C., Valls-Solé, J., 2010. European Federation of Neurological Societies/ peripheral nerve society guideline on the use of skin biopsy in the diagnosis of small fiber neuropathy. Eur. J. Neurol. 17, 903–912.

Lee, B.W., London, L., Paulauskis, J., Myers, J., Christiani, D.C., 2003. Association between human paraoxonase gene polymorphism and chronic symptoms in pesticide-exposed workers. J. Occup. Environ. Med. 45, 118–122.

Lee, S., Kim, I., Park, D., Song, J., Lee, S.G., 2019. A case of acute organic solvent poisoning during epoxy coating. Ann. Occup. Environ. Med. 31, e9.

Levin, E.C., Acharya, N.K., Han, M., Zavareh, S.B., Sedeyn, J.C., Venkataraman, V., Nagele, R.G., 2010. Brain- reactive autoantibodies are nearly ubiquitous in human sera and may be linked to pathology in the context of blood-brain barrier breakdown. Brain Res. 1345, 221–232.

Li, W.-F., Costa, L.G., Furlong, C.E., 1993. Serum paraoxonase status: a major factor in determining resistance to organophosphates. J. Toxicol. Environ. Health 40, 337–346.

Li, Z., Guan, J., Yang, X., Lin, C.-H., 2014. Source appointment of airborne particles in commercial aircraft cabin environment: contributions from outside and inside of cabin. Atmos. Environ. 89, 119–128.

Lindgren, T., Norbäck, D., Wieslander, G., 2007. Perception of cabin air quality in airline crew related to air humidification, on intercontinental flights. Indoor Air 17, 204–210.

Liyasova, M., Li, B., Schopfer, L.M., Nachon, F., Masson, P., Furlong, C.E., Lockridge, O., 2011. Exposure to tri-o-cresylphosphate detected in jet airplane passengers. Toxicol. Appl. Pharmacol. 256, 337–347.

Mackenzie Ross, S.J., Harper, A., Burdon, J., 2006. Ill health following exposure to contaminated aircraft air: psychosomatic disorder or neurological injury? J. Occup. Health. Safety-Aust.-NZ 22 (6), 521–528.

Mackenzie Ross, S.J., Clark, J.S., Harrison, V., Abraham, K.M., 2007. Cognitive impairment following exposure to organophosphate pesticides: a pilot study. J. Occup. Health Saf. Aust. NZ. 23 (2), 133–142.

Mackenzie Ross, S., 2008. Cognitive function following exposure to contaminated air on commercial aircraft: a case series of 27 pilots seen for clinical purposes. J. Nutrit. Environ. Med. 17 (2), 111–126.

Mackenzie Ross, S.J., Brewin, C.R., Curran, H.V., Furlong, C.E., Abraham-Smith, K.M., Harrison, V., 2010. Neuropsychological and psychiatric functioning in sheep farmers exposed to low levels of organophosphate pesticides. Neurotoxicol. Teratol. 32 (4), 452–459.

Mackenzie Ross, S., Harrison, V., Madeley, L., Davis, K., Abraham-Smith, K., Hughes, T., Mason, O., 2011. Cognitive function following reported exposure to contaminated air on commercial aircraft: methodological considerations fot future researchers. J. Biol. Phys. Chem. 11 (4), 180–191.

Mackenzie Ross, S., McManus, I.C., Harrison, V., Mason, O., 2013. Neurobehavioral problems following low-level exposure to organophosphate pesticides: a systematic and meta-analytic review. Crit. Rev. Toxicol. 43 (1), 21–44.

Mackenzie Ross, S., 2017. Assessing cognitive function in airline pilots: the importance of neuropsychological assessment. In: Bor, R., Eriksen, C., Oakes, M., Scragg, P. (Eds.), Pilot Mental Health Assessment and Support. A practitioner's Guide. Taylor & Francis group, Routledge.

Mangili, A., Gendreau, M.A., 2005. Transmission of infectious diseases during commercial air travel. Lancet 365, 989–996.

McGuire, S., Sherman, P., Profenna, L., Grogan, P., Sladky, J., Brown, R.A., Rowland, L., Hong, E., Patel, B., Tate, D., Kawano, E.S., Fox, P., Kochunov, P., 2013. White matter hyperintensities on MRI in high-altitude U-2 pilots. Neurology 81, 729–735.

McNeely, E., Mordukhovich, I., Tideman, S., Gale, S., Coull, B., 2018. Estimating the health consequences of flight attendant work: comparing flight attendant health to the general population in a cross-sectional study. BMC Public Health 18, 346.

Meyer-Baron, M., Blaszkewicz, M., Henke, H., Knapp, G., Muttray, A., Schäper, M., van Thriel, C., 2008. The impact of solvent mixtures on neurobehavioral performance-conclusions from epidemiological data. Neurotoxicology 29, 349–360.
Michaelis, S., 2003. A survey of health symptoms in BALPA Boeing 757 pilots. J. Occup. HealthSafety- Aust. NZ 19 (3), 253–261.
Michaelis, S., 2011. Contaminated aircraft cabin air. J. Biol. Phys. Chem. 11 (4), 132–145.
Michaelis, S., 2016. Implementation of the Requirements for the Provision of Clean Air in Crew and Passenger Compartments Using the Aircraft Bleed Air System. MSc Thesis, Cranfield University, Cranfield.
Michaelis, S., Burdon, J., Howard, C.V., 2017. Aerotoxic syndrome: a new occupational disease? Public Health Panorama 3 (2), 198–211.
Michaelis, S., Loraine, T., Howard, C.V., 2021. Ultrafine particle levels measured on board short-haul commercial passenger jet aircraft. Environ. Health 20, 89.
Montgomery, M.R., Wier, G.T., Zieve, F.J., Anders, M.W., 1977. Human intoxication following inhalation exposure to a synthetic jet lubricating oil. Clin. Toxicol. 11, 423–426.
Murawski, J.T.L., Supplee, D.S., 2008. An attempt to characterize the frequency, health impact, and operational costs of oil in the cabin and flight deck supply air on US commercial aircraft. J. ASTM Intern. 5, 5.
Mutch, E., Williams, F.M., 2006. Diazinon, chlorpyrifos and parathion are metabolised by multiple cytochromes P450 in human liver. Toxicology 224, 22–32.
National Research Council committee on air quality in passenger cabins of commercial aircraft, board on environmental studies and toxicology, 2002. The Airline Cabin Environmental Health. National Academic Press, Report.
OHRCA, 2014.www.ohrca.org/wpcontent/uploads/2014/08/final report.pdf.
Pan, J.-H., Peng, C.-Y., Lo, C.-T., Dai, C.-Y., Wang, C.-L., Chuang, H.-Y., 2017. N-hexane intoxication in a Chinese medicine pharmaceutical plant: a case report. J Med Case Reports 11, 120.
Pang, A.M., Gay, S., Yadav, R., Dolea, C., Ponce, C., Velayudhan, R., Grout, A., Fehr, J., Plenge-Boenig, A., Schlagenhauf, P., 2020. The safety and applicability of synthetic pyrethroid insecticides for aircraft disinsection: a systematic review. Travel Med. Infect. Dis. 33, 101570.
Pilkington, A., Buchanan, D., Jamal, G.A., Hansen, S., Kidd, M., Hurley, J.F., Soutar, C.A., 2001. An epidemiological study of the relations between exposure to organophosphate pesticides and indices of chronic peripheral neuropathy and neuropsychological abnormalities in sheep farmers and dippers. Occup. Environ. Med. 58, 702–710.
Polimanti, R., Piacentini, S., Manfellotto, D., Fuciarelli, M., 2012. Human genetic variation of CYP450 super-family: analysis of functional diversity in worldwide populations. Pharmacogenomics 13, 1951–1960.
Puri, V., Chaudhry, N., Tatke, M., 2007. N-hexane neuropathy in screen printers. Electromyogr. Clin. Neurophysiol. 47 (3), 145–152.
Ramsden, J.J., 2012. Contaminated aircraft cabin air: aspects of causation and acceptable risk. J. Biol. Phys. Chem. 12, 56–62.
Reinen, J., Nematollahi, L., Fidder, A., Vermeulen, N.P.E., Noort, D., Commandeur, J.N.M., 2015. Characterization of human cytochrome P450s involved in the bioactivation of tri-ortho-cresyl phosphate (ToCP). Chem. Res. Toxicol. 28, 711–721.
Reneman, L., Schagen, S.B., Mulder, M., Mutsaers, H.J., Hageman, G., de Ruiter, M.B., 2016. Cognitive impairment and associated loss in brain white microstructure in aircrew members exposed to engine oil fumes. Brain Imaging Behav. 10, 437–444.
Roig, J., Domingo, C., Burdon, J., Michaelis, S., 2021. Irritant-induced asthma caused by aerotoxic syndrome. Lung. https://doi.org/10.1007/s00408-021-00431-z.

Rosenberger, W., Beckmann, B., Wrbitzky, R., 2016. Airborne aldehydes in cabin-air of commercial aircraft: measurement by HPLC with UV absorbance detection of 2,4-dinitrophenylhydrazones. J. Chromatogr. B 1019, 117–127.
Rosenberger, W., 2018. Effect of charcoal equipped HEPA filters on cabin air quality in aircraft. A case study including smell event related in-flight measurements. Build. Environ. 143, 358–365.
Schindler, BK, Koslitz, S, Weiss, T, Broding, HC, et al., 2014. Exposure of aircraft maintenance technicians to organophosphates from hydraulic fluids and turbine oils: a pilot study. Int. J. Hyg. Environ. Health 217, 34–37.
Schindler, B.K., Weiss, T., Schütze, A., Koslitz, S., Broding, H.C., Bünger, J., Brüning, T., 2013. Occupational exposure of air crews to tricresyl phosphate isomers and organophosphate flame retardants after fume events. Arch. Toxicol. 87, 645–648.
Schopfer, L.M., Furlong, C.E., Lockridge, O., 2010. Development of diagnostics in the search for an explanation of aerotoxic syndrome. Anal. Biochem. 404, 64–74.
Schuchardt, S., Koch, W., Rosenberger, W., 2019. Cabin air quality-quantitative comparison of volatile air contaminants at different flight phases during 177 commercial flights. Build. Environ. 148, 498–507.
Shehadi, M., Jones, B., Hosni, M., 2016. Characterization of the frequency and nature of bleed air contamination events in commercial aircraft. Indoor Air 26, 478–488.
Shih, D.M., Gu, L., Xia, Y.-R., Navab, M., Li, W.-F., Hama, S., Castellani, L.W., Furlong, C.E., Costa, L.G., Fogelman, A.M., Lusis, A.J., 1998. Mice lacking serum paraoxonase are susceptible to organophosphate toxicity and atherosclerosis. Nature 394, 284–287.
Singh, B., 2004. In-flight smoke and fumes. Aviation Safety Spotlight 4, 10–13.
Solbu, K., Daae, H.L., Olsen, R., Thorud, S., Ellingsen, D.G., Lindgren, T., Bakke, B., Lundanes, E., Molander, P., 2011. Organophosphates in aircraft cabin and cockpit air-method development and measurements of contaminants. J. Environ. Monit. 13, 1393–1403.
Somers, M., 2005. Aircrew exposed to fumes on the BAe 146: an assessment of symptoms. J. Occup. Health. Safety-Aust.-NZ 21, 440–449.
Sommer, C., Geber, C., Young, P., Forst, R., 2018. Polyneuropathies. Dtsch. Ärztebl. int. 115, 83–90.
Spengler, J.D., Vallarino, J., McNeely, E., Estephan, H., 2012. In-Flight/Onboard Monitoring: ACER's Component for ASHEAE 1262, Part 2 MA. National Air Transportation Center of Excellence for Research in the Intermodal Transport Environment, Boston.
Strid, A., Smedje, G., Athanassiadid, I., Lindgren, T., Lundgren, H., Jakobsson, K., Bergman, A., 2014. Brominated flame retardant exposure of aircraft personnel. Chemosphere 116, 83–90.
Sunay, S.Z., Kayaalti, Z., Bayrak, T., Söylemezoğlu, T., 2015. Effect of paraoxonase 1 192 Q/R polymorphism on paraoxonase and acetylcholinesterase enzyme activities in a Turkish population exposed to organophosphate. Toxicol. Ind. Health 31 (12), 1061–1068.
Sutton, P.M., Vergara, X., Beckman, J., Nicas, M., Das, R., 2007. Pesticide illness among flight attendants due to aircraft disinsection. Am. J. Ind. Med. 50, 345–356.
Sykes, O.T., Walker, E., 2016. The neurotoxicology of carbon monoxide- historical perspective and review. Cortex 74, 440–448.
Tang, C.Y., Carpenter, D.M., Eaves, E.L., Ng, J., Ganeshalingam, N., Weisel, C., Qian, H., Lange, G., Fiedler, N.L., 2011. Occupational solvent exposure and brain function: an fMRI study. Environ. Health Perspect. 119 (7), 908–913.
Terry, A.V., 2012. Functional consequences of repeated organophosphate exposure: potential non-cholinergic mechanisms. Pharmacol. Ther. 134, 355–365.

Ungeheuer, F., van Pinxteren, D., Vogel, A.L., 2021. Identification and source attribution of organic compounds in ultrafine particles near Frankfurt international airport. Atmos. Chem. Phys. 21, 3763–3775.

UK, 2015. Civil Aviation Authority Guidance for Passengers and Healthcare Professionals.

van den Driessche, K.S.J., Sow, A., van Gompel, A., van Deurzen, K., 2010. Anaphylaxis in an airplane after insecticide spraying. J. Travel Med. 17 (6), 427–429.

van der Hoek, J.A.F., Verberk, M.M., Hageman, G., 2000. Criteria for solvent-induced chronic toxic encephalopathy: a systematic review. Int. Arch. Occup. Environ. Health 73, 362–368.

van Drongelen, A., van der Beek, A.J., Penders, G.B.S., Hlobil, H., Smid, T., Boot, C.R.L., 2015. Sickness absence and flight type exposure in flight crew members. Occup. Med. 65, 61–66.

van Hout, M.S.E., Schmand, B., Wekking, E.M., Deelman, B.G., 2006. Cognitive functioning in patients with suspected chronic toxic encephalopathy: evidence for neuropsychological disturbances after controlling for insufficient effort. J. Neurol. Neurosurg. Psychiatry 77, 296–303.

van Netten, C. 1998. Air quality and health effects associated with the operation of BAe 146–200 aircraft. Appl. Occup. Environ. Hyg. 13 (10), 733–739.

van Netten, C. 2005. Aircraft air quality incidents, symptoms, exposures and possible solutions. In: Proceedings of the Contaminated Air Protection Conference, London, 20–21 April 2005, pp. 243–253.

van Valen, E., van Thriel, C., Akila, R., Nilson, L.N., Bast-Petterson, R., Sainio, M., van Dijk, F., van der Laan, G., Verberk, M., Wekking, E., 2012. Chronic solvent-induced encephalopathy: European consensus of neuropsychological characteristics, assessment, and guidelines for diagnostics. Neurotoxicology 33, 710–726.

van Valen, E., Wekking, E., van Hout, M., van der Laan, G., Hageman, G., van Dijk, F., de Boer, A., Sprangers, M., 2018. Chronic solvent-induced encephalopathy: course and prognostic factors of neuropsychological functioning. Int. Arch. Occup. Environ. Health 91 (7), 843–858.

Verbeek, J., 2012. When work is related to disease, what establishes evidence for a causal relation? Saf. Health Work 3, 110–116.

Verberk, M.M., Brons, J.T., Sallé, H.J.A., 2004. Visual evoked potentials in workers with chronic solvent encephalopathy. Int. Arch. Occup. Environ. Health 77, 328–334.

Visser, I., Lavini, C., Booij, J., Reneman, L., Majoie, C., de Boer, A.G.E.M., Wekking, E.M., de Joode, E.A., van der Laan, G., van Dijk, F.J.H., Schene, A.H., den Heeten, G.J., 2008. Cerebral impairment in chronic solvent induced encephalopathy. Ann. Neurol. 63, 572–580.

Wang, D., Tao, Y., Li, Z., 1995. Toxic polyneuropathy due to flour contaminated with tricresyl phosphate in China. J. Toxicol. Clin. Toxicol. 33, 373–374.

Wang, C., Yang, X., Guan, J., Gao, K., Li, Z., 2014a. Volatile organic compounds in aircraft cabin: measurements and correlations between compounds. Build. Environ. 78, 89–94.

Wang, C., Yang, X., Guan, J., Gao, K., 2014b. Source apportionment of volatile organic compounds (VOCs) in aircraft cabin. Build. Environ. 81, 1–6.

Wei, B., Mohan, K.R., Weisel, C.P., 2012. Exposure of flight attendants to pyrethroid insecticides on commercial flights: urinary metabolite levels and implications. Int. J. Hyg. Environ. Health 215, 465–473.

Wei, B., Isukapalli, S.S., Weisel, C.P., 2013. Studying permethrin exposure in flight attendants using a physiologically based pharmacokinetic model. J. Exp. Sci. Environ. Epidemiol. 23, 416–427.

White, R.F., Proctor, S.P., 1997. Solvents and neurotoxicity. Lancet 349, 1239–1243.

White, RF, Steele, L, O'Callaghan, JP, Sullivan, K, Binns, JH, et al., 2016. Recent research on Gulf War illness and other health problems in veterans of the 1991 Gulf War: effects of toxicant exposures during deployment. Cortex 74, 449–475.
Winder, C., Balouet, J.C., 2000. Aerotoxic Syndrome: Adverse Health Effects Following Exposure to Jet Oil Mist during Commercial Flights. Bisbane, Proc Int Congress on Occup Health, pp. 196–199.
Winder, C., Balouet, J.-C., 2001. Aircrew exposure to chemicals in aircraft: symptoms of irritation and toxicity. J. Occup. Health Safety-Aust. NZ 17, 471–483.
Winder, C., Fonteyn, P., Balouet, J.-C., 2002. Aerotoxic syndrome: a descriptive epidemiological survey of aircrew exposed to in-cabin airborne contaminants. J. Occup. Health Safety-Aust. NZ 18 (4), 321–338.
Winder, C., Balouet, J.-C., 2002. The toxicity of commercial jet oils. Environ. Res. 89 (section A), 146–164.
Wolkoff, P., 2013. Indoor air pollutants in office environments: assessment of comfort, health, and performance. Int. J. Hyg. Environ. Health 216 (4), 371–394.
Wolkoff, P., Crump, D.R., Harrison, P.T.C., 2016. Pollutant exposures and health symptoms in aircrew and office workers: is there a link? Environ. Int. 87, 74–84.
World Health Organization, 2019. Handbook for the Management of Public Health Events in Air Transport: Updated with Information on Ebola Virus Disease and Middle East Respiratory Syndrome Coronavirus. accessed 2019 http://apps.who.int/iris/bitstream/10665/204628/1/9789241510165_eng.pdf.
Zhai, S., Li, Z., Zhao, B., 2014. State-space analysis of influencing factors on airborne particle concentration in aircraft cabins. Build. Environ. 74, 13–21.
Zhang, X., Wu, M., Yao, H., Yang, Y., Cui, M., Tu, Z., Stallones, L., Xiang, H., 2016. Pesticide poisoning and neurobehavioral function among workers in Jiangsu, People's Republic of China. Cortex 74, 396–404.

Further reading

Lees-Haley, P.R., Williams, C.W., 1997. Neurotoxicity of chronic low-dose exposure to organic solvents: a skeptical review. J. Clin. Psychol. 53, 699–712.
Naughton, S.X., Terry, A.V., 2018. Neurotoxicity in acute and repeated organophosphate exposure. Toxicology 408, 101–112.
Senate Rural and Regional Affairs and Transport References Committee, 2000. Parliament of the Commonwealth of Australia. Air Safety and Cabin Air Quality in the BAe 146 Aircraft, Report.

CHAPTER FIVE

Neurotoxicity of organic solvents: An update on mechanisms and effects

Christoph van Thriel[a,]* and William K. Boyes[b]

[a]Leibniz Research Centre for Working Environment and Human Factors, Department of Toxicology, Research Group: Neurotoxicology and Chemosensation, Dortmund, Germany
[b]Neurological and Endocrine Toxicology Branch, Public Health and Integrated Toxicology Division, Center for Public Health and Environmental Assessment, Office of Research and Development, Environmental Protection Agency, Research Triangle Park, NC, United States
*Corresponding author: e-mail address: thriel@ifado.de

Contents

Advances in Neurotoxicology, Volume 7
ISSN 2468-7480
https://doi.org/10.1016/bs.ant.2022.05.004

1. Introduction

1.1 Organic Solvents and Exposures

The National Institute of Occupational Safety and Health NIOSH) describes organic solvents as "... carbon-based substances capable of dissolving or dispersing one or more other substances" (https://www.cdc.gov/niosh/topics/organsolv/default.html). Organic solvents, as a group, are a diverse set of chemical structures, often applied and used as mixtures, that have among the highest volume usage of chemicals in industry, commerce, and consumer products. During the history of industrialization, organic solvents have become more frequently used in workplaces (Dick, 2006) thereby exposing workers to, at that time, unknown or at least underestimated health risks (Baker, 1994). Organic solvents have been and are still used in consumer and household products. Release into the environment, especially indoors, is very likely (Bolden et al., 2015). Indoor exposure to aromatic compounds (benzene, toluene, ethylbenzene, and xylene; abbreviated as BTEX) is typically higher than outdoor exposures due to evaporations from consumer products. Moreover, there are cases of intensive and "uncontrolled" use of organic solvents in small workshops such as nail salons (Harrichandra et al., 2020) resulting in high daily exposures for the employees. Organic solvents have among the highest volume releases into the air among chemicals tracked by the US National Emissions Inventory (U.S. Environmental Protection Agency (EPA), 2017). Thus, workers and the general population can be exposed to different concentrations of solvents for different durations.

The U.S. Clean Air Act is a piece of legislation covering both stationary and mobile sources of air pollution in the United States. Substances are categorized in the Act as either one of six Criteria Air Pollutants, including: carbon monoxide, ground level ozone, lead, nitrogen oxides, particulate matter and sulfur dioxide, or as Hazardous Air Pollutants (HAPs). The list of HAPs (also known as "Air Toxics") currently includes 187 chemicals or chemical classes, including volatile organic compounds (VOCs), many of which are common organic solvents. According to the EPA's National Emissions Inventory (NEI), over 43 million tons of VOCs were emitted into the air in 2017 (the most recent year for which data are available). The emissions of VOCs accounted for 27.7% of all national air pollutant emissions (https://gispub.epa.gov/neireport/2017/, accessed 3/9/2022). EPA uses the NEI emissions data, along with air dispersion modeling, to estimate ambient air concentrations, and then compares those values to health benchmark Reference Concentrations (RfC) for individual substances to create Hazard Quotients (HQs). By summing the HQ values established specifically for neurological hazards, a Total Neurological HQ was calculated. Although the Total Neurological HQ values were low on a national basis, there were several local census tracts where Total Neurological HQ values were greater than 1, indicating a total exposure above the levels deemed to be safe, assuming that the neurological risks of VOCs were additive (U.S. Environmental Protection Agency (EPA), 2014). Consider also, that some VOCs, such as benzene, might have had RfC values set on cancer risks or other health outcomes, but which also could contribute to the total neurological risk, making the Total Neurological HQ a potential underestimate. Over 19 million people lived in census tracts with Total Neurological HQs greater than 0.1. Some people in those regions may also have occupational exposures to organic solvents, and their atmospheric exposures could be a substantial portion of their personal exposure burden. In summary, there is a sizable volume of organic solvent vapors emitted into the air annually, and in some locations the resultant exposures may represent concerns for neurotoxicity.

Exposure to organic solvents has long been recognized as a potential occupational hazard (Hooisma et al., 1993, 1994) leading to a variety of health risks, including both acute and chronic neurotoxicity (see Fig. 1). Because of their wide variety of uses and applications, which vary by chemical, the neurotoxic health effects can also vary with dose and exposure scenarios. A schematic depiction of the types of exposures and neurotoxic outcomes, arranged approximately by exposure concentration and time, is presented in Fig. 1.

1.1.1 Organic solvents as air pollutants

High concentrations of organic solvents are usually applied as gaseous anesthetics (Wrigley et al., 1991) or in case of their abuse as an inhalant drug (Yucel et al., 2008). For abused inhalants, even if the duration of the one exposure episode is short, the health effects can be severe with repeated use. Long-term, mid-range concentration exposures that usually occur in occupational settings are linked to neurophysiological and functional perturbations that can lead to an occupational disease called chronic toxic encephalopathy, or chronic solvent encephalopathy (CSE; see Dick, 2006).

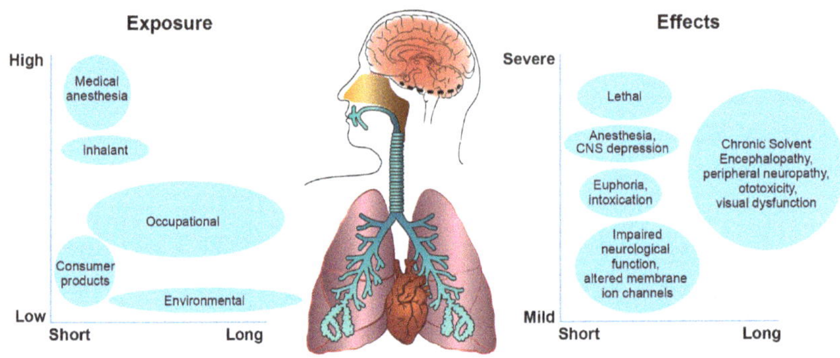

Fig. 1 Exposure to organic solvents (left side) and their effects (right side) vary markedly both as a function of concentration and time. (Please note that the concentration axis here is confounded by the potency of the substances, for example the actual inhaled concentrations of inhalation anesthetics may be lower than those experienced by severe inhalant abusers who have access to less potent substances than medical grade anesthetics. In addition, please consider "time" here to include both the duration and repetition of exposure). Like exposure profiles, the potential health effects caused by exposure also vary according to concentration and duration (or repetition) of exposure (right side). Time and dose axes are not to scale on either side of the figure and are provided only to approximate the relative exposure scenarios discussed.

1.2 Classification approaches based on molecular descriptors

Since the chemical structures (i.e., carbon-based substances) as well as the function (i.e., capable of dissolving or dispersing one or more other substances) of solvents are well described, classification approaches can be based on the chemical properties describing the various aspects and dimensions of the molecules (Parker, 1969) to group these compounds into protic, dipolar aprotic, and apolar aprotic solvent. This initial model used the dipolarity of the solvents as well as their ability to act as hydrogen bond donors as main descriptors to group them. Later, the conductor-like screening method (COSMO; Durand et al., 2011) was introduced to model the solvents as virtual conductors that surround the solute molecules. In their approach, Durand et al. (2011) grouped 153 chemicals into 10 classes with larger classes such as the aprotic dipolar solvent (e.g., ethyl acetate, cyclohexanone), the apolar solvent (e.g., n-hexane, benzene, toluene), and the amphiprotic solvents (e.g., ethanol). Based on Parker's (1969) ideas and using molecular descriptors related for instance to cavity formation, electrostatic or dispersion interaction, another mathematical approach using principal component analysis (PCA) reduced the matrix of 774 (solvents) × 100 (solvent scales) to 5 principal components (PCs) that account for 75% of the total variance

(Katritzky et al., 2005). By using these 5 PCs a total of 703 solvents were classified according to the presence of various functional groups into 11 classes of compounds. In more details, these classes were: (1) hydrocarbons, (2) halo hydrocarbons, (3) saturated, unsaturated, and cyclic ethers, (4) esters and polyesters, (5) aldehydes, ketones, and amides, (6) nitriles and nitro hydrocarbons, (7) hydroxylic compounds, (8) amines and pyridines, (9) thiols, sulfides, sulfoxides, and thiol compounds, (10) phosphorus compounds, and (11) miscellaneous. Thus handbooks addressing chemical aspects of organic solvents (Smallwood, 1996) followed this approach and some of the classes or subgroups are used there. Depending on their molecular structure, this diverse group of chemicals can be organized into subgroup such as aliphatic (straight carbon backbone) or aromatic compounds (six-carbon ring). In addition, the classification of organic solvents can also involve attached functional structures such alcohols, ethers, ketones, esters, and, of particular interest, are the chlorinated organic solvents. Some of the major industrial solvents included under these categories are listed in Table 1 (Smallwood, 1996). Organic solvents are related to several classes of organic compounds that, although not typically used as solvents, share similar physical/chemical properties, and may lead to similar effects upon exposure. Styrene, for example, is often considered to produce a solvent–like neurotoxicity even though it's primary use in industry is as a monomer for synthesis of polystyrene. These related classes of substances include some inhalation anesthetics, some inhalant drugs of abuse, ingredients of some consumer products, and components of petroleum-derived engine fuels. Some individual substances may be present in several of these categories, such as toluene, which has been a constituent of gasoline, oil-based paints, glues, printing ink and is a substance of choice for inhalant abusers.

With a focus on occupational exposures, organic solvents are used in a remarkable variety of industries, some of which are listed in Table 2. Frequent applications are cleaning and degreasing, auto or machine repair, dry cleaning, solvents in oil-based paints, inks or glues, components of cleaning solutions, and dissolution of lipid-soluble materials as in manufacture of pharmaceuticals or other chemical synthesis. In addition, the petroleum industry produces large volumes of volatile hydrocarbons that become automotive fuels, heating oils and petroleum distillate products such as white spirits, mineral spirits or kerosine. Military uses include fuels for vehicles including aircraft (JP6, JP8, etc.) and cleaning and maintenance of equipment. Some applications involve considerable dermal, as well as inhalation exposures such as cleaning and degreasing machine components.

Table 1 Commonly used organic solvents in structurally related groups.

Hydrocarbons	Alcohols	Glycol ethers	Chlorinated solvents	Ketones	Ethers	Esters	Miscellaneous solvents
n-Pentane	Methanol	Propylene glycol methyl ether	Methylene chloride	Acetone	Diethyl ether	Methyl acetate	Dimethylformamide
n-Heptane	Ethanol	Ethylene glycol methyl ether	Chloroform	Methyl ethyl ketone	Diisopropyl ether	Ethyl acetate	Dimethylacetamide
n-Octane	n-Propanol	Ethylene glycol ethyl ether	Carbon tetrachloride	Methyl isobutyl ketone	Dibutyl ether	Isopropyl acetate	Dimethlsulphoxide
n-Nonane	i-Propanol	Ethylene glycol monobutyl ether	1,2-Dichlorethane	Cyclohexanone	Methyl *tert* butyl ether	n-Butyl acetate	Sulfolane
n-Decane	n-Butanol		1,1,1-Trichloro-ethane	n-Methyl-2-pyrrolidone	1,4-Dioxane	Cellosolve acetate	Carbon disulphide
2,2,4-Trimethyl pentane	i-Butanol		Trichloroethylene	Acetophenone	Tetrahydrofuran		Acetic acid
Cyclohexane	s-Butanol		Perchloroethylene				Aniline
Benzene	n-Amyl alcohol		Monochlorobenzene				Nitrobenzene
Toluene	i-Amyl alcohol						Morpholine
Ethylbenzene	Cyclohexanol						Pyridine
Xylene (mixed isomers)	n-Octanol						2-Nitropropane
C9 Aromatics	Ethanediol						Acetonitrile
Tetralin	Diethylene glycol						Purchase
	1,2-Propenediol						Furfuraldehyde
							Phenol
							Water

Modified after Smallwood, I. M. 1996. Handbook of Organic Solvent Properties, Oxford, Butterworth-Heinemann.

Table 2 Various industries using organic solvents.

Adhesives and sealants	Inorganic chemical industry	Polymers and man-made fibers
Aerospace	Iron and steel industry	Printing industry
Asphalt compounding	Lumber and wood products	Pulp and paper
Biotechnology	Medical applications	Rubber and Plastics
Coil coating	Metal casting	Shipbuilding and ship repair
Cosmetics/personal care products	Motor vehicle assembly	Stone, clay, glass, and concrete
Dry cleaning	Organic chemical industry	Textile industry
Fabricated metal products	Paints and coatings	Transportation equipment cleaning
Food industry	Petroleum refining industry	Water transportation
Ground transportation	Pharmaceutical industry	Wood furniture

Modified after Wypych, G. 2019. 13–Solvents use in various industries. In: Wypych, G. (ed.) Handbook of Solvents (Third Edition). ChemTec Publishing.

Chlorinated hydrocarbons may also be formed as disinfection byproducts when drinking water sources contaminated with hydrocarbons and are treated with chlorine for disinfection. Among the many industrial uses of organic solvents, the tasks most often associated with neurotoxic exposures include industrial painting, construction painting, printing presses, boat construction, reinforced plastic laminating and metal degreasing (Furu et al., 2012). In summary, organic solvents are widely used for a wide variety of applications and may become unintentional environmental contaminants which raises multiple opportunities for exposures in occupational and other settings.

In addition to these "grouping" approaches (see Table 1) that are useful to structure the variety of organic solvents in the environment there have been and still are approaches to link such physicochemical properties to biological processes and effects including toxicity. These approaches will be addressed in the next section.

1.3 Physicochemical properties predicting exposures and toxicity

Due to their typically high vapor pressure organic solvents are volatile at normal temperatures, and therefore, as outlined in Fig. 1 inhalation is the most relevant route of exposure. Thus, the vapor pressure of a particular solvent is a suitable indicator in relation to airborne exposures and the risk of inhalation. However, to cross membranes and biological barriers, such as the blood brain barrier, lipophilicity is another important property. The n-Octanol/Water Partition Coefficient (K_{OW}) or the logarithmic expression of the physicochemical property log K_{ow} (or log P) is an important measure to estimate the lipophilicity of a compound. There are two OECD test guidelines (TG 117 and 123) available describing standardized methods for its assessment. These values are also relevant in other areas of toxicology, drug development or food safety as lipophilicity is a good predictor of bioavailability in biological systems (Chmiel et al., 2019), including the human body. One of the first approaches to link lipophilicity as well as other physicochemical properties of organic solvents to their neurotoxicity was described in 1996 (Cronin, 1996). Cronin used a dataset of 48 common solvents (Frantík et al., 1994) that estimated acute neurotoxic effects in mice and rats by measuring the inhibition of propagation and maintenance of the electrically evoked seizure discharge. When removing outliers Cronin (1996) showed a good correlation between the log K_{ow} and the EC_{30} for the acute neurotoxicity in both mice and rats. Recently, using the same dataset, more complex quantitative structure-toxicity/activity relationship (QSTR/QSAR) models that have been established (Basant et al., 2016). Nevertheless, in addition some other descriptors such as Shev (*E*-state indices over all non-H atoms) and Smin (the minimal E-state value in all atoms) the log K_{ow} value has still a great importance in predicting acute neurotoxicity. Based in part on evidence that momentary concentration of solvents in the brain determined their acute activity, Linakis et al. (Linakis et al., 2020) developed a high throughput model using calculated physiochemical properties of novel small lipophilic organic chemicals to predict toxicokinetics, and thereby assist rapid assessment of potential risks of a variety of exposure scenarios of different exposure concentrations and durations.

However, the mechanism underlying the neurotoxicity of the diverse group of organic solvents are by far more complex than suggested by these simple prediction models. The next section will provide an overview about the various molecular and cellular mechanisms causing acute and chronic neurotoxicity.

2. Mechanisms of solvent neurotoxicity

2.1 Background

For many years, the demonstration by Meyer and Overton that the potency of anesthetics was proportional to their lipid solubility dominated theories about anesthetic mechanisms of action and also theories about the neuroactive effects of organic solvents (Bovill, 2000). Lipid content of nerve cell membranes and the lipophilic regions of proteins residing in nerve membranes suggested a site of action for organic solvents. The leading hypothesis was that the apparently non-selective action of many structurally different solvent molecules involved partial dilution of the phospholipid bilayers of the nerve cell membrane, increasing the membrane fluidity and consequently reducing the ability to selectively partition ions necessary for neuronal polarization and depolarization (Bovill, 2000). If anesthetics and solvents act merely to increase membrane fluidity, however, they would not be expected to be stereoselective. The R(+) isomer of the anesthetic etomidate is 15 times more potent than the opposite (S-) isomer (Tomlin et al., 1998). Stereoselective potency has also been reported for isomers of isoflurane, barbiturates and steroid anesthetics on the $GABA_A$ receptor (reviewed in Bovill, 2000). Point mutations of the $GABA_A$ or glycine receptor can completely abolish the anesthetic effects (Mihic et al., 1997). Measurements of membrane fluidity or leak currents are altered only at much higher concentrations than those effecting ion channel functions (Cruz et al., 1998). The Meyer/Overton hypothesis is incompatible with these selective actions. Rather than a general effect on the membrane phospholipid bilayer, current evidence suggests that anesthetics and neuroactive organic solvents act through disrupting functions of multiple nerve membrane ion channels including notably: the $GABA_A$ receptor, the NMDA receptor, the nicotinic acetylcholine receptor (nACh), voltage-sensitive calcium channels (VSCC), and perhaps other membrane ion channels or intracellular signaling pathways. In this view, the high lipid solubility of organic solvents ensures rapid access to the lipid rich tissues of the brain but is insufficient to account for their neuroactive properties. The molecular subunit composition of ligand-gated ion channels is important for anesthetic actions (Harris et al., 1995). Reviews of the molecular actions of anesthetics can be found elsewhere (Grasshoff et al., 2006; Mashour et al., 2005). A recent review of the mechanisms of action of abused inhalant drugs provides an extensive discussion of organic solvents, and much of the mechanistic information is also relevant to occupational exposures to organic solvents (Cruz and Bowen, 2021).

2.1.1 Effects on nerve membrane ion channels

Organic solvents, like ethanol and several inhalation anesthetics, disrupt the function of several ligand- and voltage-gated nerve membrane ion channels. Toluene inhibits the functions of the glutamate neurotransmitter receptor *N*-methyl-D-aspartate (NMDA) (Bale et al., 2005b; Cruz et al., 1998, 2000). Toluene, however, had no effect on NMDA-currents when activated by the non-receptor agonist kainic acid. The deficits in NMDA receptor function were dependent on the composition of receptor subunits, NR2A and NR2B, with receptors having the NR2B form being several times more sensitive than other forms (Bale et al., 2005b, Cruz et al., 1998, 2000). Williams et al. (2005) similarly found increases in glutamatergic receptor subunits, NR1, NR2B (NMDA receptor subunits) and GluR2/3 (AMPA receptor subunits), in the medial prefrontal cortex and decreased NR1 subunits in the substantia nigra compacta (Williams et al., 2005). Like actions on the glutamate receptor, the actions of toluene on GABA receptors vary by receptor subunit composition, which also vary by brain region. The function of $GABA_A$ receptors is augmented by toluene, increasing Cl^- current when exposed in vitro (Bale et al., 2005b). In rats exposed by inhalation, toluene (8000 ppm, 30 min/d, 10 days) increased the GABAergic receptor subunit $GABA_A$ $\alpha1$ in the medial prefrontal cortex (Williams et al., 2005) and decreased $GABA_A$ $\alpha1$ in the substantia nigra. These results indicate that, in rats, toluene differentially alters glutamatergic and GABAergic receptor subunit composition in a brain region specific manner.

Nicotinic receptors are also sensitive to inhibition by organic solvents. Toluene and perchloroethylene were studied for actions on both human and rat forms of nicotinic acetylcholine receptors expressed in Xenopus oocytes in vitro or in cultured hippocampal neurons (Bale et al., 2002, 2005a). Both substances inhibited currents through the nicotinic receptors expressed with various combinations of receptor subunits. Forms of the receptor containing the $\alpha4\beta2$ and $\alpha3\beta2$ subunit combinations were more sensitive to toluene inhibition than the $\alpha4\beta4$, $\alpha3\beta4$ and $\alpha7$ receptors (Bale et al., 2002, 2005a). Perchloroethylene was more potent than toluene in inhibiting nicotinic receptor function, but there were no differences between rat and human forms of the receptors.

In PC12 phaeochromocytoma cells, a cell line that synthesizes voltage- and ligand-gated ion channels (Shafer and Atchison, 1991), toluene increased the frequency of basal neurotransmitter release in vitro, an effect mediated through increased intracellular Ca^{2+} caused by influx of Ca^{2+} through voltage-sensitive Ca^{2+} channels (VSCC) (Westerink and Vijverberg, 2002). The effect of organic solvents, including toluene, TCE and perchloroethylene on VSCC was dependent on the trans-membrane potential (Shafer et al., 2005). In whole-cell patch-clamp recordings from PC12 cells, Ca^{2+} currents were inhibited by toluene, TCE and perchloroethylene at positive transmembrane potentials, but potentiated at negative transmembrane potentials.

2.1.2 Effects on neurotransmitter systems

Toluene, like other drugs of abuse, alters brain dopamine systems (Cintra et al., 1999; Gralewicz and Dyzma, 2005; von Euler et al., 1993, 1994). Repeated exposure to toluene caused persistent changes in dopamine D2 agonist binding in the rat caudate-putamen (Hillefors-Berglund et al., 1995). Toluene inhalation (80 ppm, 6 h/d, 5d/w, 4 weeks) altered rat dopamine D2 receptor binding in striatum (von Euler et al., 1993) and impaired function in the Morris Water Maze, altered locomotor function decreased the size of cerebral cortex (von Euler et al., 2000). Repeated treatment with toluene reduced the affinity and increased the density of beta-adrenergic antagonist binding sites in prefrontal cortex of rats (Fuxe et al., 1987).

Glutamate is the primary excitatory neurotransmitter in the CNS, including in the visual pathway. Rats acutely inhaled toluene (2000 ppm) 10 min after receiving either NMDA (agonist) or MK801 (antagonist) for NMDA receptors (Bale et al., 2007). Pattern elicited visual evoked potentials (VEPs) were reduced in amplitude by toluene alone. MK801 given 10 min prior to toluene blocked the toluene-induced amplitude decrease, indicating that the toluene suppression of visual function involved NMDA receptors in some fashion.

In a simulated drug abuse exposure scenario, rats were exposed to high concentrations of toluene (800–12,000 ppm for two 15 min exposure periods) and brain regions collected afterward were evaluated using high-resolution magic angle spinning proton magnetic resonance spectroscopy (O'Leary-Moore et al., 2009). Binge toluene exposure altered several neurochemical concentrations, including decreased levels of GABA in frontal cortex and striatum, and lowered glutamate and *N*-acetyl aspartate in frontal cortex. Older animals (56 days) were more affected than adolescents (35 days).

2.1.3 Effects on neuronal networks (LTP)

Neural plasticity, the alteration of function in response to experience, is a basic property of the nervous system that can be observed at the level of neural networks. Long-Term Potentiation (LTP) refers to a persistent enhancement of synaptic transmission following high frequency stimulation and is thought to reflect the cellular processes underlying learning and memory. The induction of LTP is dependent on activation of NMDA receptors, which are inhibited by organic solvents. To test the effects of long-term exposure to organic solvents on LTP, rats were exposed to TCE (0 or 500 ppm, 6 h/d, 5 d/w, 6 months), and, after exposure terminated, brain slice preparations from visual cortex and hippocampus were assessed for paired-pulse potentiation/inhibition and LTP (Altmann et al., 2002). TCE impaired LTP and paired-pulse inhibition in the hippocampus, but not in visual cortex. No effects were observed on locomotor activity, passive avoidance, context-conditioned fear, or brainstem auditory evoked potentials. This suggests that impairment of neural plasticity in the hippocampus may be more sensitive to disruption than other neurological functions and may be the basis for reports of altered memory and cognitive function after long term occupational exposure.

2.1.4 Cytotoxicity, oxidative stress and inflammation

Direct exposure of primary cultures of rat neurons and astroglia to toluene for 60 min altered to morphological appearance of astrocytes and neurons, with astrocytes being sensitive to lower concentrations than neurons (Hansson et al., 1988).

The formation of reactive oxygen species was studied in rat cerebellar granule cells in vitro for a range of aliphatic, alicyclic and aromatic C6-C10 solvents (Dreiem et al., 2002). Dose-related increased formation of reactive oxygen indicator chemicals was observed for aromatic compounds with greater than C7 and for the other classes with C7 or greater. The aromatic solvents were generally less efficient at inducing reactive oxygen species than the other classes, and the potency correlated to octanol/water partition coefficients (LogP).

McDermott et al. studied Jurkat T cells (immortalized human T lymphocytes) and SH-SY5Y (human-derived neuroblastoma cell line) exposed in vitro for 5 days to toluene, n-hexane or methyl ethyl ketone (MEK) (McDermott et al., 2007). At micromolar concentrations of solvents in vitro resembling those of blood from workers exposed at the occupational

Threshold Limit Values (TLV), they observed membrane damage (LDH leakage), increased intracellular free calcium, and altered glutathione redox status in both cell lines. The potency of the three solvents correlated with their hydrophobicity. The changes in intracellular calcium occurred at the lowest concentrations among the measured outcomes followed by LDH leakage and then altered glutathione status.

Measures of oxidative stress in brain tissue differed by brain region and age of the rats following a single oral gavage dose (0, 0.65 or 1 g/kg) of toluene. For example, NADPH quinone oxidoreductase activity increased in frontal cortex and cerebellum of 4- and 12-month-old rats, but only in the hippocampus of 24-month-old rats (Kodavanti et al., 2011). Markers of oxidative stress were greater in rat brains following sub-chronic inhalation exposure (0, 97 or 995 ppm, 6 h/d, 5d/w, 13 weeks) than following acute inhalation exposure (1019 ppm, 6 h) (Kodavanti et al., 2015).

The role of oxidative stress in organic solvent neurotoxicity can be assessed by co-treatment with antioxidants. Kanter exposed rats to toluene (3000 ppm; 8 h/d, 6d/w, 12 weeks) or to toluene plus daily oral doses of an extract from seeds of the plant *Nigella sativa* with reported antioxidant and immunomodulatory properties (Kanter, 2008). Rats exposed to only toluene showed extensive neuropathological changes in histological sections from frontal cortex and brain stem. Damage included dark stained nuclei, shrunken cytoplasm, and increased immunostaining of caspase-3 indicative of apoptosis. These changes were reduced by cotreatment with *N. sativa* extract, suggesting that the toluene neurodegeneration involved formation of reactive oxygen species and oxidative stress triggering apoptotic neuronal degeneration.

Neuroimmune responses can be initiated by the presence of foreign microbes or cellular degradation products through activation of Tol-like receptors. Win-Shwe et al. exposed mice to toluene (0, 5, 50, 500 ppm, 6 h/d, 5 d/w, for 6 weeks) to examine the role of Tol-Like Receptor 4 (TLR4) in toluene neurotoxicity (Win-Shwe et al., 2011). Responses of wild-type mice (C3H/HeN) were compared to a TLR4-mutant strain (C3H/HeJ). In wild type mice mRNA expression in hippocampus for TLR4 was increased by 50 ppm toluene, but not 500 ppm. In TLR4 mutants, there was no increase in TLR4 mRNA expression. The expression of mRNA for heat shock protein was also increased by 50 ppm toluene in the wild type, but not the mutant mice. The results suggest a role for neuroimmune signaling via the TLR4-heat shock protein pathway in toluene neurotoxicity, although there was no explanation for the U-shaped dose-response nature of the effect.

Cruz et al. (2020) exposed rats to toluene in a simulated abuse paradigm (8000 ppm, 30 min, 2×/day, 5 days) while they were co-exposed to Minocycline, which is a potent anti-inflammatory agent that inhibits the NFκB pathway while reducing cytokine and ROS production (Cruz et al., 2020). Two days after the final exposure, brain slice preparations from pre-frontal cortex were taken. Whole-cell patch clamp recordings from layer 5 pyramidal cells showed no effect of minocycline alone, but minocycline prevented an increase in firing rate caused by toluene. Toluene exposure reduced the after- hyperpolarization current mediated through Ca^{2+}-dependent K^{+} channels, which was also prevented by minocycline. In addition, minocycline prevented toluene-induced glial cell activation. In learning tasks where training occurred after the onset of exposures including a passive avoidance task and a temporal-order object recognition task, toluene exposure significantly impaired the acquisition of both tasks, and the co-exposure to minocycline reduced or eliminated the toluene learning deficits.

2.1.5 Gene expression studies (rodent)

Ahmed et al. exposed mice to toluene (50 ppm, 6 h/d, 5d/w, 6 or 12 weeks) measured hippocampal mRNA expression of genes related to synaptic plasticity (Ahmed et al., 2007). After 12 weeks, genes related to NMDA receptor subunit 2B (NMDA NR2B) were upregulated and there was induction of CaMKIV, CREB-1, and FosB/[Delta]FosB in hippocampal tissues.

Acute toluene inhalation exposure (1000 ppm, 6 h) altered expression levels in rat brain of multiple genes associated with synaptic plasticity, GABA receptor signaling and mitochondria function. The changes in gene expression were greater after 18 than 6 h post-exposure (Hester et al., 2011). Rats exposed to toluene repeatedly (0, 10, 100, 1000 ppm, 6 h/day for 64 days), in contrast, had fewer differentially expressed genes than after acute exposure and those genes in common were not consistently altered in the same direction (Hester et al., 2012). These two experiments imply that the acute exposure to toluene produces substantial changes in brain gene expression, but following repeated exposure there is considerable tolerance at the gene expression level. This result contrasts with Kodavanti et al. (2015) described earlier, who showed greater oxidative stress in the brain after sub-chronic than acute toluene inhalation.

2.1.6 Studies in drosophila and C. elegans

The use of invertebrate species for toxicity testing has potential advantages of efficiency, reduced use of mammalian test subjects, and the ability to identify mechanisms of susceptibility in genetically sequenced strains. The sensitivity of adult animals to narcosis from acute toluene exposure was assessed in a panel of 123 genetically defined strains of *Drosophila melanogaster*, and the results mapped with a genome-wide association analysis (Bushnell et al., 2017). The analysis identified 66 candidate genes associated with phenotypic sensitivity to toluene, and 55 of those could be associated with orthologous human genes. None of the identified gene pathways associated with *Drosophila* susceptibility were orthologs of mammalian neuronal ion channels or neurotransmitter systems, but instead identified orthologs labeled as signaling, nervous system development and embryonic morphogenesis.

Drosophila larva that were exposed to BTEX showed increased apoptosis and mitosis after exposure (Adebambo et al., 2020). Using lethality as an outcome, a gene-wise analysis of larval susceptibility to xylene identified candidate genes for nervous system development, neuromuscular synaptic transmission, membrane signaling factors, carbohydrate metabolism, wing disk development, transcription factor binding sites (TFBS) and apoptotic processes.

C. elegans treated with toluene in a vapor tight chamber showed an irreversible reduction in neuron fluorescence and morphological alterations that were most evident in dopaminergic neurons, and which were associated with mitochondrial damage, reduced levels of adenosine 5'triphosphate (ATP), and mitochondrial complex I inhibition (Soares et al., 2022).

2.1.7 Limitations of in vitro studies

Currently, there is a strong push toward the expanded development and use of in vitro and alternative test species, and application of computational approaches in toxicology generally including in neurotoxicology. There is an associated movement to reduce whole animal testing for both ethical and economic reasons. In vitro toxicity testing models, although vital for evaluating mechanistic hypotheses and very promising for screening purposes, are typically very short in duration, have difficulty containing volatile materials, and typically involve single neurons or simple neural circuits that do not reflect the complex systemic functions of the intact nervous system such as learning, memory or other cognitive functions. The primary rationale for alternative test models involves reducing use of live animals in product testing, and increased speed and efficiency of chemical screening. This reasoning undervalues the benefits of verification of human observational studies, where the strength of evidence may be weakened by factors such as correlational analyses, cross-sectional designs, lack of historical exposure records, mismatched control populations and other potential experimental biases. The scientific credibility of relationships between chronic solvent exposures and neurotoxicity would be greatly advanced if there were whole animal models with high validity to support the human evidence. Unfortunately, because of the factors mentioned above, it appears likely that the dearth of valid animal models of CSE will continue indefinitely.

2.2 Summary of mechanistic studies

Organic solvents have multiple effects on the nervous system. The effects vary by dose level and duration of exposure. The solvents affect multiple molecular targets and produce multiple outcomes. The high lipid solubility of organic solvents leads to rapid delivery to the brain upon exposure, with the amount of brain penetration dependent on the compound's lipophilicity. In the brain, solvents partition to the nerve membrane phospholipid bilayer and there alter the functioning of critical membrane-resident proteins including multiple types of ligand-gated and voltage-gated ion channels. Among the ion channels altered are glutamate NMDA receptors, which are inhibited by organic solvents. Glutamate is the primary excitatory neurotransmitter in the central nervous system. The NMDA glutamate receptor is also critical for the induction of Long-Term Potentiation (LTP), a mechanism of neuronal plasticity that is theorized to represent the cellular basis of learning and memory. Organic solvents also affect $GABA_A$ receptors to potentiate their inhibitory actions. The combination of reduced glutamate excitation and increased GABA inhibitory activity is thought to cause overall CNS depression. Solvent exposures also have direct effects on nerve membranes, inducing formation of reactive oxygen species, damaging membranes, inducing cellular leakage, LDH release, increased intracellular Ca^{2+}, and through Tol-like receptors, triggering neuroimmune responses, including activation of reactive astrocytes and microglia. Solvents induce a range of gene expression changes in pathways associated with neuroplasticity, GABA receptor signaling and mitochondria function. Thus, solvent neurotoxicity is expressed acutely through multiple changes in membrane ion channels and neurotransmitter systems that vary with receptor subunit composition, brain region as well as over time and dose. In addition, prolonged oxidative stress and neuroinflammation could conceivably produce the pattern of cognitive and psychological changes reported after chronic occupational exposures to organic solvents.

3. Acute and chronic neurobehavioral toxicity in rodents and humans

The complex and diverse mechanisms described in the previous section suggest a variety of neurobehavioral and -physiological effects that indicate neurotoxicity in humans and laboratory animals. Historically, the discovery of neurotoxicity of organic solvents overlaps with the increasing use and standardization of behavioral tests that are capable to integrate all the molecular and cellular perturbations into one measure: impaired performance in neurobehavioral tests (Reiter, 1978). Accordingly, much of the knowledge about solvent neurotoxicity is based on human experimental and epidemiological studies in humans. This knowledge can be further supported by numerous studies using experimental animals including behavioral observations and more invasive measures directly in the brain.

3.1 Neurobehavioral and neurophysiological human studies

The range of human exposures, especially for workers, to organic solvents and their effects is wide and varies as a function of both the concentration and duration of exposure (Fig. 1). Carefully controlled exposure to inhalation anesthetics produces surgical anesthesia. Among inhalant drug abusers, exposure to sub-anesthetic concentrations of solvents can cause euphoria and intoxication, which with continued exposure can lead eventually to unconsciousness. Repeated high concentration inhalant abuse is associated with demonstrable neuropathology (Yucel et al., 2008). High-level, short-term exposure may be lethal through acute depression of the Central Nervous System (CNS). Lower-level acute exposures can occur from use of consumer products such as cleaning solutions and nail polishes. The effects of exposure to such products may include slight light-headedness or sensory irritation depending on the situation. Nerve membrane ion channels including GABA and glutamate, among others, are highly sensitive to organic solvents and likely underly the acute CNS dysfunction experienced after exposure. Occupational exposure to organic solvents can involve moderate to higher concentrations across years of duration. The occupational exposure to organic solvents has been linked to production of a neurological syndrome called Chronic Solvent Encephalopathy (CSE, see Sainio, 2015). This syndrome was first described in the Nordic countries (Hanninen et al., 1976) and involves a variety of diffuse cognitive, memory, mood, and affective deficits. Also associated with chronic exposure to organic solvents are sensory deficits including hearing loss (ototoxicity) and visual or olfactory dysfunction including deficits in color discrimination and sensitivity to contrast in visual patterns. Many of these issues are discussed more extensively in subsequent sections of the chapter.

There are important exceptions to the simplified depictions of exposure scenarios in Fig. 1. For example, the figure depicts environmental exposures as being low concentration and long duration. However, there are also episodic or "upset" emissions from chemical plants that may occur during start-up or shutdown of batch processes, or during plant maintenance, in which higher concentrations of emissions are released over a short period and may lead to a chemical plume exposing plant workers and nearby residents who are downwind from the site. These types of emissions may be either somewhat periodic or highly irregular, and the emission plumes can be difficult to detect because of their transient nature. Still the effects of a single large plume, or the cumulative effects of repeated exposures over time, may be significant for the workers or the exposed communities. The latter may be a particular concern because the communities residing near large chemical facilities tend to be lower income populations with limited resources or political influence and other ongoing health challenges. In addition to episodic releases, there can be large scale releases from transportation accidents such as truck, train, or tanker ship crashes, and rarely large scale environmental disasters such as the Deepwater Horizon explosion and oil spill that killed several workers outright, devastated a large segment of the Gulf of Mexico environment, and exposed many environmental species, cleanup workers and costal residents to petroleum hydrocarbons and their combustion products as the oil burned off the water surface.

3.2 Neurological effects of acute exposure

While the first evidence regarding the neurotoxic effects of organic solvents was provided by epidemiological studies of occupationally exposed workers in the 1970s (Axelson et al., 1976; Hane et al., 1977; Harkonen, 1977; Knave et al., 1978) both the mechanistic knowledge and observations in these studies, such as improved performance after a weekend, pointed to the importance of acute exposures leading to reduced performance in neurobehavioral tests or altered neurophysiological functions. One of the driving forces in this research area was the group of Francesco Gamberale from Stockholm, Sweden who used human volunteer studies to identify the acute neurotoxicity of various solvents as well as their mixtures.

3.2.1 Human volunteer studies

Using a well-designed exposure system and healthy human volunteers, the acute and dose-dependent neurotoxic effects of toluene, methyl chloroform, styrene, methylene chloride, trichloroethylene, xylene, ethylbenzene, as well as the mixture white spirit (Gamberale et al., 1975a,b, 1976, 1978; Gamberale and Hultengren, 1972, 1973, 1974) were studied in the 1970s. At these early days of neurobehavioral toxicology, the assessments of cognitive or motor performance as surrogate markers of acute solvent neurotoxicity were not standardized, and self-programmed computer tests assessing, for example, the simple reaction time of the exposed individuals were used. Nevertheless, many of these studies revealed performance deficits during high exposure conditions. But as normative data was lacking the severity of these behavioral changes were often difficulted to evaluate. Later, a comprehensive and standardized test battery, the Swedish Performance Evaluation System SPES) was developed based on these initial test systems (Iregren et al., 1996). Gamberale and Hultengren (1972) investigated the acute neurobehavioral effects of toluene in 12 young (20 to 35 years) healthy males, testing airborne concentrations of 100, 300, 500, and 700 ppm. In these early human volunteer studies, short exposure periods (20–30 min) were presented using a stair wise increase of the solvent concentration in the inspired air. During the last 15 min of the 20-min exposure periods the neurobehavioral tests were applied. Different cognitive domains such as perceptual and psychomotor speed were measured using four tests, and the results were compared to an identical test session performed under clean air conditions. The aromatic smell of toluene was masked by using menthol crystals in a container in the mouthpiece tubing. The simple and choice reaction time (RT) task was prolonged starting in the 300-ppm exposure period. During the 700-ppm period the RT-simple was reduced by 18%. The performance of a test comparable to the trail making A test, that was published in the Halstead-Reitan neuropsychology test battery and standardized decades later (Tombaugh, 2004), was reduced by 23%, compared to the clean air condition. Thus, even short-term exposures to

300 ppm toluene caused measurable performance alteration even if the volunteers were exposed under resting conditions (Gamberale and Hultengren, 1972). Styrene was another organic solvent that was also studied at these times, including additional knowledge about the importance of the systemic uptake of the solvent. Using the already described within-subject design from the toluene study, 12 healthy males were exposed to 50, 150, 250, and 350 ppm styrene applied in four continuous 30-min periods using an ascending order of the styrene exposures. During the last 20 min of the exposure periods the neurobehavioral tests were applied, and based on the results of a linked toxicokinetic study (Astrand et al., 1974), uptake was increased by light physical exercise on a bicycle ergometer. Compared to the four 30-min periods obtained during the clean air condition, the simple reaction time increased significantly and especially during the 350-ppm exposure period. These two examples illustrate that scientific knowledge about the neurobehavioral toxicity of solvents commonly used by workers dates back to the early 1970s. Later in the early 1980s neurophysiological measures were introduced into the methodology of human volunteer studies, and here the effects of ethanol and xylene, either alone or in combination were studied (Seppalainen et al., 1981). Visual (VEPs) and somatosensory (SEPs) evoked potentials as well as electroretinograms (ERGs) were recorded after a controlled 4-h exposure to 700 or 1200 mg/m^3 xylene, 0.4 or 0.8 g/kg ethanol, or the combination of the two xylene doses with 0.8 g/kg ethanol. 1200 mg/m^3 xylene as well as 1200 mg/m^3 xylene plus 0.8 g/kg ethanol reduced the N60 and P115 amplitude of the VEP. Thus, neurophysiological effects of organic solvents in human volunteers could be confirmed experimentally as early as 1981. At least for acute effects such experimental exposure studies were used to derive no-observable adverse effect concentrations (NOEAC) for styrene some years after the initial experiments (Edling and Ekberg, 1985). Such approaches included more sophisticated neuropsychological and neurophysiological methods, and the exposure duration was usually extended from 30 min to 4 to 8 h as described in various publications (Juran et al., 2012; Kobald et al., 2015; Seeber et al., 2004a; van Thriel et al., 2003).

Human volunteer studies have also been used to answer questions about the effects of the route of exposure, especially reports or subjective ratings of tiredness, sleepiness, headache, nausea, feeling of intoxication, and dizziness. Such reports might be related to the chemosensory effects of organic solvents where the odor or sensory irritation might trigger such subjective effects. Using eight combinations of toluene (1.5 and 4 mg/min), trichloroethylene (1.5 and 4 mg/min), and n-hexane (0.3 and 1.0 mg/min) applied for 60 min into the stomach via a feeding tube by adding the solvents to a commercial tube-feeding product (Baelum, 1999). The subsequent systemic exposures of the eight male volunteers were estimate by using the exhaled concentrations of the solvents. Thereby, exposures equivalents to airborne concentrations of approximately 120 ppm (toluene), 230 ppm (trichloroethylene), and 12 ppm (*n*-hexane) could be generated. Neither alone, nor in combination did these non-inhalation exposures cause acute, neurotoxic symptoms. However, one shortcoming of this study was the lack of neurobehavioral testing that could have shown the neurotoxicity of the internal dose that might have reached the brain. Another "uptake" aspect related to organic solvents has also been described more precisely in human volunteer studies. *N*-Methyl-2-pyrrolidone (NMP) was introduced as a less-toxic substitute for dichloromethane in paint strippers (van Thriel et al., 2007). In a human volunteer study (Bader et al., 2008) focusing on the dermal uptake of NMP from the vapor phase, 16 volunteers were exposed to 80 mg/m^3 NMP for 8 h either as inhalational plus dermal or dermal-only conditions. For this dermal-only exposure, the volunteers were equipped with face shield masks that covered the whole face and supplied activated carbon filtered air. By means of this experimental variation Bader et al. (2008) could estimate that dermal uptake from the vapor phase can account for 33–42% of the total body burden of NMP. Other organic solvents sharing some physicochemical probertites of NMP might also bear the risk of high dermal uptake from the vapor phase. However, more details about pharmacokinetics and pharmacodynamics and their role in the context of solvent neurotoxicity will be described in the sections about animal studies.

Another way that human volunteer studies can inform about the neurotoxic risks of solvent mixtures was recently shown in an experiment comparing white spirits with high and low aromatic content (Juran et al., 2014). As toluene and xylene showed high acute neurotoxic potential in the experiments summarized above, these aromatic hydrocarbons have been removed from commercial mixtures marked as white spirit. Thus, the neurotoxic potency of these products is supposed to be lower than that of products containing up to 19% aromatics. Using five computerized neurobehavioral tests, challenging the cognitive functions of sustained attention (vigilance), response inhibition, response shifting, divided attention and working memory the authors aim to capture even mild performance alterations. Both products, aromatized white spirit (aWS) and a dearomatized mixture (daWS), were used in concentrations of 100 and 300 mg/m^3. The exposure duration was 4 h, and the uptake was not systematically increased via physical activity. The results revealed no evidence that aWS in relatively low concentration and in resting volunteers had a stronger potency than daWS to impair performance different cognitive functions.

Due to the increased knowledge about the neurobiological mechanism underlying the acute interaction of organic solvents with excitatory, inhibitory and modulating neurotransmitter systems, state-of-the-art methodology from clinical and cognitive neuroscience, such as non-invasive brain stimulation (Nitsche and Paulus, 2000), can be integrated into human volunteer studies (Yavari et al., 2018). This study has already been described in the section about mechanisms, as neurobiological key players underlying neuronal plasticity (Kandel, 2001) have been directly manipulated via transcranial direct current stimulation (tDCS). This study not only contributed to a better mechanistic understanding, it also showed that the sophisticated neurophysiological methods might be more sensitive than neurobehavioral measures when searching for small neurotoxic effects of solvents.

3.2.2 Inhalant abuse

While the previous section addresses controlled acute exposures under medial surveillance, this section will focus on organic solvents as drugs of abuse. Products containing volatile intoxicating solvents are inexpensive, readily available, and legal in many places. Inhalant abuse has become very common in lower opportunity youth in impoverished communities worldwide (Cruz and Bowen, 2021). Repeated episodes of inhalant abuse have been shown to produce frank neuropathology, making this an urgent public health concern. The adverse neurophysiological, pathological and biological consequences of inhalant abuse, mainly among children and adolescents, were described in various publications (Balster, 1998; Lubman et al., 2008; Yucel et al., 2008, 2010). Balster (1998) first described important neurobiological mechanisms underlying the addictive behavior of the abuse of toluene or trichloroethane including reinforcing effects similar to addiction to ethanol, barbiturates or benzodiazepines. Later, experiments using brain slices from male Sprague–Dawley rats (Riegel et al., 2007) showed that toluene directly increased the release of dopamine in neurons located in the nucleus accumbens that are elements of neural circuits mediating arousal, motivation, and reinforcement. However, almost mM concentrations were necessary to produce these effects. The concentrations that are typically inhaled by abusers can only roughly be estimated, but are likely in the range of about 1000ppm (Svenson et al., 2022). In his review, Yucel et al. (2008) showed severe neuropathological changes in toluene-abusing adolescents including extensive white matter abnormalities and cerebral atrophy observed by MRI in various brain areas including the corpus callosum. Such findings were confirmed in a larger cohort study (Al-Hajri and Del Bigio, 2010) including neuropathological staining and autopsy samples. Other neuroimaging methods such as functional-Single-Photon Emission Computed Tomography (SPECT) or Magnetic Resonance Spectroscopy (MRS) were reviewed by Yucel et al. (2008) and showed multi-focal impairment of perfusion in bilateral frontal, parietal cortex, basal ganglia and thalami. One MRS study revealed decreased *N*-acetylaspartate (NAA) peak (indicating neuronal loss), and a mildly increased choline (Cho) peak (could be related to acetylcholine levels) in an abuser. The Myoinositol (mI) peak, which is used as marker for Alzheimer's disease (AD, Voevodskaya et al., 2019) was also increased in the abuser (Aydin et al., 2003). However, a summary review by Yucel et al. (2008) showed that most common cognitive and motor functions were severely impaired. The impairments in solvent abusers were sometimes stronger than those observed in cocaine and alcohol using populations (Rosenberg et al., 2002). Working memory and executive functions were most affected in comparison to other drug abusers. Thus, the high-dose and additive exposures to organic solvents are associated with severe brain damage and loss of cognitive functioning. For more information, including dose-response relationships, the reviews by Yucel et al. (2008) and Cruz and Bowen (2021) are recommended.

3.3 Neurological effects of chronic exposure

The mechanisms and the results from the human volunteer studies suggest that repeated exposure such as lifetime occupational exposures might bear the risk of severe and irreversible effects on structure and function of the central and peripheral nervous system. In workers such effects of chronic exposure were either investigated in the context of an occupational disease caused by organic solvents or in epidemiological studies searching for safe exposure levels that can be used to establish Occupational Exposure Levels (OEL).

3.3.1 Occupational diseases: Chronic toxic encephalopathy (CTE) or chronic solvent encephalopathy (CSE)

Even though occupational exposures might have never reached the exposure levels described in the previous sections, another clinically relevant neurotoxic syndrome was described as early as the 1970s (Hanninen et al., 1976). The terms used to describe the syndrome in the early studies included psycho-organic syndrome (POS), presenile syndrome, toxic dementia, or the painters' syndrome (see review by Errebo-Knudsen and Olsen, 1986). The syndrome later was called Chronic Toxic Encephalopathy (CTE) or more precisely Chronic Solvent Encephalopathy (CSE). The CSE syndrome was initially described in populations of painters in Scandanavia, The Netherlands, Germany (Hooisma et al., 1993; Triebig, 2000) and other northern European countries (Mikkelsen, 1980). The initial reports faced criticism of a lack of substantiation of such severe intellectual impairments. As a result, neuropsychological tests were introduced (Errebo-Knudsen and Olsen, 1986), which also became controversial due to other shortcomings of the initial studies including small sample sizes, mis-matched control populations, cross sectional designs, lack of historical, or even current exposure sampling, and potential incentives for subjects to under-perform in testing due to generous workman compensation policies. Moreover, there was no standardized assessment of symptoms, neurological impairments, neurophysiological or neuroimaging techniques, or neuropsychological testing. The standardization of the neurobehavioral/ -psychological assessment was addressed some years later by several authors (Anger et al., 1994, 2000, 2001; Iregren et al., 1996; Letz, 1991; Proctor et al., 2000; Rohlman et al., 2003). Many of these approaches, especially the WHO neurobehavioral core test battery (WHO-NCTB; see review by Anger et al., 2000) used an assembly of tests that were originally developed in intelligence research assessing fluid aspects of intelligence (Russell, 1980) as measured with some subtests from the Wechsler Adult Intelligence Scale (WAIS). Many of the early studies used a combined approach for the clinical evaluation of suspected solvent-induced toxic effects and also addressing aspects of the severity of the disease. A comprehensive overview in 1997 included classification about the severity of acute and chronic organic mental disorders caused by solvents (White and Proctor, 1997). Moreover, the classification of CSE strongly depends on the presence of objectified cognitive impairment by neuropsychological test (see van Valen et al., 2012 for more details). In the WHO classification type II includes the criteria "reduced CNS function" and the Raleigh criteria described in type 2B that symptoms such as "…difficulty in concentration, impairment of memory…" should be "…accompanied by objective evidence of impairment" (van Valen et al., 2012). Thus, the diagnosis of Chronic Solvent Encephalopathy (CSE) for workers chronically exposed to organic solvents depended heavily on performance on neuropsychological tests without hard physical signs or evidence of cerebral pathology. In some cases, there is a potential motive for patients to be diagnosed with CSE if it involved litigation or claims for financial compensation. Therefore, the issue of inadequate effort on neuropsychological testing was a concern. Van Hout et al., (van Hout et al., 2006) included tests designed to identify insufficient effort, which were several memory tasks that appeared to be difficult yet could be performed even by patients with severe brain damage, and eliminated subjects who appeared to show insufficient effort on testing. Evaluating the remaining participants who showed good testing efforts, the solvent exposed painters had significantly more psychological complaints and performed worse than the control group on tests of information processing speed, memory, and learning.

In the early 2010s the Dutch solvent team (van der Hoek et al., 2001) took the initiative and the European consensus of neuropsychological characteristics, assessment, and guidelines for diagnostics of CSE was prepared in several round table discussions (van Valen et al., 2012). Based on an extensive amount of preliminary work done by researchers in Europe studying CSE (see Table 2 in Van Valen et al., 2012) and neuropsychological domains and tests that have been used demonstrate solvent use dependent differences (see Table 3 in Van Valen et al., 2012), six recommendations were provided for the neuropsychological assessment of CSE (see Fig. 2).

These recommendations were meant to be adjusted to the legal framework of the respective country, where the occupational disease should be diagnosed with adjustments for the locally available standardized (e.g., normative data; see recommendation 2) neuropsychological tests. The link of these neuropsychological changes to neurophysiological effects could also be shown, for instance in a Dutch neuroimaging study including CSE patients (Visser et al., 2008) showing reduced striatal D2 receptor binding in CSE patents by means of single-photon emission computed tomographic (SPECT). There were also abnormalities in frontostriatothalamic (FST) circuitry as confirmed by fractional anisotropy (FA) maps derived from MRI scans. Thus, with these considerations, the proposed strategy for the valid assessment of CSE can be used worldwide.

Recommendation 1. Cognitive domains.:
Assessment of the following domains and associated sub-domains/related aspects is recommended for the neuropsychological examination of CSE patients:

1. Attention
   a. Processing speed
   b. Complex attention
2. Memory
   a. Immediate recall
   b. Delayed recall
   c. Recognition
3. Fine motor performance
   a. Motor speed
   b. Dexterity
4. Concept formation and reasoning
   a. Verbal
   b. Non-verbal
5. Construction

Recommendation 2. Estimation of the severity of cognitive impairment.:
A method of estimating a rating the severity of the overall cognitive impairment in a patient is recommended. According to the proposed method, all of the test variables are converted into impairment scores. Test results that are lower than 16th percentile and that are higher than or equal to the 5th percentile* are recoded as 0.5 'impairment points'. Scores that are lower than the 5th percentile are recoded as 1 'impairment point'. The impairment scores of the tests are averaged to obtain an overall impairment score for each sub-domain/related aspects of the domain. The sub-domain/related aspect impairment scores are subsequently averaged to obtain an impairment score for the entire domain, with the exception of domain 5, which should be treated as a related aspect of domain 4. Finally, the impairment scores of all of the domains are averaged to yield a total impairment score that ranges between 0 and 1.
* Percentile refers to the distribution of test results in a non-selected population of the same age, gender and education. Test results <the 16th percentile are equivalent to normally distributed test results that are lower than −1 SD, −1 *z*-score, T40 or the lower half of stanine 3. Test results <the 5th percentile are equivalent to normally distributed test results that are lower than −1.65 SD, −1.65 *z*-score, T32 or stanine 1.

Recommendation 3. Assessment of symptoms.:
During the diagnostic process, symptoms should be evaluated separately from cognitive functioning. In addition to collecting anamnestic information about symptoms during the neuropsychological interview, questionnaires that have been specifically designed to assess symptoms that are relevant to solvent effects should be used.

Recommendation 4. Differential diagnosis.:
CSE may coexist with another pathological condition and general neuropsychological assessment alone cannot be used to establish which condition is the main contributor to the neuropsychological impairment. However, if neuropsychological impairment persists after the successful treatment of a reversible etiological condition, CSE should be suspected. Treatable conditions that affect cognition should therefore be treated prior to neuropsychological assessment when possible.

Recommendation 5. Reliability and validity of test performance.:
The neuropsychological assessment should include at least one test to assess sub-optimal performance, and implicit signs of sub-optimal test performance should be documented. In cases of suspected sub-optimal performance, poor test results must be considered inconclusive.

Recommendation 6. Retesting.:
Patients for whom the diagnostic results are inconclusive may be retested. We recommend allowing a minimum interval of one year to elapse prior to retesting. Stable or slightly improved neuropsychological function after the cessation of solvent exposure is to be expected, whereas pronounced decline suggests the presence of another disease.

Fig. 2 Recommendations for the standardized assessment and evaluation of cognitive impairments in the neuropsychological diagnosis of CSE (van Valen et al., 2012).

As organic solvents are indispensable in many workplaces, the safe use of them should be guaranteed by scientifically derived Occupational Exposure Values. These recommendations should strongly rely on human data and various epidemiological studies addressing the risk of neurotoxicity of occupational exposures to various solvents. Neurobehavioral approaches have been and still are a strong pillar in risk assessment (Lucchini et al., 2005).

3.3.2 Occupational exposures and Occupational exposure limits

While the early studies and the diagnosis of CSE usually dealt with high exposures that also acutely could cause neurobehavioral effects (see previous section) the neurobehavioral methodology developed during these times should be sensitive enough to capture even mild effects (Rohlman et al., 2008) that are not considered to be an adverse health outcome (Ankley et al., 2010). Several reviews and meta-analysis focusing on single solvents have been published. Due its wide occupational application, toluene is probably the most frequently studied solvent. A meta-analytic summary about neurobehavioral test results in cross-sectional studies showed small and mainly insignificant effects for the ten studies that could be included (Meyer-Baron, 2005). In the studies that could be included to calculate effect sizes (Cohens *d*) for six cognitive tests that have been repeatably used in the epidemiological studies, the average exposures of the exposed workers were estimated between 33 and 89 ppm. This cross-sectional approach has often been criticized as (a) covariates such as age, premorbid intelligence, alcohol consumption, depression, etc. (see CSE section) might not been considered, and (b) only associations but not causality could be investigated. Thus, longitudinal studies or estimated of individual exposures are required to investigate causality more directly or by providing clear dose-response relationships. In the case of toluene, both approaches have been combined (Seeber et al., 2004b, Wameling et al., 2000) to derive a No-Observed Adverse Effect Level (NOAEL) based on epidemiological studies. Seeber et al. (2004b) showed that rotogravure printers (initial sample size 333 participants) that were tested neuropsychologically every year over a total study period of 5 years showed no decline in various cognitive and motor tests. Moreover, high- and low-exposed workers did not differ dose-dependently at any time of the four assessments. Moreover, an individual lifetime weighted average exposure (LWAE) estimating the past exposure based on job exposure matrices and airborne toluene concentrations provided by the companies, had occasionally a significant impact on the performance parameters. The stepwise regression analyses (forward, using LWAE first) yielded proportions of explained variances by the LWAE between 1 and 3% when testing the influence of life-time exposures for the four assessments individually. Thus, in contrast to the impact of age or education no persistent effect of occupational toluene exposure could be confirmed. On average, the LWAE of the workers was around 45 ppm and thus, the German OEL was set to 50 ppm as this was consider as a human NOAEL for chronic exposures. Nevertheless, other counties used different studies and risk assessment approaches and thus, the OELs (time-weighted average TWA, 8 h) vary between 20 and 200 ppm (https://limitvalue.ifa.dguv.de/).

Examples of other solvents for which epidemiological studies supported the risk assessment include trichloroethylene (Kaneko et al., 1997), where carcinogenicity was also a concern as some solvents are known carcinogens (e.g., benzene), or styrene (Hall and Rebert, 1994; Rebert and Hall, 1994). These reviews were helpful to draw the attention of neurotoxicologists to the most sensitive endpoint as in the case of styrene effects of the visual system (see Sensory Neurotoxicity section).

As outlined in the introduction, solvents are often used as mixtures. Their composition is often unknown and might change over time. For the group of ketones, a systematic review (Noraberg and Arlien-Soborg, 2000) described 54 studies investigating neurotoxic effects of ketones such as acetone (ACE), methyl ethyl ketone (MEK) and methyl isobutyl ketone (MIBK). The summarized results are heterogeneous with respect to possible neurotoxic interactions. Possible synergistic interactions of the ketones were reported in 12 studies, only two studies reported neurotoxic potentiation after acute short-term combined exposure to human volunteers, and when excluding studies using n-hexane or 2,5-hexanedione (2,5-HD) in their mixtures, eight studies reported synergistic interactions. In their review, the authors concluded that some of the results of combined exposure, especially those obtained in animals or human volunteers, cannot be extrapolated to occupational situations and that the real risk of occupational exposures to ketone mixtures needs more research to be conclusively answered. Another example of mixtures are hydrocarbon fuels where animal and human data have been reviewed (Ritchie et al., 2001), again without a clear picture of possible neurotoxic interactions.

The impact of occupational exposures to solvent mixtures was addressed in a meta-analysis (Meyer-Baron et al., 2008). Here, 46 epidemiological studies (published between 1977 and 2004) investigating mainly painters, printers, and glue users (e.g., floor layers) were identified using a standardized search strategy. Of those studies, 17 papers reported concentrations of the individual constituents of mixtures and thereby allowed the calculation of an Exposure Index (EI). The EI was calculated if concentrations of constituents were reported. Based on OEL recommendations the EI was defined as follows: $EI = conc_1/OEL_1 + conc_2\ /OEL_2 \ldots\ conc_N\ /\ OEL_N$. This procedure was necessary as the EI is a relative measure and is comparable among studies only when the same limit value is used for all calculations. According to the rationale of the calculation an $EI = 1$ means here that based on limit setting in 2004, adverse health effects are unlikely. The EIs calculated from the concentrations reported in the included studies ranged from 0.07 to 14.86. In addition to the EIs the individual effect sizes (Cohens *d*) were calculated for each neuropsychological test that was repeatedly used in the studies. The overall effect for every neuropsychological test was calculated as a weighted mean of the individual results by using a random effects model (d_{RE}) accounting for potential additional variability between the studies that cannot be explained by chance. In general, negative overall effects in accordance with the hypothesis of a negative impact of solvent exposure were obtained for 43 out of 48 tests. Their magnitude ranged from $d_{RE} = -0.01$ to -0.64, and 12 of these d_{RE} were statistically significant. One example was the simple reaction time task that was used in 26 studies/ samples with an $d_{RE} = -0.34$ (CI: −0.15 - -0.53). Thus, this "cross-sectional" approach confirmed the negative impact of exposures to solvent mixtures on cognitive and motor performance. However, when using the EI or the reported exposure duration as quantitative indicator of the cumulative exposure, the results were not consistent across the different tests. While larger effect sizes are associated with higher EIs for the digit symbol, the reverse tendency applies to the Benton visual retention task. Also, for exposure duration different trends were observed and differences between painters and workers from other jobs for these "dose-response" relationships were observed. The lack of clear exposure–effect relationships was discussed extensively by the authors (Meyer-Baron et al., 2008) and the central conclusion was the lack of rigorous exposure descriptions in epidemiological studies that allows a more precise estimation of the chronic exposure of the individuals included in such epidemiological studies. Again, the neurobehavioral deficits repeatedly obtained in the epidemiolocal studies summarized in this meta-analysis could be supported by neuroimaging results obtained in painters, drywall tapers, and carpenters (Tang et al., 2011). The workers exposed to solvent mixtures performed worse in a working memory task and their corresponding fMRI activation maps showed compensatory activations of brain areas that were not activated in controls.

3.3.3 Other exposures

The 2010 Deepwater Horizon oil disaster spilled an approximated 5 million barrels of crude oil into the Gulf of Mexico and spurned a massive cleanup effort involving thousands of oil spill response and cleanup workers (Kwok et al., 2017). Those workers were exposed to a mixture of hydrocarbon components of the crude oil including inhalation of volatile components and skin contact during the cleanup efforts. The air concentrations of total hydrocarbons, however, were lower for the Gulf cleanup workers than in many occupational settings, providing an opportunity to evaluate the effects of lower-level hydrocarbon exposures on neurobehavioral functions. Quist et al. (Quist et al., 2019) identified approximately 3000 participants who were divided into four exposure categories based on job classifications with total airborne hydrocarbon concentrations of <0.30 ppm, 0.30–0.99 ppm, 1.0–2.99 ppm, ≥3.00 ppm. In addition, workers were classified by job duties in a hierarchy based on anticipated likelihood of exposure. Neurobehavioral testing consisted of a computerized Behavioral Assessment and Research System. After confounders were considered, there was only limited evidence of an association between total hydrocarbon exposure and neurobehavioral performance. An association was observed for poorer performance on a digit-span test between the highest and lowest exposure category by job classification, land cleanup, decontamination and operations workers showed slower response times than administration workers. The associations were stronger among male than female workers. Overall, the worker exposed to total hydrocarbon concentrations greater than 1 ppm performed worse on several behavioral testing domains including digit span, symbol digit, and continuous performance tests.

As a component of the Agricultural Health Study involving longitudinal assessments of licensed pesticide applicators in the US states of Iowa and North Carolina, participants were asked to self-report use of organic solvents, including gasoline or petroleum distillates. Use could include mixing pesticides, cleaning hands or equipment or other activities. Regardless of pesticide use, the history of using solvents, petroleum distillates or gasoline was associated with increased depressive symptoms on the Center for Epidemiologic Studies Depression Scale (Siegel et al., 2017). These factors and a couple of other comorbidities need to be considered in the differential diagnosis of CSE.

3.4 Summary human studies

For many individual solvents the available experimental and epidemiolocal studies provided clear evidence for their neurobehavioral toxicity but only for a few solvents, e.g., toluene the results are comprehensive enough to derive an OEL solely based on human data. However, during the last decades almost no epidemiolocal studies among solvent exposed workers have been published. The reason for this lack of publications is unknown and it can only be speculated that (a) the exposure conditions in industrialized countries has been reduced to the existing scientific knowledge, or (b) the exposure in other parts of the world is not sufficiently monitored and their neurotoxic effects are unknown and unpublished. In addition, the number of cases of CSE in European countries has declined over the last two decades, likely due to increased surveillance of at-risk professions, substitution of less toxic alternatives, engineering controls, and reduced indoor spraying of oil-based paints (Kaukiainen et al., 2004; Spee et al., 2012). It might seem that solvents are no longer a health risk for workers. This cannot be concluded from the previous sections, however, as the reduced solvent exposures in Europe and North America has not necessarily been the case in less developed countries. There may still be an underdiagnosis of CSE cases in developed countries as well (Furu et al., 2012). Even among workers without a diagnosis of CSE, those exposed chronically to solvents have greater numbers of symptoms related to memory and concentration than do non-exposed workers (Furu et al., 2018). In addition, the neurobehavioral toxicity of solvent mixtures is not fully understood. These interactions have been addressed systematically in human volunteer studies but only with simple mixtures that do not represent the workplace situation. To improve this, more research is needed that also includes the neurotoxic mechanisms of the single solvent to build prediction models that use convergent or divergent molecular or cellular targets (see section on mechanisms) as relevant modulators of the effects of mixtures (Ankley et al., 2010).

3.5 Neurobehavioral and neurophysiological rodent studies

3.5.1 Animal studies acute exposure

There are numerous studies spanning many decades of scientific literature reporting the acute effects of exposure to organic solvents in experimental animals. This early literature has been reviewed elsewhere (Arlien-Søborg, 2000; Benignus, 1981; Evans and Balster, 1991). There are recent reviews of animal models of solvent abuse, including toluene and other organic solvents (Bowen, 2006; Cruz and Bowen, 2021). These animal models typically employ exposure concentrations much higher than workplace levels, and short term or repeated exposure paradigms that limit applicability to occupational health. Never-the-less these studies are rich data sources regarding the cellular and molecular effects of solvents that also could be relevant to the lower concentration, longer duration occupational exposures. There are experimental design factors, however, that limit the usefulness of some of laboratory animal studies for occupational health including:

- Inappropriate routes of exposure, i.e., anything other than inhalation, which is the primary route of occupational exposure. Time course and metabolism are very different for inhalation and other routes of exposure. The rate of absorption, distribution, time to peak effect and elimination may be slower following oral or intraperitoneal (i.p.) administration than by inhalation. With repeated i.p. or oral routes of exposure there is increased likelihood of painful irritation to the site of administration or systemic toxicity (e.g., liver, kidney, gastrointestinal) contributing to outcomes, and first-pass metabolism in the liver prior to transport to CNS may result in lower CNS concentrations of the parent compound and greater exposure to metabolites than occurs via inhalation.
- Testing neurological function after the end of an inhalation exposure session rather than during exposure when testing for acute effects. Many volatile organic compounds are rapidly taken up during exposure and rapidly excreted after exposure terminates. Experimental designs that exposure animals in typical exposure chambers, then turn off the exposure source, blow down the chamber concentration to remove the animals and then transport them to behavioral testing chambers means that by the time behavioral testing begins the animals have a considerably lower tissue concentration of the test substance than during the actual exposure. Further, the tissue concentration continues to decrease during the course of the behavioral test session.
- Inadequate exposure systems. Static inhalation systems have been used in which the animals are placed in a sealed system, such as a bell jar, with the source of exposure being something like a pool of liquid solvent under a mesh floor, which evaporates to generate the vapor exposure. These systems are unstable for other than short term exposures because the source of exposure may be depleted over time reducing the air concentrations, and presence of animals in a closed system may cause buildup of CO_2 or ammonia over time.
- Failure to monitor inhalation systems for air quality including concentrations of the test material, and also factors such as temperature, humidity and concentrations of CO_2 or ammonia.

Overall, a general pattern of behavioral changes has emerged from many studies of the effects of acute organic solvent exposure on animal behavior. At low concentrations and early in the exposure protocol, a phase of excitation and increased activity may be observed (e.g., Moser and Balster, 1981; Warren et al., 2000). This excitatory phase is somewhat dependent on the test substance and the behavioral paradigm, but it is often observed. Subsequently, and at higher concentrations, the excitatory phase is followed by reduced activity levels, behavioral suppression, impaired task performance, and CNS depression (e.g., Moser and Balster, 1981, Warren et al., 2000) .

It is important to link the tissue concentrations of organic solvents to their effects. The effects of acute exposure to 1,1,1-trichloroethane on schedule-controlled behavior (Warren et al., 1998) and motor activity (Warren et al., 2000) were correlated to both the blood and brain concentrations of the test compound. Response rates increased at low tissue concentrations and decreased at high tissue concentrations. Both peak tissue concentrations and area under the curve (AUC) tissue concentrations were well correlated with behavior. In these experiments, however, all the exposure sessions lasted the same amount of time (100 min). For exposures lasting the same duration, the peak concentration and the area under the curve at perfectly correlated, and the experimental design cannot distinguish between these exposure metrics.

When the behavioral effects of acute toluene inhalation exposure were measured at numerous times while the animals were being exposed to toluene, and brain concentrations were estimated at the times of assessment using a pharmacokinetic model, it was possible to differentiate between peak concentration and the area under the curve (Bushnell et al., 2007). In this case, the momentary brain concentration at the time of testing, and no other dose metric, correlated well with behavioral impairment (Fig. 3).

When rats were exposed to constant or fluctuating concentrations of toluene over 7.5 h (total dose 10,000 or 20,000 ppmh), and tested after exposure, the effects on behavioral performance differed by exposure pattern and not according to concentrations of toluene in blood or brain (Lammers et al., 2005). These results suggested that the changing brain concentration was more a determinant of effects than the tissue concentration itself. The fluctuating exposure concentrations and post-exposure behavioral assessment, however, complicate an interpretation.

Rats were exposed acutely to trichloroethylene (TCE) vapors and visual evoked potentials were measured periodically before, during and after exposure (Boyes et al., 2005). The amplitude of the potentials recorded from visual cortex decreased during exposure and recovered after exposure was terminated (Fig. 4). A physiologically-based pharmacokinetic model (PBPK) model (Simmons et al., 2002) was used to estimate the brain concentration of TCE at the moments of electrophysiological recordings.

(A)

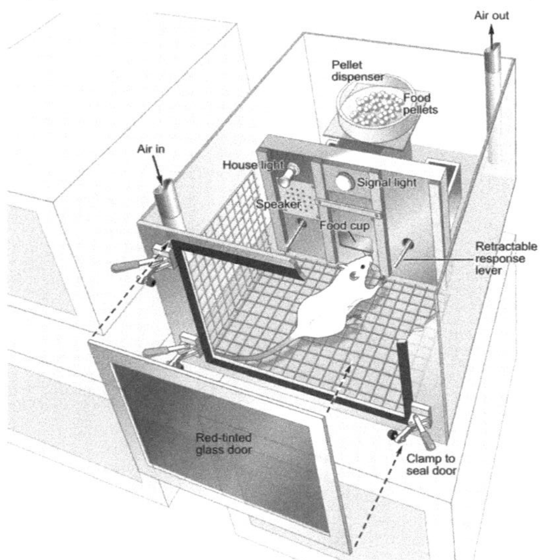

(B)

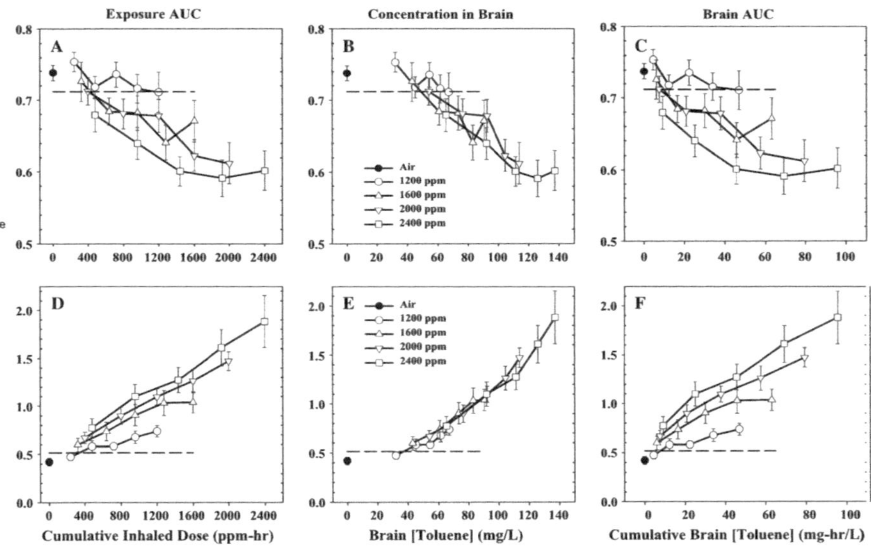

Fig. 3 (A) Diagram of the operant system within an inhalation exposure chamber and rats performing a food-reinforced signal detection/ attention task while exposed to toluene vapors. Source: (Bushnell et al., 2007). (B) Task performance measures include accuracy (P(correct) top row) and response time (*sec*) (bottom row) are plotted against three potential measures of exposure including the Area Under the Curve (AUC) of the chamber air (left side), concentration of toluene in the brain at the time of testing as predicted using a physiologically-based pharmacokinetic model (middle), and the AUC of brain concentration (right side). The concentration of toluene in the brain was a more consistent predictor of performance accuracy and response time than the other two measures at all exposure concentrations and times tested. Source: (Bushnell et al., 2007).

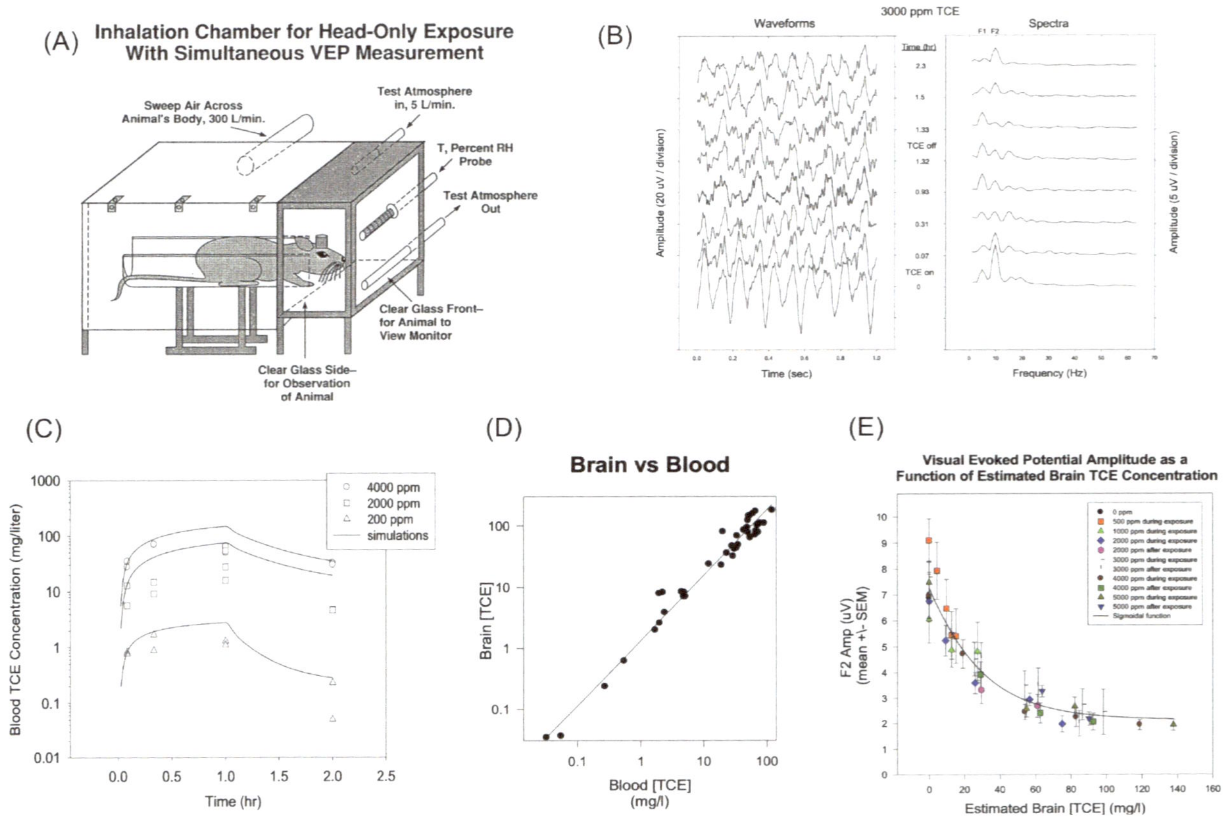

Fig. 4 See figure legend on next page.

Across a range of air concentrations and exposure durations, the predicted momentary brain concentration at the time of recording predicted the magnitude of the deficit in brain function (Boyes et al., 2005). Additional experiments showed that for other volatile organic solvents, including toluene and perchloroethylene, there was a similar relationship between momentary concentration of the organic solvent in the brain and a measured deficit in brain neurophysiology (Boyes et al., 2007, 2009).

The data in Fig. 4 illustrate that momentary concentration of TCE in the brain is an accurate predictor of the deficit in electrophysiological function both during exposure and afterward. In addition, the succession between panels C–E in which the vertical axis of panel C becomes the horizontal axis of panel D, and in turn the vertical axis of panel D becomes the horizontal axis of panel E. This sequence becomes a quantitative exposure-dose-response model in which external exposure conditions determine the absorbed dose (panel C), the absorbed dose determines the target tissue dose (panel D), and the target tissue dose determines the adverse outcome (panel E).

Benignus et al. (2007) estimated toluene molar concentrations in brain tissue using physiologically based pharmacokinetic modeling for humans

Fig. 4 Recordingvisual evoked potentials (VEP) while rats are concurrently exposed to organic solvent vapors. (A) Diagram of recording apparatus in which awake restrained rats with previously implanted electrodes over visual cortex are exposed to organic solvent vapors while watching a video monitor (not pictured) through a clear glass front plate (Boyes et al., 2003). (B) Left side. Visual evoked potential (VEP) waveforms recorded to a visual pattern that modulated (on/off) at 5 Hz and produced a strong response in visual cortex at twice the response rate. Right side. Amplitude spectrum of the corresponding VEP waveform showing a predominate response amplitude at twice the stimulation rate (F2). The waveform and spectrum presented at the bottom of each panel were recorded while the rats breathed clean air prior to the onset of exposure to TCE. Time proceeds upward during the course of exposure to 3000 ppm TCE, and then after the TCE was turned off and rats returned to breathing clean air. During exposure to TCE the VEP waveform structure became disorganized and the F2 amplitude decreased, which then recovered after returning the clean air (Boyes et al., 2005). (C) Concentrations of TCE measured in rat blood, and the predictions of the physiologically-based pharmacokinetic model during and after a 1-h exposure session to different air concentrations of TCE (Simmons et al., 2002). (D) Measured TCE concentrations in blood and brain of rats sacrificed at various exposure concentrations and times and a linear regression fit to the data. (E) VEP F2 amplitude recorded across a range of exposure concentrations and times, and during or after exposure plotted against model-predicted brain concentration at the time of recording (Boyes et al., 2005).

and rats that were tested behaviorally or neurophysiologically during the exposure period (Benignus et al., 2007). Behavioral tasks included choice reaction time for human volunteers and food-motivated signal detection task and shock-motivated avoidance in rats. The electrophysiological measure was pattern-elicited VEPs in rats. Behavioral and electrophysiological measures were converted to a common effect scale as percentage of the maximum magnitude of effect for each dependent variable. When this was done, human choice reaction times and rat VEP amplitudes were of equivalent sensitivity to brain toluene concentration, and more sensitive than food-motivated signal detection accuracy. The lease sensitive measure was shock-motivated avoidance behavior. Thus, the sensitivity of the task depended on the strength of the motivational force. Where there was a strong motivational force, such as shock avoidance, it was very difficult for toluene exposure to disrupt performance. The least/unmotivated tasks of human choice reaction time and rat visual evoked potentials were disrupted at the same low toluene concentrations, indicated that there were no significant differences among the species in sensitivity to toluene.

Expanding this analysis to data from other organic solvents including trichloroethylene (TCE), 1,1,1-trichloroethane (TCA), and perchloroethylene (PERC) along with toluene (TOL), Benignus et al. (2009) modeled brain molar concentrations (Benignus et al., 2009) of each material based on exposure concentration and time parameters (Fig. 5). When this was done, there was no statistically significant difference between the effects of the four solvents. Therefore, the data could be collapsed across the four solvents revealing tight predictive functions between the molar brain concentration and performance on the various tasks (Fig. 5, Panel B). Again, the motivational force driving each task appeared to be a strong determinant of the sensitivity to solvent exposure.

Comparing across the panels of Fig. 5 suggests (1) the four solvents were not different in potency on a molar basis, (2) tasks in humans and rats differ in sensitivity to disruption based on motivational contingencies to perform the tasks, (3) rats and human tissues are not different in sensitivity at the neurotransmitter receptor level, and (4) the molar concentrations which alter behavior and whole animal physiology (panels A and B) are similar to those which disrupt neurotransmitter functions (Panel C) (note that $1000\,\mu M = 1\,mM$).

Expressing dose levels of solvents as molar concentrations in the brain at the time of testing also enables comparisons to invitro mechanistic studies of the effects of solvents on membrane ion channels (Fig. 5, Panel C).

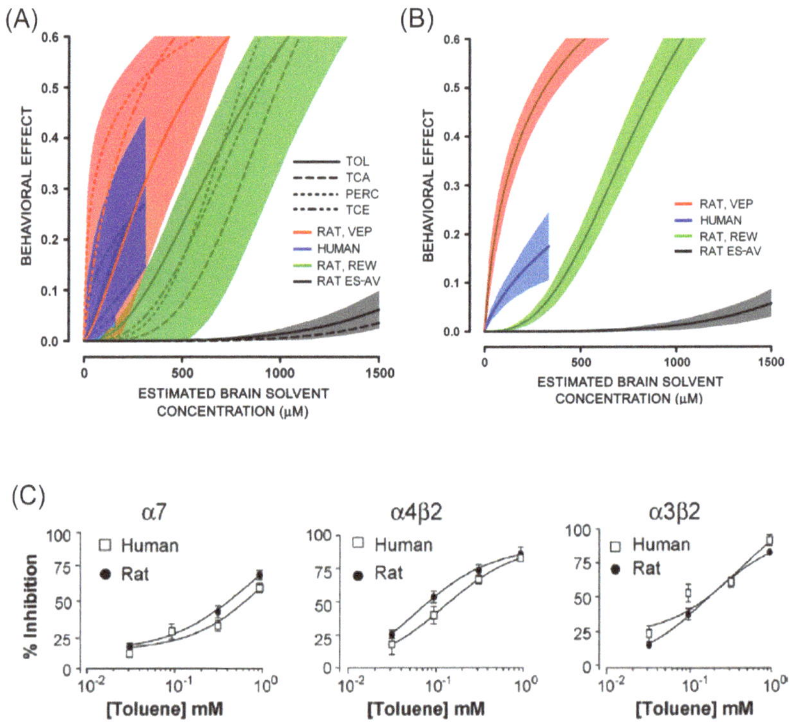

Fig. 5 Rat and human sensitivity to organic solvents on behavioral and electrophysiological measures and at cellular and system levels. (A) Dose response curves fit to literature data for rat behavior: schedule controlled operant behavior for food reward (REW) and shock motivated escape avoidance (ES_AV), rat visual evoked potential (VEP) and human choice reaction time data (Human) for four organic solvents (toluene (TOL), 1,1,1-trichloroethane (TCA), perchloroethylene (PERC) and trichloroethylene (TCE). Effects were transformed to proportion of maximal effect to fit on a common axis. Dose is expressed as the micromolar concentration of the solvent in the brain at the time of testing, as estimated using physiologically based pharmacokinetic modeling (Benignus et al., 2009). (B) The data from Panel A were collapsed across solvents since no statistically significant differences were found between potency of the four solvents when dose was expressed as brain micromolar concentration (Benignus et al., 2009). (C) The effects of toluene expressed as millimolar concentration to inhibit the function of three subtypes of acetylcholine receptors expressed in *Xenopus* oocytes. Rat and human versions of each receptor were tested. The IC_{50} values differed across receptor subtypes, but not between rat and human versions of the receptor. *Panel C: Data from Bale, A. S., Meacham, C. A., Benignus, V. A., Bushnell, P. J., Shafer, T. J. 2005a. Volatile organic compounds inhibit human and rat neuronal nicotinic acetylcholine receptors expressed in xenopus oocytes. Toxicol. Appl. Pharmacol., 205, 77–88; Bushnell, P. J., Oshiro, W. M., Samsam, T. E., Benignus, V. A., Krantz, Q. T., Kenyon, E. M. 2007. A dosimetric analysis of the acute behavioral effects of inhaled toluene in rats. Toxicol. Sci., 99, 181–9.*

Toluene has been shown to modulate the function of *N*-methyl-D-aspartate (NMDA), gamma-aminobutyric acid (GABA), glycine, and nicotinic acetylcholine receptors at micromolar concentrations (Bale et al., 2002; Cruz et al., 1998, 2000). Both human and rat forms of neuronal nicotinic receptors when expressed in *Xenopus oocytes* show differential sensitivity to toluene based on the makeup of the receptor subunits, but the rat and human forms do not differ from each other in sensitivity to toluene (Fig. 4, Panel C). This suggests that there are no significant differences between rat and human sensitivity to organic solvents at the cellular level. In addition, it is important to observe that the dose ranges over which function is impaired are similar for the nerve membrane receptors and the whole animal/human measures of neurophysiology and behavior.

Repeated or prolonged exposure to organic solvents leads to tolerance. For example, when exposure durations extended to 24 h tolerance to the behavioral effects of toluene was observed which was beyond what could be attributed to metabolic induction reducing the effective brain concentrations (Oshiro et al., 2011). The tolerance to prolonged toluene exposure apparently reflected neural plasticity and behavioral adaptations which enabled the animals to continue task performance despite on-going presence of toluene in the brain.

3.5.2 Animal studies chronic exposure

In contrast to the effects of acute solvent exposure, the effects of low-level chronic exposure involving cognitive effects and memory dysfunction, which have been extensively described in occupational settings, have not been reliably replicated in animal models. Although there have been multiple attempts to assess neurological function following sub-chronic or chronic exposure of laboratory animals, some of which have been positive and others negative, to date no widely accepted animal model of CSE has emerged. The lack of successful animal models of CSE could be attributed to multiple factors, including the duration of exposure: human cases may take many years of exposure to emerge whereas animal experiments typically do not extend the exposures beyond a few weeks or months; and the difficulty of assessing the kind of diffuse cognitive and affective dysfunction in animal models that have been reported in exposed workers.

The US National Institute for Occupational Safety and Health (NIOHS) issued a Current Intelligence Bulletin on Organic Solvent Neurotoxicity (National Institute of Occupational Safety and Health (NIOSH), 1987) recognizing the concern for neurotoxicity for workers exposed to organic solvents. NIOSH recognized hazards of both acute and chronic exposures, the latter of which at their most severe included irreversible deterioration of intellect and memory accompanied by CNS structural damage. The NIOSH bulletin included a list of research needs, among them was the establishment of an adequate animal model to predict chronic neurobehavioral neurotoxicity. NIOSH also recognized a need for research to test chemicals and structural analogues to provide the ability to predict neurotoxicity.

The US Environmental Protection Agency (EPA) in 1993 attempted to exercise the authority provided under the Toxic Substances Control Act (TSCA) to address some of the issues for chronic solvent neurotoxicity through The Multi-Substance Test Rule (U.S. Environmental Protection Agency (EPA), 1993). The rule required manufacturers of 10 volatile organic solvents to undertake toxicity testing that included assessment of cognitive functions and screening level tests for neurotoxicity. The substances named in the rule included: acetone, *n*-methyl acetate, 1-butanol, n-butyl acetate, diethyl ether, 2-ethoxyethanol, ethyl acetate, isobutyl alcohol, methyl isobutyl ketone, and tetrahydrofuran. These substances were selected because of their high production volume, and workplace or consumer exposures. The test required sub-chronic (90 day) inhalation exposure and evaluation by schedule-controlled operant behavior (SCOB). According to the test rule the experimental animals were to be trained to perform multi-schedule operant tasks under variable ratio- and interval-based reinforcement schedules prior to the onset of sub-chronic inhalation exposure, and the testing phase post-exposure evaluated primarily the ability of the animals to perform the previously learned tasks under the variable reinforcement schedules. Therefore, the testing did not evaluate the effects of solvents on the learning phase of the process, but rather the ability to perform previously learned tasks (Moser et al., 2000).

The EPA received extensive criticisms on the Multi-substance Test Rule from the regulated industries, represented primarily by the Chemical Manufacturers Association (CMA), but also expressed elsewhere (e..g. Maurissen, 1994), regarding the necessity of testing the individual substances, and the expense, lack of facilities and logistical difficulty of organizing inhalation exposures along with SCOB testing. Subsequent to a lawsuit, the EPA and CMA entered into a consent decree revoking The Multi-Substance Test Rule (U.S. Environmental Protection Agency (EPA), 1995). SCOB testing would no longer be required, other than for two substances (acetone and methyl isobutyl ketone) where testing had already begun. According to the consent decree, the remaining substances would undergo testing with the established TSCA neurotoxicity screening battery (motor activity, functional observational battery, and neuropathology) which have no direct measures of cognitive function or memory. As it turned out, however, data from schedule-controlled behavioral experiments were eventually provided from industry-supported labs for several of the test rule substances. Reports were published for acetone (Christoph et al., 2003), amyl acetate (Gill et al., 2000), isobutanol (Li et al., 1999), methyl isobutyl ketone (David et al., 1999), and tetrahydrofuran (Malley et al., 2001). Almost uniformly, there were no or limited effects of sub-chronic inhalation on the scheduled-controlled operant performance. Among the solvents tested, significant differences from control were observed only for acetone and ethyl acetate and, according to the authors of the reports, the pattern of results did not reflect adverse neurotoxic effects (Li et al., 1999).

In retrospect, it is not clear that the testing paradigms of the Multi-Substance Test Rule addressed neuropsychological processes analogous to the types of diffuse confusion, mood changes, and learning and memory problems expressed by painters and other exposed workers considered to have CSE. After exposure, the test animals had only to perform previously learned tasks since the training phase of the experiments preceded exposure. In addition, there was no requirement for testing of positive control substances, such as toluene or trichloroethylene, widely reported to produce CSE. Therefore, there was no proof-of-concept demonstration that the proposed testing paradigm was sensitive to the neurotoxic effects of concern. Although no positive controls were required, two of the test-rule compounds, diethyl ether (ether) and isobutyl alcohol (isobutanol), were among substances associated with CSE in a list compiled by Schaumburg and Spencer (Schaumburg and Spencer, 2000). That list, however, was merely compounds associated with CSE in occupational settings where multiple exposures may have occurred, exposures histories may have been undocumented or there may have been other shortcomings in the strength of evidence as noted elsewhere.

EPA initiated another extensive set of toxicity testing experiments on hydrocarbon vapors under the authority of the Clean Air Act, which includes provisions concerning the constituents and emissions from automotive sources. At that time, there was an emphasis on developing oxygenated additives to automobile fuels that would enhance combustion efficiency and reduce overall emissions (U.S. Environmental Protection Agency (EPA), 1994, Henley et al., 2014, Swick et al., 2014). EPA entered into a consent agreement with fuel manufacturers coordinated by the American Petroleum Institute (API) to test the potential toxicity of several proposed oxygenate compounds. Test materials included baseline gasoline vapor condensate (BGVC) and BGVC combined with methyl t-butyl ether (G/MTBE), ethanol (G/EtOH), t-amyl ethyl ether (G/TAME), ethyl t-butyl ether (G/ETBE), diisopropyl ether (G/DIPE), or t-butyl alcohol (G/TBA) at concentrations of 0, 2000, 10,000, 20,000 mg/m^3, 5 days/week for 13 weeks. The Baseline Gasoline Vapor Condensate contained 131 identified chemical components with volume percentages split among paraffins (76.6%), olefins (13.8%), cycloparaffins (5.9%) and aromatics (3.7%) (Henley et al., 2014). An unusual component of this testing program derived from stipulations in the Clean Air Act acknowledging that gasoline itself is a toxic material, and that new gasoline formulations such as those with new oxygenate substances be no more toxic than baseline gasoline without the additives. The testing program therefore compared the effects of exposure to baseline gasoline with those of baseline gasoline plus the oxygenates, and there was no clean air control condition and no exposures to the neat oxygenates themselves. The toxicity testing included adult and developing organisms, reproductive toxicity and special add-on testing for both immune and neurotoxicity. The neurotoxicity testing did not specify schedule-controlled behavior as had the multi-Substance Test Rule, but instead included the standard neurotoxicity screening battery (Motor activity, FOB and neuropathology) along with the additional requirement for measuring Glial Fibrillary Acidic Protein (GFAP). GFAP is a protein expressed in astrocytic glial cells that increases in concentration during reactive gliosis in response to neuronal injury. The results of these experiments showed no substantial indications of neurotoxicity after the termination of exposures in behavioral, neuropathological or neurochemical assays except for BGVC plus ethanol, which showed low level increases in GFAP in the cerebellum that did not show a clear dose-response pattern (Henley et al., 2014; O'Callaghan et al., 2014). Although the statistical comparisons were against a BGVC alone group, there were no overall indications of neurotoxicity in behavior, neuropathology or GFAP levels. Subsequently, another test rule using the same testing protocols evaluated the potential toxicity of baseline gasoline plus isobutanol against baseline gasoline alone, and again failed to show substantial neurotoxicity (U.S. Environmental Protection Agency (EPA), 2018). Therefore, these studies of sub-chronic exposures to high concentrations of gasoline hydrocarbon vapors alone or with the specified oxygenate compounds did not identify an animal model of CSE.

No changes in behavior that persisted after the termination of the exposure period were observed following sub-chronic inhalation exposure to toluene (Pryor et al., 1983), trichloroethylene (Kulig, 1987; Oshiro et al., 2004), or perchloroethylene (Mattsson et al., 1998). Following sub-chronic exposure to 2000 ppm 1,1,1-trichloroethane there was a slight decrease of forelimb grip strength, but no changes beyond the exposure period in a Functional Observational Battery (FOB), a battery of electrophysiological tests, or histopathology (Mattsson et al., 1993). In one study, rats exposed to toluene (600 ppm, 24 h/d, 50 days) showed fewer errors than controls on a radial arm task (Miyagawa et al., 1995). Despite not observing changes in baseline behavioral performance, 20 days of exposure to TCE caused a persistent change in sensitivity to d-amphetamine (Oshiro et al., 2004) and 16 weeks of toluene exposure altered brain catecholamine synthesis (Soulage et al., 2004). In addition, subtle changes in visual electrophysiology were seen following 30 days exposure to toluene (Dyer et al., 1984) and 13 weeks of exposure to perchloroethylene (Mattsson et al., 1998). Clear indications of ototoxicity were seen following exposure to toluene (Mattsson et al., 1990; Pryor et al., 1983) and TCE (Rebert et al., 1991), which are discussed in subsequent sections.

4. Sensory neurotoxicity

4.1 Auditory effects

Many organic solvents damage the auditory system. This was first demonstrated by Gordon Pryor and Chuck Rebert in laboratory animals using behavioral tests with auditory cues (Pryor et al., 1983) and brainstem auditory evoked potentials (Rebert et al., 1983, 1991). Histopathology of the cochlea showed damage primarily to apical microvilli of outer hair cells (Johnson, 1993; Johnson and Canlon, 1994). In addition to the cochlear outer hair cells, damage has been observed to the spiral ganglion cells by trichloroethylene (Fechter et al., 1998) and styrene (Lataye et al., 2001). The Dieters cells, which are support cells for the cochlear outer hair cells, also undergo apoptotic cell death induced by styrene (Chen et al., 2007; Chen and Henderson, 2009). Older animals were more susceptible to cochlear damage than young animals (Campo et al., 2003). These animal experiments have often employed relatively high concentrations out of the range of most workplace settings. It soon became clear, however, that ototoxicity could also being detected in workers occupationally exposed to solvents including toluene and styrene (Morata et al., 1997, 2002).

4.1.1 Frequency specificity

Early animal studies reported that the hearing loss from exposure to toluene was greater at high frequencies (Pryor et al., 1983). High frequency hearing loss is common for other ototoxic substances such as the aminoglycosides (Jiang et al., 2017). However, these early studies of organic solvents tested at a frequency range appropriate for humans (up to ~16 Khz) where rats can hear much higher frequencies (up to ~80 KHz). When an acoustic range appropriate for rats was tested, the pattern of hearing loss was greater at between 8 and 20 KHz (Campo et al., 1997; Crofton et al., 1994; Crofton and Zhao, 1993; Jaspers et al., 1993; Sullivan et al., 1988), which is mid-range for rats. A midfrequency hearing loss is unusual among classic ototoxic chemicals and suggests a novel mode of action, although a clear mode of action for solvent ototoxicity has yet to be established.

4.1.2 Structure-activity

Not all solvents cause ototoxic effects. Pryor (Pryor, 1995) provided a list of positive and negative ototoxic compounds based on inhalation exposure to rats and testing auditory thresholds with behavioral and electrophysiological assays (Table 3). It is interesting that among the isomers of xylene, only *p*-xylene, and not *o*-xylene and *m*-xylene, was ototoxic. This isomeric specificity was observed using behavioral tasks and electrophysiology, and confirmed independently with morphological cochlear cytograms (Maguin et al., 2006).

Structure-activity relationships were explored for a series of 21 aromatic solvents (Gagnaire and Langlais, 2005). The experimental evidence of Gagnaire and Langlais (2005) was based on rats treated orally with 8.47 mmol/Kg by oral gavage 5 d/w for 2 weeks, and afterward evaluated by morphological evaluation of the cochlea. In contrast, evidence from the Pryor and Rebert group was based repeated inhalation exposure. The combined determinations from Pryor (1995) and Gagnaire and Langlais (2005) are presented in Table 3. For those compounds evaluated by both groups, there is complete agreement as to which were ototoxic, and which were not. The ototoxic solvents were active by either inhalation or oral routes of exposure, although it is hard to equate tissue concentrations across exposure routes. Delivery to the cochlea appears to be blood-borne rather the via the cerebrospinal fluid or the inner ear fluid (Campo et al., 1999).

Gagnaire and Langlais (2005) observed that the ototoxic aromatic organic solvents differed widely in potency, and their activity (ototoxic or not) and potency of the ototoxic substances was unrelated to LogP.

The specificity of action among isomers of xylene is additional evidence that the ototoxicity is related to selective molecular targets rather than being a simple function of lipophilicity. The exact structural features making the ototoxic solvents active, and the molecular target sites in the ear, have yet to be determined. The ototoxic effects of toluene, ethylbenzene and three isomers of xylene were determined in rats by measuring distortion product otoacoustic emissions (DPOAEs). The ototoxic potency of the substances was uncorrelated to their lipophilocity (LogP) and also uncorrelated to the fluidity of brain lipid microsomes as determined by solid state Nuclear Magnetic Resonance (NMR) spectra, indicating that brain lipid fluidity was not associated with the ototoxic effects of these organic solvents. Further, the potency differences between isomers of xylene indicated that stereospecific interactions determined the toxic effects rather than dissolution of lipid membranes (Wathier et al., 2016).

4.1.3 Co-exposures to solvents

Ototoxic solvents, including chlorobenzene, styrene, toluene, trichloroethylene and mixed xylenes when administered in any set of pairs were dose-additive for ototoxicity, implying and that one could substitute for another and they likely act through common mechanisms (Rebert et al., 1993, 1995). Ototoxicity of toluene was enhanced by doses of acetyl salicylic acid which were not toxic by themselves (Johnson, 1992). Ototoxicity was also observed after exposure to solvent mixtures such as jet fuel JP-8 (Fechter et al., 2007a, 2010). These results suggest that workplaces with exposure to multiple ototoxic solvents or mixed solvents should consider the summed potency of individual ototoxic components rather than evaluate each material separately for occupational safety.

4.1.4 Co-exposure to noise and solvents

Noise-induced hearing loss is one of the most common occupational injuries. The ototoxicity from exposure to organic solvents is enhanced by co-exposures to noise. This has been demonstrated in both animal models (Campo et al., 2003; Fechter et al., 2007a, 2007b; Johnson et al., 1990; Muijser et al., 2000; Sliwińska-Kowalska et al., 2003) and occupational settings (Jacobsen et al., 1993; Morata et al., 1993, 2002; Yang et al., 2016). Notably, a multicenter study of workers exposed to styrene while making reinforced fiberglass involved 1620 workers in Finland, Sweden and Poland (Morata et al., 2011). This study showed that workers with styrene exposure were more likely to exhibit poor audiometric thresholds, and those workers co-exposed to styrene and noise were worse than those exposed only to styrene. The hearing loss from co-exposures to ototoxic organic solvents and noise produces greater hearing loss than exposure to equivalent levels of either stressor alone.

4.1.5 Reviews

Lawton et al. reviewed seven studies concerning potential ototoxicity of occupational styrene exposure (Lawton et al., 2006). They concluded that the results were equivocal with four studies failing to show an effect of styrene exposure on hearing thresholds, with other studies showing a relationship that was enhanced by concurrent noise exposure. They found the positive studies compromised by experimental design and statical analysis shortcomings. One year later, Johnson (Johnson, 2007) reviewed nine studies concerning the relationship between occupational exposure to styrene and hearing loss including the same studies reviewed previously by Lawton et al. (2006). Johnson determined that seven of the nine studies evaluated showed a relationship between occupational styrene exposure and effects on the auditory system (Johnson, 2007). The studies assessed in these reviews were cross sectional in design, several had small sample sizes, relied on retrospective exposure records based on company records, had inconsistent use of biomonitoring of styrene metabolites, had co-exposures to other compounds than styrene, had an unclear history of use of personal protective equipment for either vapor inhalation or for noise exposure, and used different methods to assess toxicity to the auditory and vestibular systems. These limitations lead to differential conclusions among literature reviewers and illustrated the need for better occupational studies.

In yet another review, this time including both styrene and toluene and considering both animal and human studies, the authors concluded that it was clear from animal studies that both substances were ototoxic at high dose level, but the occupational evidence was equivocal (Hoet and Lison, 2008). It is clear from higher dose short term exposure animal experiments that exposure to styrene and noise both damage the cochlea and vestibular apparatus and cause hearing loss. The issues for occupational exposure were determining what exposure concentrations were safe, determining the effect of years of cumulative exposures, what is the interaction with noise, and how can protective measures be determined.

A systematic review and meta-analysis identified 15 studies including a total of 7530 workers suitable for including to evaluate the risks of hearing loss from exposures to only organic solvents mixtures and to organic solvent mixtures with co-exposure to noise (Hormozi et al., 2017). In the combined data set, the odds ratio for hearing loss from occupational exposures to organic solvents was 2.05 (95% confidence interval (CI) 1.44–2.9) and for co-exposures to solvents and noise the OR for nearing loss was 2.95 (CI = 2.1–4.17). The authors conclude that the accumulated data provide evidence of the increased risk of hearing loss among workers exposed to organic solvents, even at low exposure levels, and that risk was exacerbated by co-exposure to noise. More recently, another systematic review and meta-analysis of data from 71,865 workers in China found an increased risk of high frequency hearing loss from noise exposure (OR = 1.95) was further enhanced by co-exposure to chemicals including organic solvents (OR = 2.36) (Zhou et al., 2020). Note that these two independent meta-analyses derived remarkably similar odds ratios for co-exposure to noise and ototoxic chemicals including organic solvents that indicate between a 2–3 times greater risk of suffering hearing loss from solvent and noise co-exposure over non-exposed workers.

4.2 Vision (color and contrast)

Visual impairments have been detected repeatedly following visual function testing of workers occupationally exposed to organic solvents. Among the most common visual changes are increased errors in color chip arrangement tests that are indicative of a lower ability to discriminate along the blue-yellow color axis, and lower sensitivity to visual contrast. Among the solvents reported to produce such changes are styrene (Campagna et al., 1995, 1996; Castillo et al., 2001; Gong et al., 2002; Iregren et al., 2005), toluene (Campagna et al., 2001; Cavalleri et al., 2000; Muttray et al., 1997), benzene (Lee et al., 2007) perchloroethylene (Cavalleri et al., 1994), n-hexane (Issever et al., 2002), xylene (Lee et al., 2013), and a variety of mixed solvent exposures including among painters (Allam et al., 2018), workers in printing shops (Böckelmann et al., 2005; Boeckelmann and Pfister, 2003; Braun et al., 1989), electronics manufacturing (Broadwell et al., 1995; Mergler et al., 1991), petrochemical plants (Indhushree et al., 2016), drycleaners (Jiménez Barbosa et al., 2015), and gasoline filling stations (Costa et al., 2012; de Oliveira et al., 2017; Oliveira et al., 2018). Non-occupational exposures to perchloroethylene, either in residents of neighborhoods with nearby dry-cleaning shops (Altmann et al., 1995), or in buildings with co-located dry cleaners have been reported to cause visual function loss among the residents, including both adults and children (Schreiber et al., 2002; Storm et al., 2011). In these studies, color vision problems typically first appear as subtle blue/yellow errors, that progress over time and with continued exposures may eventually involve red/green errors as well. Deficits in contrast sensitivity typically appear as a higher threshold (i.e., lower sensitivity) to detecting subtle differences in shades of gray in visual patterns. In both cases subjects may be unaware of their visual disturbances, which only become evident on testing these specific functions.

Both the perception of color and contrast reflect primarily neural processing in the retina and central visual structures as opposed to the ocular components responsible for focusing images on the retina. Congenital defects or absence of the gene for middle wavelength sensitive (green or m-cones) or long-wavelength (red or l-cones) are common, and about 8% of human males have congenital red-green color blindness. Females, in contrast, having two copies of both genes rarely have congenital red-green color blindness and the rate of congenital color blindness in females is approximately 1 in 200. The gene for the s-cone is on chromosome 7, meaning that congenital defects in blue/yellow perception are very rare. For this reason, blue/yellow color confusion errors, when observed, and typically considered to be acquired rather than congenital in origin. There are several standard tests for color perception, and the one most commonly used for occupational surveys is the Lanthony D15-d procedure which requires the subject to arrange a sequence of desaturated colored chips of equal luminance that can be distinguished only by subtle gradients in saturation of different hues. The pattern of errors in chip arrangement is indicative of either red/green or blue/yellow deficits. Performance on the Lanthony D15-d is dependent on testing conditions, in particular the luminance and spectral color of lamps used to illuminate the test chips, which may contribute to variability of CCI scores across experiments (Geller and Hudnell, 1997).

Visual contrast is also encoded initially in the outer retina. Visual contrast is expressed as the luminance difference between the light and dark portions of a visual pattern, adjusted for the overall mean luminance. Visual contrast thresholds are the minimum amount of contrast detectable, and it is common to express contrast sensitivity and the inverse of contrast thresholds so that higher values of contrast sensitivity reflect superior perceptual function. Visual contrast is encoded as the differential luminance stimulation of the center and surround portions of receptive fields, which translates into the firing rates of retinal bipolar cells and is transmitted to retinal ganglion cells, and then via the optic nerve to visual centers of the brain. Because retinal bipolar receptive fields differ in size, contrast sensitivity is often expressed as a function of pattern size, measured as cycles of a repeated pattern/degree of visual angle (cpd). Thus, both color and contrast perception are initially encoded by retinal neural networks and transmitted on to the visual centers of the brain.

4.2.1 Animal studies

One of the issues contributing to uncertainty regarding long-term exposure to organic solvents and visual deficits has been a relative lack of animal models. Although there are multiple demonstrations of changes in visual function in animal models from acute organic solvent exposures (discussed previously) there have been very limited reports of persistent changes in visual functions of animals following prolonged exposure. Vettori et al. (2000) showed that the retinas of rats exposed to 300 ppm styrene for 12 weeks had a loss of dopaminergic amacrine cells. In another experimental study, rats were exposed by inhalation (6 h/d, 5 d/w) to 0, 10, 100 or 1000 ppm toluene for 13 weeks (Boyes et al., 2016). Although no significant changes in visual evoked potentials were observed, the rats exposed to 1000 ppm had significantly reduced B-wave amplitudes in dark-adapted electrograms (ERG) recorded approximately 4 weeks after termination of exposure. Importantly, rats tested approximately 1 year after exposure to 1000 ppm toluene still exhibited reduced amplitudes of ERG B-waves. The ERG B-wave is generated primarily by the retinal bipolar cells. These data demonstrated that under experimentally controlled conditions, it was possible to demonstrate persistent changes in retinal function following organic solvent exposure. Alternatively, it is also possible that changes in the central visual components of the CNS contribute to these effects (Paallysaho et al., 2007).

4.2.2 Systematic reviews and meta-analysis

Several critical reviews of the literature linking organic solvent exposures to visual disfunction are available (Fox and Boyes, 2019; Iregren et al., 2002; Lomax et al., 2004; Mergler, 1995). Among these reviews of the literature, only Lomax et al. (2004) raised concerns regarding methodological limitations associated with linking color vision impairments to organic solvent exposures, with the exception of styrene, for which they found the weight of evidence sufficient, however they considered the nature of the deficit to be minimal and the exposure concentrations and durations causing effects to be uncertain.

In addition to the review articles listed above, there have been three reviews involving a quantitative meta-analysis of the literature on color-vision loss and exposure to organic solvents, in particular for styrene (Benignus et al., 2005; Choi et al., 2017; Paramei et al., 2004), In the first of these, Paramei et al. (2004) note that, although 13 of the 15 studies noted a positive association between organic solvent exposure and increased CCI scores, their meta-analysis revealed large heterogeneity between studies preventing the conclusion of an effect of solvents on color vision. Benignus et al. (2005) found 6 studies suitable for meta-analysis and converted CCI to effect magnitude as percent of baseline. Styrene exposures, when not reported directly, were estimated from biometric data. A statistically significant relationship was observed between cumulative styrene exposure (ppm-years) and increased CCI scores. Choi et al. (2017) performed a systematic review and updated meta-analysis of occupational styrene-induced color vision impairments. Of 15 studies reviewed in depth, data from eight studies met their criteria for meta-analysis. Color confusion scores were log-transformed to achieve more normal distributions and to reduce the mean-variance relationship. The pooled data for meta-analysis included 352 styrene-exposed workers and 355 control subjects. The analysis revealed only low-to-moderate heterogeneity among the studies and indicated a statistically significant increase color-confusion error scores associated with styrene exposure. Whether solvent-induced changes in visual function derive from deficits the retina or central visual processing, it is clear that they are among the most sensitive of neurotoxic changes from chronic solvent exposures.

4.3 Olfaction

As described in the section about physicochemical properties, organic solvents share an important feature of odorants (Doleman et al., 1998); they are volatile under normal ambient conditions and can activate olfactory receptor neurons even in low concentrations (van Thriel et al., 2006). Some of them have a characteristic odor that can be found in their descriptive name, such as aromatic hydrocarbons. However, as outlined by van Thriel et al. (2006), many compounds that are classified as organic solvents activate trigeminal chemoreceptors in higher concentrations thereby causing sensory irritation. This particular type of irritation and the transition to more severe tissue irritation might change the cellular and molecular composition of the intranasal respiratory and olfactory epithelium (Bruning et al., 2014). Thereby, the sense of smell might be affected in workers occupationally exposed to higher concentrations of organic solvents. Mergler and Beauvais (1992) were among the first who described this health effect in human volunteers. A sample of five volunteers were exposed to toluene (50 ppm), xylene (40 ppm), or an additive mixture of the two for 7 h on three consecutive days/ week (Mergler and Beauvais, 1992). The olfactory perception threshold (OPT) for toluene was six-fold reduced after these exposures, while the OPT remained unchanged for PM-carbinol (phenylethyl methyl ethyl carbinol) a commonly used test odorant in OPT assessment. This shift in the OPT was (a) substance or substance family depended, and (b) reversible as all volunteers returned to pre-exposure values within a couple of days. Here, receptor-specific mechanisms of adaptation might be the underlying mechanism, and such OPT shifts are probably not an adverse health effect. Among occupationally exposed workers a series of studies confirmed such specific olfactory detection threshold impairments after styrene exposures (Dalton et al., 2003, 2007; Lees et al., 2003). At suprathreshold levels, olfaction normally serves several biologically relevant functions and behaviors (Stevenson, 2010). The olfactory abilities of odor identification or discrimination are used in clinical assessment of chemosensory dysfunctions (Doty, 2018). Metal toxicity as a consequence of occupational exposures has been reported impair olfaction (Antunes et al., 2007). In the case of organic solvents, impaired olfactory function as a neurotoxic consequence has long been neglected (Gobba, 2003). This review described among others the sensory neurotoxicity of solvents, and studies reporting impaired olfaction were less common than vision, hearing, touch, and taste. A comprehensive overview focusing on olfactory toxicity (Gobba, 2006) extended the knowledge about long-term occupational exposures and reduced olfaction. Again, the studies focusing on organic solvents and their mixture were a minority compared to other exposures such as metals. In summary, the olfactory system will be affected by various environmental exposure to volatiles, fumes, and particles but occupational exposures to organic solvents will contribute to this overall impact of environment pollutants of olfaction. Again, the exact neuropathological mechanisms underlying this adverse effect of organic solvents are not fully understood.

5. Peripheral neuropathy

One of the best documented neurotoxic effects of repeated exposure to some, but not all, organic solvents is peripheral neuropathy in both exposed humans and experimental animal models featuring the degeneration of sensory and motor nerve axons in the peripheral nervous system, and the ascending spinal sensory nerves. This is associated with a loss of sensory-motor function that progresses from distal to proximal regions of the extremities. The best-known examples of this are *n*-hexane and methyl-*n*-butyl ketone that act through a common metabolite, the γ-diketone 2,5-hexanedione. Additional substances producing similar damage to peripheral nervous system axons are carbon disulfide and acrylamide. The mechanism of action producing peripheral neuropathy is thought to involve protein adduction and cross-linking axonal neuroproteins required for axonal transport. For critical reviews of this topic please see (Graham et al., 1995; Spencer, 2020; Spencer and Chen, 2021). Another theory includes adduction to critical synaptic proteins (LoPachin and Gavin, 2015; LoPachin and Lehning, 1997).

6. Developmental neurotoxicity

The developing nervous system can be particularly sensitive to chemical exposure, and developmental exposure to organic solvents can occur prenatally through pregnant workers exposed occupationally, postnatally through atmospheric or household sources, or from children in the workforce. In addition, adolescents may become abusers of inhalants containing toluene or other intoxicating solvents. Children of pregnant solvent abusers experience high concentrations of exposure in utero during critical stages of brain development and the associated physical and neurodevelopmental disorder in the offspring has been labeled "fetal solvent syndrome," which is analogous to fetal alcohol syndrome (Khattak et al., 1999; Mcmartin et al., 1998). Neurodevelopmental effects in offspring are a concern following maternal occupational exposure (Laslo-Baker et al., 2004; Till et al., 2001a). Working children exposed to solvents are at risk (Saddik et al., 2009; Scott and Pocock, 2021). Developmental solvent exposures may result in cognitive, motor and sensory abnormalities (Bowen et al., 2005, 2007; Laslo-Baker et al., 2004; Till et al., 2001a,b, 2003, 2005). Prenatal exposure altered neurogenesis and migration of rat cortical neurons (Gospe and Zhou, 2000). This is a complex topic, and the reader is referred to recent review articles for a thorough assessment (Brown Dzubow et al., 2010; Cruz and Bowen, 2021).

Additional evidence could now be derived from in vitro assays that are thought to enhance the risk assessment for developmental neurotoxicity (DNT) testing (Sachana et al., 2019). As an example, in neurospheres constructed from neonatal rat cortical neuronal progenitor cells, 0.25 μM concentrations of TCE for 7 or 14 days reduced neuronal proliferation and induced apoptosis and necrosis (Salama et al., 2018). However, volatile organic solvents may be difficult to contain many in in vitro systems, and they also might have affinity for plastics commonly used in in vitro test systems, leading to uncertainty regarding the tested concentrations.

7. Conclusions

This chapter summarized many aspects of the existing scientific knowledge on solvent neurotoxicity with a special focus on occupational exposures. The scientific literature is comprehensive and diverse giving the impression that the neurotoxicity of solvents is a well-studied research area. We could provide information from "simple" interactions of solvents with specific neurotransmitter receptors (e.g., NMDAR) to complex changes in the human brain. Nevertheless, we could identify some knowledge gaps especially with respect to the neurotoxicity of mixtures and the transition from acute to chronic changes in the brain. We could describe the relevance of the momentary concentration of the solvent in the target organ for the neurobehavioral and -physiological changes that are acutely induced in humans and laboratory animals. We can only speculate that a repeated perturbation of molecular pathways involved in neuronal plasticity will cause neuropathological changes that underly the cognitive impairments in CSE patients or chronically exposed workers. Here, more coordinated research approaches combining mechanistic in vitro *and* in vivo (rodent) studies with human studies exploiting the whole repertoire of cognitive neurosciences would be necessary to close these gaps. Thereby, the safety and health of workers exposed to solvent could be further improved.

Acknowledgments

The authors thank Markku Sainio, Heidi Furu, Timothy Shafer, and Kelly Carstens for helpful comments on a previous version of the manuscript.

Disclaimer

This manuscript has been reviewed by the Center for Public Health and Environmental Assessment, United States Environmental Protection Agency and approved for publication. Approval does not signify that the contents necessarily reflect the views and policies of the Agency nor does mention of trade names or commercial products constitute endorsement or recommendation for use.

References

Adebambo, T.H., Fox, D.T., Otitoloju, A.A., 2020. Toxicological study and genetic basis of BTEX susceptibility in Drosophila melanogaster. Front. Genet. 11, 594179.

Ahmed, S., Tin Tin Win, S., Yamamoto, S., Tsukahara, S., Kunugita, N., Arashidani, K., Fujimaki, H., 2007. Increased hippocampal mRNA expression of neuronal synaptic plasticity related genes in mice chronically exposed to toluene at a low-level human occupational-exposure. Neurotoxicology 28, 168–174.

Al-Hajri, Z., Del Bigio, M.R., 2010. Brain damage in a large cohort of solvent abusers. Acta Neuropathol. 119, 435–445.

Allam, H.K., Soliman, S., Wasfy, T., Ghoneim, A., Serag, Y., Sembajwe, G., 2018. The neuro-ophthalmological effects related to long-term occupational exposure to organic solvents in painters. Toxicol. Ind. Health 34, 91–98.

Altmann, L., Neuhann, H.F., Kramer, U., Witten, J., Jermann, E., 1995. Neurobehavioral and neurophysiological outcome of chronic low-level tetrachloroethene exposure measured in neighborhoods of dry cleaning shops. Environ. Res. 69, 83–89.

Altmann, L., Welge, P., Mensing, T., Lilienthal, H., Voss, B., Wilhelm, M., 2002. Chronic exposure to trichloroethylene affects neuronal plasticity in rat hippocampal slices. Environ. Toxicol. Pharmacol. 12, 157–167.

Anger, W.K., Letz, R., Chrislip, D.W., Frumkin, H., Hudnell, K., Russo, J.M., Chappell, W., Hutchinson, L., 1994. Neurobehavioral test methods for environmental health studies of adults. Neurotoxicol. Teratol. 16, 489–497.

Anger, W.K., Liang, Y.X., Nell, V., Kang, S.K., Cole, D., Bazylewicz-Walczak, B., Rohlman, D.S., Sizemore, O.J., 2000. Lessons learned--15 years of the WHO-NCTB: a review. Neurotoxicology 21, 837–846.

Anger, W.K., Rohlman, D.S., Storzbach, D., 2001. Neurobehavioral testing in humans. Curr. Protoc. Toxicol. (Chapter 11, Unit11 7).

Ankley, G.T., Bennett, R.S., Erickson, R.J., Hoff, D.J., Hornung, M.W., Johnson, R.D., Mount, D.R., Nichols, J.W., Russom, C.L., Schmieder, P.K., Serrano, J.A., Tietge, J.E., Villeneuve, D.L., 2010. Adverse outcome pathways: a conceptual framework to support ecotoxicology research and risk assessment. Environ. Toxicol. Chem. 29, 730–741.

Antunes, M.B., Bowler, R., Doty, R.L., 2007. San Francisco/Oakland Bay bridge welder study: olfactory function. Neurology 69, 1278–1284.

Arlien-Søborg, P., 2000. Solvent Neurotoxicity. CRC Press Inc, Boca Raton, Florida.

Astrand, I., Kilbom, A., Ovrum, P., Wahlberg, I., Vesterberg, O., 1974. Exposure to styrene. I. Concentration in alveolar air and blood at rest and during exercise and metabolism. Work Environ. Health 11, 69–85.

Axelson, O., Hane, M., Hogstedt, C., 1976. A case-referent study on neuropsychiatric disorders among workers exposed to solvents. Scand. J. Work Environ. Health 2, 14–20.

Aydin, K., Ozmen, M., Tatli, B., Sencer, S., 2003. Single-voxel MR spectroscopy and diffusion-weighted MRI in two patients with l-2-hydroxyglutaric aciduria. Pediatr. Radiol. 33, 872–876.

Bader, M., Wrbitzky, R., Blaszkewicz, M., Schaper, M., Van Thriel, C., 2008. Human volunteer study on the inhalational and dermal absorption of N-methyl-2-pyrrolidone (NMP) from the vapour phase. Arch. Toxicol. 82, 13–20.

Baelum, J., 1999. Acute symptoms during non-inhalation exposure to combinations of toluene, trichloroethylene, and n-hexane. Int. Arch. Occup. Environ. Health 72, 408–410.

Baker, E.L., 1994. A review of recent research on health effects of human occupational exposure to organic solvents. A critical review. J. Occup. Med. 36, 1079–1092.

Bale, A.S., Smothers, C.T., Woodward, J.J., 2002. Inhibition of neuronal nicotinic acetylcholine receptors by the abused solvent, toluene. Br. J. Pharmacol. 137, 375–383.

Bale, A.S., Meacham, C.A., Benignus, V.A., Bushnell, P.J., Shafer, T.J., 2005a. Volatile organic compounds inhibit human and rat neuronal nicotinic acetylcholine receptors expressed in xenopus oocytes. Toxicol. Appl. Pharmacol. 205, 77–88.

Bale, A.S., Tu, Y., Carpenter-Hyland, E.P., Chandler, L.J., Woodward, J.J., 2005b. Alterations in glutamatergic and gabaergic ion channel activity in hippocampal neurons following exposure to the abused inhalant toluene. Neuroscience 130, 197–206.

Bale, A.S., Jackson, M.D., Krantz, Q.T., Benignus, V.A., Bushnell, P.J., Shafer, T.J., Boyes, W.K., 2007. Evaluating the NMDA-glutamate receptor as a site of action for toluene, in vivo. Toxicol. Sci. 98, 159–166.

Balster, R.L., 1998. Neural basis of inhalant abuse. Drug Alcohol Depend. 51, 207–214.

Basant, N., Gupta, S., Singh, K.P., 2016. Predicting the acute neurotoxicity of diverse organic solvents using probabilistic neural networks based QSTR modeling approaches. Neurotoxicology 53, 45–52.

Benignus, V.A., 1981. Health effects of toluene: a review. Neurotoxicology 2, 567–588.

Benignus, V.A., Geller, A.M., Boyes, W.K., Bushnell, P.J., 2005. Human neurobehavioral effects of long-term exposure to styrene: a meta-analysis. Environ. Health Perspect. 113, 532–538.

Benignus, V.A., Boyes, W.K., Kenyon, E.M., Bushnell, P.J., 2007. Quantitative comparisons of the acute neurotoxicity of toluene in rats and humans. Toxicol. Sci. 100, 146–155.

Benignus, V.A., Bushnell, P.J., Boyes, W.K., Eklund, C., Kenyon, E.M., 2009. Neurobehavioral effects of acute exposure to four solvents: meta-analyses. Toxicol. Sci. 109, 296–305.

Böckelmann, I., Pfister, E.A., Peters, B., 2005. Influence of chronic exposure to organic solvent mixtures on contrast sensitivity in silk-screen printers: matched-pair analysis. Environ. Toxicol. Pharmacol. 19, 505–510.

Boeckelmann, I., Pfister, E.A., 2003. Influence of occupational exposure to organic solvent mixtures on contrast sensitivity in printers. J. Occup. Environ. Med. 45, 25–33.

Bolden, A.L., Kwiatkowski, C.F., Colborn, T., 2015. New look at BTEX: are ambient levels a problem? Environ. Sci. Technol. 49, 5261–5276.

Bovill, J.G., 2000. Mechanisms of anaesthesia: time to say farewell to the Meyer-Overton rule. Curr. Opin. Anaesthesiol. 13, 433–436.

Bowen, S.E., 2006. Increases in amphetamine-like discriminative stimulus effects of the abused inhalant toluene in mice. Psychopharmacology (Berl) 186, 517–524.

Bowen, S.E., Batis, J.C., Mohammadi, M.H., Hannigan, J.H., 2005. Abuse pattern of gestational toluene exposure and early postnatal development in rats. Neurotoxicol. Teratol. 27, 105–116.

Bowen, S.E., Mohammadi, M.H., Batis, J.C., Hannigan, J.H., 2007. Gestational toluene exposure effects on spontaneous and amphetamine-induced locomotor behavior in rats. Neurotoxicol. Teratol. 29, 236–246.

Boyes, W.K., Bercegeay, M., Ali, J.S., Krantz, T., Mcgee, J., Evans, M., Raymer, J.H., Bushnell, P.J., Simmons, J.E., 2003. Dose-based duration adjustments for the effects of inhaled trichloroethylene on rat visual function. Toxicol. Sci. 76, 121–130.

Boyes, W.K., Bercegeay, M., Krantz, T., Evans, M., Benignus, V., Simmons, J.E., 2005. Momentary brain concentration of trichloroethylene predicts the effects on rat visual function. Toxicol. Sci. 87, 187–196.
Boyes, W.K., Bercegeay, M., Krantz, Q.T., Kenyon, E.M., Bale, A.S., Shafer, T.J., Bushnell, P.J., Benignus, V.A., 2007. Acute toluene exposure and rat visual function in proportion to momentary brain concentration. Toxicol. Sci. 99, 572–581.
Boyes, W.K., Bercegeay, M., Oshiro, W.M., Krantz, Q.T., Kenyon, E.M., Bushnell, P.J., Benignus, V.A., 2009. Acute perchloroethylene exposure alters rat visual-evoked potentials in relation to brain concentrations. Toxicol. Sci. 108, 159–172.
Boyes, W.K., Bercegeay, M., Degn, L., Beasley, T.E., Evansky, P.A., Mwanza, J.C., Geller, A.M., Pinckney, C., Nork, T.M., Bushnell, P.J., 2016. Toluene inhalation exposure for 13 weeks causes persistent changes in electroretinograms of long-Evans rats. Neurotoxicology 53, 257–270.
Braun, C.M.J., Daigneault, S., Gilbert, B., 1989. Color discrimination testing reveals early printshop solvent neurotoxicity better than a neuropsychological test battery. Arch. Clin. Neuropsychol. 4, 1–13.
Broadwell, D.K., Darcey, D.J., Hudnell, H.K., Otto, D.A., Boyes, W.K., 1995. Work-site clinical and neurobehavioral assessment of solvent-exposed microelectronics workers. Am. J. Ind. Med. 27, 677–698.
Brown Dzubow, R., Makris, S., Siegel Scott, C., Barone, S., Jr., 2010. Early lifestage exposure and potential developmental susceptibility to tetrachloroethylene. Birth Defects Res. B Dev. Reprod. Toxicol. 89, 50–65.
Bruning, T., Bartsch, R., Bolt, H.M., Desel, H., Drexler, H., Gundert-Remy, U., Hartwig, A., Jackh, R., Leibold, E., Pallapies, D., Rettenmeier, A.W., Schluter, G., Stropp, G., Sucker, K., Triebig, G., Westphal, G., Van Thriel, C., 2014. Sensory irritation as a basis for setting occupational exposure limits. Arch. Toxicol. 88, 1855–1879.
Bushnell, P.J., Oshiro, W.M., Samsam, T.E., Benignus, V.A., Krantz, Q.T., Kenyon, E.M., 2007. A dosimetric analysis of the acute behavioral effects of inhaled toluene in rats. Toxicol. Sci. 99, 181–189.
Bushnell, P.J., Ward, W.O., Morozova, T.V., Oshiro, W.M., Lin, M.T., Judson, R.S., Hester, S.D., Mckee, J.M., Higuchi, M., 2017. Editor's highlight: genetic targets of acute toluene inhalation in Drosophila melanogaster. Toxicol. Sci. 156, 230–239.
Campagna, D., Mergler, D., Huel, G., Belanger, S., Truchon, G., Ostiguy, C., Drolet, D., 1995. Visual dysfunction among styrene-exposed workers. Scand. J. Work Environ. Health 21, 382–390.
Campagna, D., Gobba, F., Mergler, D., Moreau, T., Galassi, C., Cavalleri, A., Huel, G., 1996. Color vision loss among styrene-exposed workers neurotoxicological threshold assessment. Neurotoxicology 17, 367–373.
Campagna, D., Stengel, B., Mergler, D., Limasset, J.C., Diebold, F., Michard, D., Huel, G., 2001. Color vision and occupational toluene exposure. Neurotoxicol. Teratol. 23, 473–480.
Campo, P., Lataye, R., Cossec, B., Placidi, V., 1997. Toluene-induced hearing loss: a mid-frequency location of the cochlear lesions. Neurotoxicol. Teratol. 19, 129–140.
Campo, P., Loquet, G., Blachère, V., Roure, M., 1999. Toluene and styrene intoxication route in the rat cochlea. Neurotoxicol. Teratol. 21, 427–434.
Campo, P., Pouyatos, B., Lataye, R., Morel, G., 2003. Is the aged rat ear more susceptible to noise or styrene damage than the young ear? Noise Health 5, 1–18.
Castillo, L., Baldwin, M., Sassine, M.P., Mergler, D., 2001. Cumulative exposure to styrene and visual functions. Am. J. Ind. Med. 39, 351–360.
Cavalleri, A., Gobba, F., Paltrinieri, M., Fantuzzi, G., Righi, E., Aggazzotti, G., 1994. Perchloroethylene exposure can induce colour vision loss. Neurosci. Lett. 179, 162–166.

Cavalleri, A., Gobba, F., Nicali, E., Fiocchi, V., 2000. Dose-related color vision impairment in toluene-exposed workers. Arch. Environ. Health 55, 399–404.

Chen, G.D., Henderson, D., 2009. Cochlear injuries induced by the combined exposure to noise and styrene. Hear. Res. 254, 25–33.

Chen, G.D., Chi, L.H., Kostyniak, P.J., Henderson, D., 2007. Styrene induced alterations in biomarkers of exposure and effects in the cochlea: mechanisms of hearing loss. Toxicol. Sci. 98, 167–177.

Chmiel, T., Mieszkowska, A., Kempińska-Kupczyk, D., Kot-Wasik, A., Namieśnik, J., Mazerska, Z., 2019. The impact of lipophilicity on environmental processes, drug delivery and bioavailability of food components. Microchem. J. 146, 393–406.

Choi, A.R., Braun, J.M., Papandonatos, G.D., Greenberg, P.B., 2017. Occupational styrene exposure and acquired dyschromatopsia: A systematic review and meta-analysis. Am. J. Ind. Med. 60, 930–946.

Christoph, G.R., Malley, L.A., Stadler, J.C., 2003. Subchronic inhalation exposure to acetone vapor and scheduled controlled operant performance in male rats. Inhal. Toxicol. 15, 781–798.

Cintra, A., Aguirre, J.A., Andbjer, B., Finnman, U.B., Hagman, M., Agnati, L.F., Hoglund, C., Moller, A., Fuxe, K., 1999. Subchronic toluene exposure in low concentrations produces signs of reduced dysfunction in the 6-hydroxydopamine lesioned nigrostriatal dopaminergic system of the rat. Neurosci. Lett. 274, 5–8.

Costa, T.L., Barboni, M.T., Moura, A.L., Bonci, D.M., Gualtieri, M., De Lima Silveira, L.C., Ventura, D.F., 2012. Long-term occupational exposure to organic solvents affects color vision, contrast sensitivity and visual fields. PLoS One 7, e42961.

Crofton, K.M., Zhao, X., 1993. Mid-frequency hearing loss in rats following inhalation exposure to trichloroethylene: evidence from reflex modification audiometry. Neurotoxicol. Teratol. 15, 413–423.

Crofton, K.M., Lassiter, T.L., Rebert, C.S., 1994. Solvent-induced ototoxicity in rats: an atypical selective mid-frequency hearing deficit. Hear. Res. 80, 25–30.

Cronin, M.T.D., 1996. Quantitative structure-activity relationship (QSAR) analysis of the acute sublethal neurotoxicity of solvents. Toxicol. In Vitro 10, 103–110.

Cruz, S.L., Bowen, S.E., 2021. The last two decades on preclinical and clinical research on inhalant effects. Neurotoxicol. Teratol. 87, 106999.

Cruz, S.L., Mirshahi, T., Thomas, B., Balster, R.L., Woodward, J.J., 1998. Effects of the abused solvent toluene on recombinant N-methyl-D-aspartate and non-N-methyl-D-aspartate receptors expressed in xenopus oocytes. J. Pharmacol. Exp. Ther. 286, 334–340.

Cruz, S.L., Balster, R.L., Woodward, J.J., 2000. Effects of volatile solvents on recombinant N-methyl-D-aspartate receptors expressed in xenopus oocytes. Br. J. Pharmacol. 131, 1303–1308.

Cruz, S.L., Armenta-Reséndiz, M., Carranza-Aguilar, C.J., Galván, E.J., 2020. Minocycline prevents neuronal hyperexcitability and neuroinflammation in medial prefrontal cortex, as well as memory impairment caused by repeated toluene inhalation in adolescent rats. Toxicol. Appl. Pharmacol. 395, 114980.

Dalton, P., Cowart, B., Dilks, D., Gould, M., Lees, P.S., Stefaniak, A., Emmett, E., 2003. Olfactory function in workers exposed to styrene in the reinforced-plastics industry. Am. J. Ind. Med. 44, 1–11.

Dalton, P., Lees, P.S., Gould, M., Dilks, D., Stefaniak, A., Bader, M., Ihrig, A., Triebig, G., 2007. Evaluation of long-term occupational exposure to styrene vapor on olfactory function. Chem. Senses 32, 739–747.

David, R.M., Bernard, L.G., Banton, M.I., Tyler, T.R., Topping, D.C., Gill, M.W., O'donoghue, J.L., 1999. The effect of repeated methyl iso-butyl ketone vapor exposure on schedule-controlled operant behavior in rats. Neurotoxicology 20, 583–593.

De Oliveira, A.R., Campos Neto, A.D.A., Bezerra De Medeiros, P.C., De Andrade, M.J.O., Dos Santos, N.A., 2017. Chronic organic solvent exposure changes visual tracking in men and women. Front. Neurosci. 11.
Dick, F.D., 2006. Solvent neurotoxicity. Occup. Environ. Med. 63 (221–6), 179.
Doleman, B.J., Severin, E.J., Lewis, N.S., 1998. Trends in odor intensity for human and electronic noses: relative roles of odorant vapor pressure vs. molecularly specific odorant binding. Proc. Natl. Acad. Sci. U. S. A. 95, 5442–5447.
Doty, R.L., 2018. Measurement of chemosensory function. World J. Otorhinolaryngol. Head Neck Surg. 4, 11–28.
Dreiem, A., Myhre, O., Fonnum, F., 2002. Relationship between lipophilicity of C6-10 hydrocarbon solvents and their ROS-inducing potency in rat cerebellar granule cells. Neurotoxicology 23, 701–709.
Durand, M., Molinier, V., Kunz, W., Aubry, J.M., 2011. Classification of organic solvents revisited by using the COSMO-RS approach. Chemistry 17, 5155–5164.
Dyer, R.S., Muller, K.E., Janssen, R., Barton, C.N., Boyes, W.K., Benignus, V.A., 1984. Neurophysiological effects of 30 day chronic exposure to toluene in rats. Neurobehav. Toxicol. Teratol. 6, 363–368.
Edling, C., Ekberg, K., 1985. No acute behavioural effects of exposure to styrene: a safe level of exposure? Br. J. Ind. Med. 42, 301–304.
Errebo-Knudsen, E.O., Olsen, F., 1986. Organic solvents and presenile dementia (the painters' syndrome). A critical review of the Danish literature. Sci. Total Environ. 48, 45–67.
Evans, E.B., Balster, R.L., 1991. CNS depressant effects of volatile organic solvents. Neurosci. Biobehav. Rev. 15, 233–241.
Fechter, L.D., Liu, Y., Herr, D.W., Crofton, K.M., 1998. Trichloroethylene ototoxicity: evidence for a cochlear origin. Toxicol. Sci. 42, 28–35.
Fechter, L.D., Gearhart, C., Fulton, S., Campbell, J., Fisher, J., Na, K., Cocker, D., Nelson-Miller, A., Moon, P., Pouyatos, B., 2007a. JP-8 jet fuel can promote auditory impairment resulting from subsequent noise exposure in rats. Toxicol. Sci. 98, 510–525.
Fechter, L.D., Gearhart, C., Fulton, S., Campbell, J., Fisher, J., Na, K., Cocker, D., Nelson-Miller, A., Moon, P., Pouyatos, B., 2007b. Promotion of noise-induced cochlear injury by toluene and ethylbenzene in the rat. Toxicol. Sci. 98, 542–551.
Fechter, L.D., Gearhart, C.A., Fulton, S., 2010. Ototoxic potential of JP-8 and a Fischer-Tropsch synthetic jet fuel following subacute inhalation exposure in rats. Toxicol. Sci. 116, 239–248.
Fox, D.A., Boyes, W.K., 2019. Toxic responses of the cornea, retina, and central visual system. In: KLAASSEN, C.D. (Ed.), Casarett & Doull's Toxicology the Basic Science of Poisons, ninth Edition. McGraw Hill Education, New York.
Frantík, E., Hornychová, M., Horváth, M., 1994. Relative acute neurotoxicity of solvents: isoeffective air concentrations of 48 compounds evaluated in rats and mice. Environ. Res. 66, 173–185.
Furu, H., Sainio, M., Hyvärinen, H.K., Akila, R., Bäck, B., Uuksulainen, S., Kaukiainen, A., 2012. Detecting chronic solvent encephalopathy in occupations at risk. Neurotoxicology 33, 734–741.
Furu, H., Sainio, M., Hyvärinen, H.-K., Kaukiainen, A., 2018. Work ability score of solvent-exposed workers. Int. Arch. Occup. Environ. Health 91, 559–569.
Fuxe, K., Martire, M., Von Euler, G., Agnati, L.F., Hansson, T., Andersson, K., Gustafsson, J.A., Harfstrand, A., 1987. Effects of subacute treatment with toluene on cerebrocortical alpha- and beta-adrenergic receptors in the rat. Evidence for an increased number and a reduced affinity of beta-adrenergic receptors. Acta Physiol. Scand. 130, 307–311.
Gagnaire, F., Langlais, C., 2005. Relative ototoxicity of 21 aromatic solvents. Arch. Toxicol. 79, 346–354.

Gamberale, F., Hultengren, M., 1972. Toluene exposure II. Psychophysiological functions. Work Environ. Health 9, 131–139.
Gamberale, F., Hultengren, M., 1973. Methylchloroform exposure II. Psychophysiological functions. Work Environ. Health 10, 82–92.
Gamberale, F., Hultengren, M., 1974. Exposure to styrene. II. Psychological functions. Work Environ. Health 11, 86–93.
Gamberale, F., Annwall, G., Hultengren, M., 1975a. Exposure to methylene chloride. II. Psychological functions. Scand. J. Work Environ. Health 1, 95–103.
Gamberale, F., Annwall, G., Hultengren, M., 1975b. Exposure to white spirit. II. Psychological functions. Scand. J. Work Environ. Health 1, 31–39.
Gamberale, F., Annwall, G., Olson, B.A., 1976. Exposure to trichloroethylene III. Psychological functions. Scand. J. Work Environ. Health 2, 220–224.
Gamberale, F., Annwall, G., Hultengren, M., 1978. Exposure to xylene and ethylbenzene. III. Effects on central nervous functions. Scand. J. Work Environ. Health 4, 204–211.
Geller, A.M., Hudnell, H.K., 1997. Critical issues in the use and analysis of the Lanthony desaturate color vision test. Neurotoxicol. Teratol. 19, 455–465.
Gill, M.W., Tyler, T.R., Beyrouty, P.C., 2000. Subchronic inhalation neurotoxicity study of amyl acetate in rats. J. Appl. Toxicol. 20, 463–469.
Gobba, F., 2003. Occupational exposure to chemicals and sensory organs: a neglected research field. Neurotoxicology 24, 675–691.
Gobba, F., 2006. Olfactory toxicity: long-term effects of occupational exposures. Int. Arch. Occup. Environ. Health 79, 322–331.
Gong, Y.Y., Kishi, R., Katakura, Y., Tsukishima, E., Fujiwara, K., Kasai, S., Satoh, T., Sata, F., Kawai, T., 2002. Relation between colour vision loss and occupational styrene exposure level. Occup. Environ. Med. 59, 824–829.
Gospe Jr., S.M., Zhou, S.S., 2000. Prenatal exposure to toluene results in abnormal neurogenesis and migration in rat somatosensory cortex. Pediatr. Res. 47, 362–368.
Graham, D.G., Amarnath, V., Valentine, W.M., Pyle, S.J., Anthony, D.C., 1995. Pathogenetic studies of hexane and carbon disulfide neurotoxicity. Crit. Rev. Toxicol. 25, 91–112.
Gralewicz, S., Dyzma, M., 2005. Organic solvents and the dopaminergic system. Int. J. Occup. Med. Environ. Health 18, 103–113.
Grasshoff, C., Drexler, B., Rudolph, U., Antkowiak, B., 2006. Anaesthetic drugs: linking molecular actions to clinical effects. Curr. Pharm. Des. 12, 3665–3679.
Hall, T.A., Rebert, C.S., 1994. Review of neurotoxicity of styrene in humans. Vet. Hum. Toxicol. 36, 249.
Hane, M., Axelson, O., Blume, J., Hogstedt, C., Sundell, L., Ydreborg, B., 1977. Psychological function changes among house painters. Scand. J. Work Environ. Health 3, 91–99.
Hanninen, H., Eskelinen, L., Husman, K., Nurminen, M., 1976. Behavioral effects of long-term exposure to a mixture of organic solvents. Scand. J. Work Environ. Health 2, 240–255.
Hansson, E., Von Euler, G., Fuxe, K., Hansson, T., 1988. Toluene induces changes in the morphology of astroglia and neurons in striatal primary cell cultures. Toxicology 49, 155–163.
Harkonen, H., 1977. Relationship of symptoms to occupational styrene exposure and to the findings of electroencephalographic and psychological examinations. Int. Arch. Occup. Environ. Health 40, 231–239.
Harrichandra, A., Roelofs, C., Pavilonis, B., 2020. Occupational exposure and ventilation assessment in new York City nail salons. Ann. Work Expo. Health 64, 468–478.

Harris, R.A., Mihic, S.J., Dildy-Mayfield, J.E., Machu, T.K., 1995. Actions of anesthetics on ligand-gated ion channels: role of receptor subunit composition. FASEB J. 9, 1454–1462.

Henley, M., Letinski, D.J., Carr, J., Caro, M.L., Daughtrey, W., White, R., 2014. Health assessment of gasoline and fuel oxygenate vapors: generation and characterization of test materials. Regul. Toxicol. Pharmacol. 70, S13–S17.

Hester, S.D., Johnstone, A.F.M., Boyes, W.K., Bushnell, P.J., Shafer, T.J., 2011. Acute toluene exposure alters expression of genes in the central nervous system associated with synaptic structure and function. Neurotoxicol. Teratol. 33, 521–529.

Hester, S.D., Johnstone, A.F.M., Boyes, W.K., Bushnell, P.J., Shafer, T.J., 2012. Transcriptional responses in rat brain associated with sub-chronic toluene inhalation are not predicted by effects of acute toluene inhalation. Neurotoxicol. Teratol. 34, 530–533.

Hillefors-Berglund, M., Liu, Y., Von Euler, G., 1995. Persistent, specific and dose-dependent effects of toluene exposure on dopamine D2 agonist binding in the rat caudate-putamen. Toxicology 100, 185–194.

Hoet, P., Lison, D., 2008. Ototoxicity of toluene and styrene: state of current knowledge. Crit. Rev. Toxicol. 38, 127–170.

Hooisma, J., Hänninen, H., Emmen, H.H., Kulig, B.M., 1993. Behavioral effects of exposure to organic solvents in Dutch painters. Neurotoxicol. Teratol. 15, 397–406.

Hooisma, J., Hänninen, H., Emmen, H.H., Kulig, B.M., 1994. Symptoms indicative of the effects of organic solvent exposure in Dutch painters. Neurotoxicol. Teratol. 16, 613–622.

Hormozi, M., Ansari-Moghaddam, A., Mirzaei, R., Dehghan Haghighi, J., Eftekharian, F., 2017. The risk of hearing loss associated with occupational exposure to organic solvents mixture with and without concurrent noise exposure: A systematic review and meta-analysis. Int. J. Occup. Med. Environ. Health 30, 521–535.

Indhushree, R., Monica, R., Coral, K., Angayarkanni, N., Punitham, R., Subburathinam, B.M., Krishnakumar, R., Santanam, P.P., 2016. Visual functions of workers exposed to organic solvents in petrochemical industries. Indian J. Occup. Environ. Med. 20, 133–137.

Iregren, A., Gamberale, F., Kjellberg, A., 1996. SPES: a psychological test system to diagnose environmental hazards. Swedish performance evaluation system. Neurotoxicol. Teratol. 18, 485–491.

Iregren, A., Andersson, M., Nylen, P., 2002. Color vision and occupational chemical exposures. II. Visual functions in non-exposed subjects. Neurotoxicology 23, 735–745.

Iregren, A., Johnson, A.-C., Nylén, P., 2005. Low-level styrene exposure and color vision in Swedish styrene workers. Environ. Toxicol. Pharmacol. 19, 511–516.

Issever, H., Malat, G., Sabuncu, H.H., Yuksel, N., 2002. Impairment of colour vision in patients with n-hexane exposure-dependent toxic polyneuropathy. Occup. Med. (Lond) 52, 183–186.

Jacobsen, P., Hein, H.O., Suadicani, P., Parving, A., Gyntelberg, F., 1993. Mixed solvent exposure and hearing impairment: an epidemiological study of 3284 men. The Copenhagen male study. Occup. Med. 43, 180–184.

Jaspers, R.M., Muijser, H., Lammers, J.H., Kulig, B.M., 1993. Mid-frequency hearing loss and reduction of acoustic startle responding in rats following trichloroethylene exposure. Neurotoxicol. Teratol. 15, 407–412.

Jiang, M., Karasawa, T., Steyger, P.S., 2017. Aminoglycoside-induced Cochleotoxicity: A review. Front. Cell. Neurosci. 11, 308.

Jiménez Barbosa, I.A., Boon, M.Y., Khuu, S.K., 2015. Exposure to organic solvents used in dry cleaning reduces low and high level visual function. PLoS One 10, e0121422.

Johnson, A.C., 1992. Auditory sensitivity in rats exposed to toluene and/or acetyl salicylic acid. Neuroreport 3, 1141–1144.

Johnson, A.C., 1993. The ototoxic effect of toluene and the influence of noise, acetyl salicylic acid, or genotype. A study in rats and mice. Scand. Audiol. Suppl. 39, 1–40.

Johnson, A.C., 2007. Relationship between styrene exposure and hearing loss: review of human studies. Int. J. Occup. Med. Environ. Health 20, 315–325.

Johnson, A.C., Canlon, B., 1994. Progressive hair cell loss induced by toluene exposure. Hear. Res. 75, 201–208.

Johnson, A.C., Nylén, P., Borg, E., Höglund, G., 1990. Sequence of exposure to noise and toluene can determine loss of auditory sensitivity in the rat. Acta Otolaryngol. 109, 34–40.

Juran, S.A., Van Thriel, C., Kleinbeck, S., Schaper, M., Falkenstein, M., Iregren, A., Johanson, G., 2012. Neurobehavioral performance in human volunteers during inhalation exposure to the unpleasant local irritant cyclohexylamine. Neurotoxicology 33, 1180–1187.

Juran, S.A., Johanson, G., Ernstgard, L., Iregren, A., Van Thriel, C., 2014. Neurobehavioral performance in volunteers after inhalation of white spirits with high and low aromatic content. Arch. Toxicol. 88, 1127–1140.

Kandel, E.R., 2001. The molecular biology of memory storage: a dialogue between genes and synapses. Science 294, 1030–1038.

Kaneko, T., Wang, P.Y., Sato, A., 1997. Assessment of the health effects of trichloroethylene. Ind. Health 35, 301–324.

Kanter, M., 2008. Protective effects of Nigella sativa on the neuronal injury in frontal cortex and brain stem after chronic toluene exposure. Neurochem. Res. 33, 2241.

Katritzky, A.R., Fara, D.C., Kuanar, M., Hur, E., Karelson, M., 2005. The classification of solvents by combining classical QSPR methodology with principal component analysis. J. Phys. Chem. A 109, 10323–10341.

Kaukiainen, A., Riala, R., Martikainen, R., Akila, R., Reijula, K., Sainio, M., 2004. Solvent-related health effects among construction painters with decreasing exposure. Am. J. Ind. Med. 46, 627–636.

Khattak, S., K-Moghtader, G., Mcmartin, K., Barrera, M., Kennedy, D., Koren, G., 1999. Pregnancy outcome following gestational exposure to organic SolventsA prospective controlled study. JAMA 281, 1106–1109.

Knave, B., Olson, B.A., Elofsson, S., Gamberale, F., Isaksson, A., Mindus, P., Persson, H.E., Struwe, G., Wennberg, A., Westerholm, P., 1978. Long-term exposure to jet fuel. II. A cross-sectional epidemiologic investigation on occupationally exposed industrial workers with special reference to the nervous system. Scand. J. Work Environ. Health 4, 19–45.

Kobald, S.O., Wascher, E., Blaszkewicz, M., Golka, K., Van Thriel, C., 2015. Neurobehavioral and neurophysiological effects after acute exposure to a single peak of 200 ppm toluene in healthy volunteers. Neurotoxicology 48, 50–59.

Kodavanti, P.R., Royland, J.E., Richards, J.E., Besas, J., Macphail, R.C., 2011. Toluene effects on oxidative stress in brain regions of young-adult, middle-age, and senescent Brown Norway rats. Toxicol. Appl. Pharmacol. 256, 386–398.

Kodavanti, P.R., Royland, J.E., Moore-Smith, D.A., Besas, J., Richards, J.E., Beasley, T.E., Evansky, P., Bushnell, P.J., 2015. Acute and subchronic toxicity of inhaled toluene in male long-Evans rats: oxidative stress markers in brain. Neurotoxicology 51, 10–19.

Kulig, B.M., 1987. The effects of chronic trichloroethylene exposure on neurobehavioral functioning in the rat. Neurotoxicol. Teratol. 9, 171–178.

Kwok, R.K., Engel, L.S., Miller, A.K., Blair, A., Curry, M.D., Jackson, W.B., Stewart, P.A., Stenzel, M.R., Birnbaum, L.S., Sandler, D.P., 2017. The GuLF STUDY: A prospective study of persons involved in the Deepwater horizon oil spill response and clean-up. Environ. Health Perspect. 125, 570–578.

Lammers, J.H.C.M., Van Asperen, J., De Groot, D., Rijcken, W.R.P., 2005. Behavioural effects and kinetics in brain in response to inhalation of constant or fluctuating toluene concentrations in the rat. Environ. Toxicol. Pharmacol. 19, 625–634.

Laslo-Baker, D., Barrera, M., Knittel-Keren, D., Kozer, E., Wolpin, J., Khattak, S., Hackman, R., Rovet, J., Koren, G., 2004. Child neurodevelopmental outcome and maternal occupational exposure to solvents. Arch. Pediatr. Adolesc. Med. 158, 956–961.

Lataye, R., Campo, P., Barthelemy, C., Loquet, G., Bonnet, P., 2001. Cochlear pathology induced by styrene. Neurotoxicol. Teratol. 23, 71–79.

Lawton, B.W., Hoffmann, J., Triebig, G., 2006. The ototoxicity of styrene: a review of occupational investigations. Int. Arch. Occup. Environ. Health 79, 93–102.

Lee, E.-H., Eum, K.D., Cho, S.-I., Cheong, H.-K., Paek, D.M., 2007. Acquired dyschromatopsia among petrochemical industry workers exposed to benzene. Neurotoxicology 28, 356–363.

Lee, E.H., Paek, D., Kho, Y.L., Choi, K., Chae, H.J., 2013. Color vision impairments among shipyard workers exposed to mixed organic solvents, especially xylene. Neurotoxicol. Teratol. 37, 39–43.

Lees, P.S., Stefaniak, A., Emmett, E.A., Dalton, P., 2003. Exposure assessment for study of olfactory function in workers exposed to styrene in the reinforced-plastics industry. Am. J. Ind. Med. 44, 12–23.

Letz, R., 1991. Use of computerized test batteries for quantifying neurobehavioral outcomes. Environ. Health Perspect. 90, 195–198.

Li, A.A., Thake, D.C., Kaempfe, T.A., Branch, D.K., O'donnell, P., Speck, F.L., Tyler, T.R., Faber, W.D., Jasti, S.L., Ouellette, R., Banton, M.I., 1999. Neurotoxicity evaluation of rats after subchronic inhalation exposure to isobutanol. Neurotoxicology 20, 889–900.

Linakis, M.W., Sayre, R.R., Pearce, R.G., Sfeir, M.A., Sipes, N.S., Pangburn, H.A., Gearhart, J.M., Wambaugh, J.F., 2020. Development and evaluation of a high throughput inhalation model for organic chemicals. J. Expo. Sci. Environ. Epidemiol. 30, 866–877.

Lomax, R.B., Ridgway, P., Meldrum, M., 2004. Does occupational exposure to organic solvents affect colour discrimination? Toxicol. Rev. 23, 91–121.

Lopachin, R.M., Gavin, T., 2015. Toxic neuropathies: mechanistic insights based on a chemical perspective. Neurosci. Lett. 596, 78–83.

Lopachin, R.M., Lehning, E.J., 1997. The relevance of axonal swellings and atrophy to gamma-diketone neurotoxicity: a forum position paper. Neurotoxicology 18, 7–22.

Lubman, D.I., Yucel, M., Lawrence, A.J., 2008. Inhalant abuse among adolescents: neurobiological considerations. Br. J. Pharmacol. 154, 316–326.

Lucchini, R., Albini, E., Benedetti, L., Alessio, L., 2005. Neurobehavioral science in hazard identification and risk assessment of neurotoxic agents- -what are the requirements for further development? Int. Arch. Occup. Environ. Health 78, 427–437.

Maguin, K., Lataye, R., Campo, P., Cossec, B., Burgart, M., Waniusiow, D., 2006. Ototoxicity of the three xylene isomers in the rat. Neurotoxicol. Teratol. 28, 648–656.

Malley, L.A., Christoph, G.R., Stadler, J.C., Hansen, J.F., Biesemeier, J.A., Jasti, S.L., 2001. Acute and subchronic neurotoxicological evaluation of tetrahydrofuran by inhalation in rats. Drug Chem. Toxicol. 24, 201–219.

Mashour, G.A., Forman, S.A., Campagna, J.A., 2005. Mechanisms of general anesthesia: from molecules to mind. Best Pract. Res. Clin. Anaesthesiol. 19, 349–364.

Mattsson, J.L., Gorzinski, S.J., Albee, R.R., Zimmer, M.A., 1990. Evoked potential changes from 13 weeks of simulated toluene abuse in rats. Pharmacol. Biochem. Behav. 36, 683–689.

Mattsson, J.L., Albee, R.R., Lomax, L.G., Beekman, M.J., Spencer, P.J., 1993. Neurotoxicologic examination of rats exposed to 1,1,1-trichloroethane vapor for 13 weeks. Neurotoxicol. Teratol. 15, 313–326.

Mattsson, J.L., Albee, R.R., Yano, B.L., Bradley, G., Spencer, P.J., 1998. Neurotoxicologic examination of rats exposed to 1,1,2,2-tetrachloroethylene (perchloroethylene) vapor for 13 weeks. Neurotoxicol. Teratol. 20, 83–98.

Maurissen, J.P.J., 1994. Operant methodology: an evaluation in the context of the multisubstance rule for solvent neurotoxicity testing. In: O'donoghue, B.W.A.J. (Ed.), Neurobehavioral Toxicology: Analysis and Interpretation. Raven Press Ltd., New York.

Mcdermott, C., Allshire, A., Pelt, V., F. N. a. M., Heffron, J. J. A., 2007. Sub-chronic toxicity of low concentrations of industrial volatile organic pollutants in vitro. Toxicol. Appl. Pharmacol. 219, 85–94.

Mcmartin, K.I., Chu, M., Kopecky, E., Einarson, T.R., Koren, G., 1998. Pregnancy outcome following maternal organic solvent exposure: A meta-analysis of epidemiologic studies. Am. J. Ind. Med. 34, 288–292.

Mergler, D., 1995. Behavioral neurophysiology: quantitative measures of sensory toxicity. In: CHANG, L.W., SLIKKER, W.J. (Eds.), Neurotoxicology Approaches and Methods. Academic Press, San Diego, CA.

Mergler, D., Beauvais, B., 1992. Olfactory threshold shift following controlled 7-hour exposure to toluene and/or xylene. Neurotoxicology 13, 211–215.

Mergler, D., Huel, G., Bowler, R., Frenette, B., Cone, J., 1991. Visual dysfunction among former microelectronics assembly workers. Arch. Environ. Health 46, 326–334.

Meyer-Baron, M., 2005. A meta-analytical approach to neurobehavioural effects of occupational toluene exposure. Environ. Toxicol. Pharmacol. 19, 651–657.

Meyer-Baron, M., Blaszkewicz, M., Henke, H., Knapp, G., Muttray, A., Schäper, M., Van Thriel, C., 2008. The impact of solvent mixtures on neurobehavioral performance: conclusions from epidemiological data. Neurotoxicology 29, 349–360.

Mihic, S.J., Ye, Q., Wick, M.J., Koltchine, V.V., Krasowski, M.D., Finn, S.E., Mascia, M.P., Valenzuela, C.F., Hanson, K.K., Greenblatt, E.P., Harris, R.A., Harrison, N.L., 1997. Sites of alcohol and volatile anaesthetic action on GABA(A) and glycine receptors. Nature 389, 385–389.

Mikkelsen, S., 1980. A cohort study of disability pension and death among painters with special regard to disabling presenile dementia as an occupational disease. Scand. J. Soc. Med. Suppl. 16, 34–43.

Miyagawa, M., Honma, T., Sato, M., 1995. Effects of subchronic exposure to toluene on working and reference memory in rats. Neurotoxicol. Teratol. 17, 657–664.

Morata, T.C., Dunn, D.E., Kretschmer, L.W., Lemasters, G.K., Keith, R.W., 1993. Effects of occupational exposure to organic solvents and noise on hearing. Scand. J. Work Environ. Health 19, 245–254.

Morata, T.C., Fiorini, A.C., Fischer, F.M., Colacioppo, S., Wallingford, K.M., Krieg, E.F., Dunn, D.E., Gozzoli, L., Padrão, M.A., Cesar, C.L., 1997. Toluene-induced hearing loss among rotogravure printing workers. Scand. J. Work Environ. Health 23, 289–298.

Morata, T.C., Johnson, A.C., Nylen, P., Svensson, E.B., Cheng, J., Krieg, E.F., Lindblad, A.C., Ernstgård, L., Franks, J., 2002. Audiometric findings in workers exposed to low levels of styrene and noise. J. Occup. Environ. Med. 44, 806–814.

Morata, T.C., Sliwinska-Kowalska, M., Johnson, A.C., Starck, J., Pawlas, K., Zamyslowska-Szmytke, E., Nylen, P., Toppila, E., Krieg, E., Pawlas, N., Prasher, D., 2011. A multicenter study on the audiometric findings of styrene-exposed workers. Int. J. Audiol. 50, 652–660.

Moser, V.C., Balster, R.L., 1981. The effects of acute and repeated toluene exposure on operant behavior in mice. Neurobehav. Toxicol. Teratol. 3, 471–475.

Moser, V.C., Bowen, S.E., Li, A.A., Sette, W.S., Weisenburger, W.P., 2000. Cognitive evaluation: is it needed in neurotoxicity screening? Symposium presented at the annual behavioral toxicology society meeting, may 1999. Neurotoxicol. Teratol. 22, 785–798.

Muijser, H., Lammers, J.H., Kullig, B.M., 2000. Effects of exposure to trichloroethylene and noise on hearing in rats. Noise Health 2, 57–66.

Muttray, A., Wolff, U., Jung, D., Konietzko, J., 1997. Blue-yellow deficiency in workers exposed to low concentrations of organic solvents. Int. Arch. Occup. Environ. Health 70, 407–412.

National Institute of Occupational Safety and Health (NIOSH), 1987. Organic Solvent Neurotoxicity. Center for Disease Control and Prevention (CDC).

Nitsche, M.A., Paulus, W., 2000. Excitability changes induced in the human motor cortex by weak transcranial direct current stimulation. J. Physiol. 527 (Pt 3), 633–639.

Noraberg, J., Arlien-Soborg, P., 2000. Neurotoxic interactions of industrially used ketones. Neurotoxicology 21, 409–418.

O'callaghan, J.P., Daughtrey, W.C., Clark, C.R., Schreiner, C.A., White, R., 2014. Health assessment of gasoline and fuel oxygenate vapors: neurotoxicity evaluation. Regul. Toxicol. Pharmacol. 70, S35–S42.

O'leary-Moore, S.K., Galloway, M.P., Mcmechan, A.P., Irtenkauf, S., Hannigan, J.H., Bowen, S.E., 2009. Neurochemical changes after acute binge toluene inhalation in adolescent and adult rats: A high-resolution magnetic resonance spectroscopy study. Neurotoxicol. Teratol. 31, 382–389.

Oliveira, A.R., Campos Neto, A.A., De Andrade, M.J.O., De Medeiros, P.C.B., Dos Santos, N.A., 2018. Organic solvent exposure and contrast sensitivity: comparing men and women. Braz. J. Med. Biol. Res. 51, e6568.

Oshiro, W.M., Krantz, Q.T., Bushnell, P.J., 2004. A search for residual behavioral effects of trichloroethylene (TCE) in rats exposed as young adults. Neurotoxicol. Teratol. 26, 239–251.

Oshiro, W.M., Kenyon, E.M., Gordon, C.J., Bishop, B., Krantz, Q.T., Ford, J., Bushnell, P.J., 2011. Extrapolating the acute behavioral effects of toluene from 1- to 24-h exposures in rats: roles of dose metric and metabolic and behavioral tolerance. Toxicol. Sci. 123, 180–192.

Paallysaho, J., Nasanen, R., Mantyjarvi, M., Kaukiainen, A., Sainio, M., 2007. Colour vision defects in occupational chronic solvent encephalopathy. Hum. Exp. Toxicol. 26, 375–383.

Paramei, G.V., Meyer-Baron, M., Seeber, A., 2004. Impairments of colour vision induced by organic solvents: A meta-analysis study. Neurotoxicology 25, 803–816.

Parker, A.J., 1969. Protic-dipolar aprotic solvent effects on rates of bimolecular reactions. Chem. Rev. 69, 1–32.

Proctor, S.P., Letz, R., White, R.F., 2000. Validity of a computer-assisted neurobehavioral test battery in toxicant encephalopathy. Neurotoxicology 21, 703–714.

Pryor, G.T., 1995. Solvent-induced neurotoxicity: effects and mechanisms. In: CHANG, L.W., DYER, R.S. (Eds.), Handbook of Neurotoxicity. Marcel Dekker, Inc., New York, NY.

Pryor, G.T., Dickinson, J., Howd, R.A., Rebert, C.S., 1983. Transient cognitive deficits and high-frequency hearing loss in weanling rats exposed to toluene. Neurobehav. Toxicol. Teratol. 5, 53–57.

Quist, A.J.L., Rohlman, D.S., Kwok, R.K., Stewart, P.A., Stenzel, M.R., Blair, A., Miller, A.K., Curry, M.D., Sandler, D.P., Engel, L.S., 2019. Deepwater horizon oil spill exposures and neurobehavioral function in GuLF study participants. Environ. Res. 179, 108834.

Rebert, C.S., Hall, T.A., 1994. The neuroepidemiology of styrene: a critical review of representative literature. Crit. Rev. Toxicol. 24 (Suppl), S57–106.

Rebert, C.S., Sorenson, S.S., Howd, R.A., Pryor, G.T., 1983. Toluene-induced hearing loss in rats evidenced by the brainstem auditory-evoked response. Neurobehav. Toxicol. Teratol. 5, 59–62.
Rebert, C.S., Day, V.L., Matteucci, M.J., Pryor, G.T., 1991. Sensory-evoked potentials in rats chronically exposed to trichloroethylene: predominant auditory dysfunction. Neurotoxicol. Teratol. 13, 83–90.
Rebert, C.S., Boyes, W.K., Pryor, G.T., Svensgaard, D.J., Kassay, K.M., Gordon, G.R., Shinsky, N., 1993. Combined effects of solvents on the rat's auditory system: styrene and trichloroethylene. Int. J. Psychophysiol. 14, 49–59.
Rebert, C.S., Schwartz, R.W., Svendsgaard, D.J., Pryor, G.T., Boyes, W.K., 1995. Combined effects of paired solvents on the rat's auditory system. Toxicology 105, 345–354.
Reiter, L., 1978. An introduction to neurobehavioral toxicology. Environ. Health Perspect. 26, 5–7.
Riegel, A.C., Zapata, A., Shippenberg, T.S., French, E.D., 2007. The abused inhalant toluene increases dopamine release in the nucleus accumbens by directly stimulating ventral tegmental area neurons. Neuropsychopharmacology 32, 1558–1569.
Ritchie, G.D., Still, K.R., Alexander, W.K., Nordholm, A.F., Wilson, C.L., Rossi 3rd, J., Mattie, D.R., 2001. A review of the neurotoxicity risk of selected hydrocarbon fuels. J. Toxicol. Environ. Health B Crit. Rev. 4, 223–312.
Rohlman, D.S., Gimenes, L.S., Eckerman, D.A., Kang, S.K., Farahat, F.M., Anger, W.K., 2003. Development of the behavioral assessment and research system (BARS) to detect and characterize neurotoxicity in humans. Neurotoxicology 24, 523–531.
Rohlman, D.S., Lucchini, R., Anger, W.K., Bellinger, D.C., Van Thriel, C., 2008. Neurobehavioral testing in human risk assessment. Neurotoxicology 29, 556–567.
Rosenberg, N.L., Grigsby, J., Dreisbach, J., Busenbark, D., Grigsby, P., 2002. Neuropsychologic impairment and MRI abnormalities associated with chronic solvent abuse. J. Toxicol. Clin. Toxicol. 40, 21–34.
Russell, E.W., 1980. Fluid and crystallized intelligence: effects of diffuse brain damage on the WAIS. Percept. Mot. Skills 51, 121–122.
Sachana, M., Bal-Price, A., Crofton, K.M., Bennekou, S.H., Shafer, T.J., Behl, M., Terron, A., 2019. International regulatory and scientific effort for improved developmental neurotoxicity testing. Toxicol. Sci. 167, 45–57.
Saddik, B., Williamson, A., Black, D., Nuwayhid, I., 2009. Neurobehavioral impairment in children occupationally exposed to mixed organic solvents. Neurotoxicology 30, 1166–1171.
Sainio Sr., M.A., 2015. Neurotoxicity of solvents. Handb. Clin. Neurol. 131, 93–110.
Salama, M.M., El-Naggar, D.A., Abdel-Rahman, R.H., Elhak, S. a. G., 2018. Toxic effects of trichloroethylene on rat Neuroprogenitor cells. Front. Pharmacol. 9.
Schaumburg, H.H., Spencer, P.S., 2000. Organic solvent mixtures. In: SPENCER, P.S., SCHAUMBURG, H.H., LUDOLPH, A.C. (Eds.), Experimental and Clinical Neurotoxicology, Second Edition. Oxford University Press, New York.
Schreiber, J.S., Hudnell, H.K., Geller, A.M., House, D.E., Aldous, K.M., Force, M.S., Langguth, K., Prohonic, E.J., Parker, J.C., 2002. Apartment residents' and day care workers' exposures to tetrachloroethylene and deficits in visual contrast sensitivity. Environ. Health Perspect. 110, 655–664.
Scott, N.B., Pocock, N.S., 2021. The health impacts of hazardous chemical exposures among child Labourers in low- and middle-income countries. Int. J. Environ. Res. Public Health 18.
Seeber, A., Blaszkewicz, M., Golka, K., Hallier, E., Kiesswetter, E., Schaper, M., Van Thriel, C., 2004a. Neurobehavioral effects of experimental exposures to low levels of styrene. Toxicol. Lett. 151, 183–192.

Seeber, A., Schaper, M., Zupanic, M., Blaszkewicz, M., Demes, P., Kiesswetter, E., Van Thriel, C., 2004b. Toluene exposure below 50 ppm and cognitive function: a follow-up study with four repeated measurements in rotogravure printing plants. Int. Arch. Occup. Environ. Health 77, 1–9.
Seppalainen, A.M., Savolainen, K., Kovala, T., 1981. Changes induced by xylene and alcohol in human evoked potentials. Electroencephalogr. Clin. Neurophysiol. 51, 148–155.
Shafer, T.J., Atchison, W.D., 1991. Transmitter, ion channel and receptor properties of pheochromocytoma (PC12) cells: a model for neurotoxicological studies. Neurotoxicology 12, 473–492.
Shafer, T.J., Bushnell, P.J., Benignus, V.A., Woodward, J.J., 2005. Perturbation of voltage-sensitive Ca2+ channel function by volatile organic solvents. J. Pharmacol. Exp. Ther. 315, 1109–1118.
Siegel, M., Starks, S.E., Sanderson, W.T., Kamel, F., Hoppin, J.A., Gerr, F., 2017. Organic solvent exposure and depressive symptoms among licensed pesticide applicators in the agricultural health study. Int. Arch. Occup. Environ. Health 90, 849–857.
Simmons, J.E., Boyes, W.K., Bushnell, P.J., Raymer, J.H., Limsakun, T., Mcdonald, A., Sey, Y.M., Evans, M.V., 2002. A physiologically based pharmacokinetic model for trichloroethylene in the male long-Evans rat. Toxicol. Sci. 69, 3–15.
Sliwińska-Kowalska, M., Zamyslowska-Szmytke, E., Szymczak, W., Kotylo, P., Fiszer, M., Wesolowski, W., Pawlaczyk-Luszczynska, M., 2003. Ototoxic effects of occupational exposure to styrene and co-exposure to styrene and noise. J. Occup. Environ. Med. 45, 15–24.
Smallwood, I.M., 1996. Handbook of Organic Solvent Properties. Butterworth-Heinemann, Oxford.
Soares, M.V., Mesadri, J., Gonçalves, D.F., Cordeiro, L.M., Franzen Da Silva, A., Obetine Baptista, F.B., Wagner, R., Dalla Corte, C.L., Soares, F.A.A., Ávila, D.S., 2022. Neurotoxicity induced by toluene: in silico and in vivo evidences of mitochondrial dysfunction and dopaminergic neurodegeneration. Environ. Pollut. 298, 118856.
Soulage, C., Perrin, D., Cottet-Emard, J.M., Pequignot, J., Dalmaz, Y., Pequignot, J.M., 2004. Central and peripheral changes in catecholamine biosynthesis and turnover in rats after a short period of ozone exposure. Neurochem. Int. 45, 979–986.
Spee, T., Van Valen, E., Van Duivenbooden, C., Van Der Laan, G., 2012. A screening programme on chronic solvent-induced encephalopathy among Dutch painters. Neurotoxicology 33, 727–733.
Spencer, P.S., 2020. Neuroprotein targets of γ-diketone metabolites of aliphatic and aromatic solvents that induce central-peripheral axonopathy. Toxicol. Pathol. 48, 411–421.
Spencer, P.S., Chen, X., 2021. The role of protein adduction in toxic neuropathies of exogenous and endogenous origin. Toxics 9, 98.
Stevenson, R.J., 2010. An initial evaluation of the functions of human olfaction. Chem. Senses 35, 3–20.
Storm, J.E., Mazor, K.A., Aldous, K.M., Blount, B.C., Brodie, S.E., Serle, J.B., 2011. Visual contrast sensitivity in children exposed to tetrachloroethylene. Arch. Environ. Occup. Health 66, 166–177.
Sullivan, M.J., Rarey, K.E., Conolly, R.B., 1988. Ototoxicity of toluene in rats. Neurotoxicol. Teratol. 10, 525–530.
Svenson, D.W., Davidson, C.J., Thakur, C., Bowen, S.E., 2022. Acute exposure to abuse-like concentrations of toluene induces inflammation in mouse lungs and brain. J. Appl. Toxicol.
Swick, D., Jaques, A., Walker, J.C., Estreicher, H., 2014. Gasoline toxicology: overview of regulatory and product stewardship programs. Regul. Toxicol. Pharmacol. 70, S3–S12.
Tang, C.Y., Carpenter, D.M., Eaves, E.L., Ng, J., Ganeshalingam, N., Weisel, C., Qian, H., Lange, G., Fiedler, N.L., 2011. Occupational solvent exposure and brain function: an fMRI study. Environ. Health Perspect. 119, 908–913.

Till, C., Koren, G., Rovet, J.F., 2001a. Prenatal exposure to organic solvents and child neurobehavioral performance. Neurotoxicol. Teratol. 23, 235–245.

Till, C., Westall, C.A., Rovet, J.F., Koren, G., 2001b. Effects of maternal occupational exposure to organic solvents on offspring visual functioning: a prospective controlled study. Teratology 64, 134–141.

Till, C., Rovet, J.F., Koren, G., Westall, C.A., 2003. Assessment of visual functions following prenatal exposure to organic solvents. Neurotoxicology 24, 725–731.

Till, C., Westall, C.A., Koren, G., Nulman, I., Rovet, J.F., 2005. Vision abnormalities in young children exposed prenatally to organic solvents. Neurotoxicology 26, 599–613.

Tombaugh, T.N., 2004. Trail making test A and B: normative data stratified by age and education. Arch. Clin. Neuropsychol. 19, 203–214.

Tomlin, S.L., Jenkins, A., Lieb, W.R., Franks, N.P., 1998. Stereoselective effects of etomidate optical isomers on gamma-aminobutyric acid type A receptors and animals. Anesthesiology 88, 708–717.

Triebig, G., 2000. Solvent related chronic encephalopathy. Int. Arch. Occup. Environ. Health 73, 361.

U.S. Environmental Protection Agency (EPA), 1993. Multi-Substance Rule for the Testing of Neurotoxicity. Federal Register, Washington DC.

U.S. Environmental Protection Agency (EPA), 1994. Part 79 - Registration of Fuels and Fuel Additives. Code of Federal Register (CFR), Washington DC.

U.S. Environmental Protection Agency (EPA), 1995. Revocation of Final Multi-Substance Rule for the Testing of Neurotoxicity. Federal Register, Washington D.C.

U.S. Environmental Protection Agency (EPA), 2014. *National air Toxics Assessment. 2014 NATA: Assessment Results* [Online]. Available: https://www.epa.gov/national-air-toxics-assessment/2014-nata-assessment-results. (Accessed 03/09 2022).

U.S. Environmental Protection Agency (EPA), 2017. *2017 National Emissions Inventory Report* [Online]. EPA, Washington DC. Available: https://gispub.epa.gov/neireport/2017. (Accessed 03/09 2022).

U.S. Environmental Protection Agency (EPA), 2018. Registration of Isobutanol as a Gasoline Additive: Opportunity for Public Comment. Federal Register, Washington DC.

Van Der Hoek, J.A., Verberk, M.M., Van Der Laan, G., Hageman, G., 2001. Solvent-induced chronic encephalopathy; the 'solvent team' project. Ned. Tijdschr. Geneeskd. 145, 256–260.

Van Hout, M.S., Schmand, B., Wekking, E.M., Deelman, B.G., 2006. Cognitive functioning in patients with suspected chronic toxic encephalopathy: evidence for neuropsychological disturbances after controlling for insufficient effort. J. Neurol. Neurosurg. Psychiatry 77, 296–303.

Van Thriel, C., Kiesswetter, E., Blaszkewicz, M., Golka, K., Seeber, A., 2003. Neurobehavioral effects during experimental exposure to 1-octanol and isopropanol. Scand. J. Work Environ. Health 29, 143–151.

Van Thriel, C., Schaper, M., Kiesswetter, E., Kleinbeck, S., Juran, S., Blaszkewicz, M., Fricke, H.H., Altmann, L., Berresheim, H., Bruning, T., 2006. From chemosensory thresholds to whole body exposures-experimental approaches evaluating chemosensory effects of chemicals. Int. Arch. Occup. Environ. Health 79, 308–321.

Van Thriel, C., Blaszkewicz, M., Schaper, M., Juran, S.A., Kleinbeck, S., Kiesswetter, E., Wrbitzky, R., Stache, J., Golka, K., Bader, M., 2007. Chemosensory effects during acute exposure to N-methyl-2-pyrrolidone (NMP). Toxicol. Lett. 175, 44–56.

Van Valen, E., Van Thriel, C., Akila, R., Nilson, L.N., Bast-Pettersen, R., Sainio, M., Van Dijk, F., Van Der Laan, G., Verberk, M., Wekking, E., 2012. Chronic solvent-induced encephalopathy: European consensus of neuropsychological characteristics, assessment, and guidelines for diagnostics. Neurotoxicology 33, 710–726.

Vettori, M.V., Corradi, D., Coccini, T., Carta, A., Cavazzini, S., Manzo, L., Mutti, A., 2000. Styrene-induced changes in amacrine retinal cells: an experimental study in the rat. Neurotoxicology 21, 607–614.
Visser, I., Lavini, C., Booij, J., Reneman, L., Majoie, C., De Boer, A.G., Wekking, E.M., De Joode, E.A., Van Der Laan, G., Van Dijk, F.J., Schene, A.H., Den Heeten, G.J., 2008. Cerebral impairment in chronic solvent-induced encephalopathy. Ann. Neurol. 63, 572–580.
Voevodskaya, O., Poulakis, K., Sundgren, P., Van Westen, D., Palmqvist, S., Wahlund, L.O., Stomrud, E., Hansson, O., Westman, E., Swedish Bio, F.S.G., 2019. Brain myoinositol as a potential marker of amyloid-related pathology: A longitudinal study. Neurology 92, e395–e405.
Von Euler, G., Ögren, S.O., Li, X.M., Fuxe, K., Gustafsson, J.Å., 1993. Persistent effects of subchronic toluene exposure on spatial learning and memory, dopamine-mediated locomotor activity and dopamine D2 agonist binding in the rat. Toxicology 77, 223–232.
Von Euler, G., Ogren, S.O., Eneroth, P., Fuxe, K., Gustafsson, J.A., 1994. Persistent effects of 80 ppm toluene on dopamine-regulated locomotor activity and prolactin secretion in the male rat. Neurotoxicology 15, 621–624.
Von Euler, M., Pham, T.M., Hillefors, M., Bjelke, B., Henriksson, B., Von Euler, G., 2000. Inhalation of low concentrations of toluene induces persistent effects on a learning retention task, beam-walk performance, and cerebrocortical size in the rat. Exp. Neurol. 163, 1–8.
Wameling, A., Schaper, M., Kunert, J., Blaszkewicz, M., Van Thriel, C., Zupanic, M., Seeber, A., 2000. Individual toluene exposure in rotary printing: increasing accuracy of estimation by linear models based on protocols of daily activity and other measures. Biometrics 56, 1218–1221.
Warren, D.A., Reigle, T.G., Muralidhara, S., Dallas, C.E., 1998. Schedule-controlled operant behavior of rats during 1,1,1-trichloroethane inhalation: relationship to blood and brain solvent concentrations. Neurotoxicol. Teratol. 20, 143–153.
Warren, D.A., Bowen, S.E., Jennings, W.B., Dallas, C.E., Balster, R.L., 2000. Biphasic effects of 1,1,1-trichloroethane on the locomotor activity of mice: relationship to blood and brain solvent concentrations. Toxicol. Sci. 56, 365–373.
Wathier, L., Venet, T., Thomas, A., Nunge, H., Bonfanti, E., Cosnier, F., Parietti-Winkler, C., Campo, P., Tsan, P., Bouguet-Bonnet, S., Gansmüller, A., 2016. Membrane fluidity does not explain how solvents act on the middle-ear reflex. Neurotoxicology 57, 13–21.
Westerink, R.H.S., Vijverberg, H.P.M., 2002. Toluene-induced, Ca2+−dependent vesicular catecholamine release in rat PC12 cells. Neurosci. Lett. 326, 81–84.
White, R.F., Proctor, S.P., 1997. Solvents and neurotoxicity. Lancet 349, 1239–1243.
Williams, J.M., Stafford, D., Steketee, J.D., 2005. Effects of repeated inhalation of toluene on ionotropic GABA A and glutamate receptor subunit levels in rat brain. Neurochem. Int. 46, 1–10.
Win-Shwe, T.-T., Kunugita, N., Yoshida, Y., Fujimaki, H., 2011. Role of hippocampal TLR4 in neurotoxicity in mice following toluene exposure. Neurotoxicol. Teratol. 33, 598–602.
Wrigley, S.R., Fairfield, J.E., Jones, R.M., Black, A.E., 1991. Induction and recovery characteristics of desflurane in day case patients: a comparison with propofol. Anaesthesia 46, 615–622.
Yang, H.Y., Shie, R.H., Chen, P.C., 2016. Hearing loss in workers exposed to epoxy adhesives and noise: a cross-sectional study. BMJ Open 6, e010533.
Yavari, F., Van Thriel, C., Nitsche, M.A., Kuo, M.F., 2018. Effect of acute exposure to toluene on cortical excitability, neuroplasticity, and motor learning in healthy humans. Arch. Toxicol. 92, 3149–3162.

Yucel, M., Takagi, M., Walterfang, M., Lubman, D.I., 2008. Toluene misuse and long-term harms: a systematic review of the neuropsychological and neuroimaging literature. Neurosci. Biobehav. Rev. 32, 910–926.

Yucel, M., Zalesky, A., Takagi, M.J., Bora, E., Fornito, A., Ditchfield, M., Egan, G.F., Pantelis, C., Lubman, D.I., 2010. White-matter abnormalities in adolescents with long-term inhalant and cannabis use: a diffusion magnetic resonance imaging study. J. Psychiatry Neurosci. 35, 409–412.

Zhou, J., Shi, Z., Zhou, L., Hu, Y., Zhang, M., 2020. Occupational noise-induced hearing loss in China: a systematic review and meta-analysis. BMJ Open 10, e039576.

CHAPTER SIX

Identifying and preventing the neurotoxic effects of pesticides

Diane S. Rohlman[a,*], James R. Olson[b,c], Ahmed A. Ismail[d,e], Matthew R. Bonner[b], Gaafar Abdel Rasoul[e], and Olfat Hendy[f]

[a]Department of Occupational and Environmental Health, University of Iowa, College of Public Health, Iowa City, IA, United States
[b]Department of Epidemiology and Environmental Health, State University of New York at Buffalo, Buffalo, NY, United States
[c]Department of Pharmacology and Toxicology, State University of New York at Buffalo, Buffalo, NY, United States
[d]Kansas Department of Health and Environment, Topeka, KS, United States
[e]Community, Environmental, and Occupational Medicine Department, Faculty of Medicine, Menoufia University, Shebin El-Kom, Egypt
[f]National Liver Institute, Menoufia University, Shebin El-Kom, Egypt
*Corresponding author: e-mail address: diane-rohlman@uiowa.edu

Contents

Advances in Neurotoxicology, Volume 7
ISSN 2468-7480
https://doi.org/10.1016/bs.ant.2022.05.001

1. Introduction

Pesticides are agents used to control pests and they are widely used in agriculture, vector control, landscaping, and to control weed and insect infestations in homes. Much attention has focused on the agricultural use of pesticides, most likely due to the use of more toxic products and their widespread and frequent application. They have been shown to increase crop and livestock yields, expand storage time for food, eliminate manual labor, control the spread of invasive species, and protect humans and livestock from vector borne diseases (Cooper and Dobson, 2007; Cooper et al., 2007). These benefits all have downstream implications on human health and well-being, the economy, and the environment (Aktar et al., 2009). It is because of these benefits that pesticides will continue to be used throughout the world. However, pesticides are also associated with adverse health effects. Workers, their family members, including children, and community members living near agricultural fields are exposed through multiple pathways. Workers are exposed when pesticides are applied in fields or greenhouses, during the manufacturing of pesticides, and through extermination activities. Family and community members are exposed through their diet and drinking water, off-target pesticide application or drift, para-occupational or take-home exposure, and through the residential use of pesticides. Therefore, there is a critical need to understand the impact of pesticide exposure on health and to focus on the prevention of exposure.

Prior to the 1940s, pesticides included compounds such as arsenic, copper, sulfur, or botanical products such as pyrethrum and tobacco. In the 1940s, dichloro-diphenyl-trichloroethane or DDT was introduced as the first modern synthetic insecticide. DDT is an organochlorine insecticide and has been widely used in many countries. In addition to controlling insect-borne diseases such as malaria and typhus, DDT has been used for insect control in agricultural and residential settings. Along with DDT, other synthetic insecticides (organophosphates, carbamates, pyrethroids, and neonicotinoids) and herbicides were also developed. Over time the types of pesticides being used has evolved due to environmental concerns, pest resistance, or concerns about human or animal health effects. The goal is to find the safest pesticide for humans, animals, and the environment that effectively controls the pest at an affordable price.

This chapter presents background information on the types of pesticides, their toxicity, and mechanisms of action. It also provides an overview of how pesticides are used in agriculture, the pathways of exposure, and methods to control pesticide exposure. Understanding these factors is critical to understand how to prevent exposure to pesticides and the subsequent neurotoxic effects.

2. Pesticide products

The US Environmental Protection Agency (EPA) defines pesticides as "any substance or mixture of substances intended for preventing, destroying, repelling, or mitigating any pest" (U.S. Environmental Protection Agency, 2021i). It applies to a range of products designed to control specific pests (e.g., insecticides, rodenticides, herbicides, larvicide), including fumigants and disinfectants designed to control microorganisms. Pesticide products consist of both active (the chemical used to control the pest) and inert ingredients (e.g., solvents, adjuvants). Inert ingredients are added to improve the effectiveness of the active ingredient. For example, they may assist with the absorption of the active ingredient into the soil, expand the shelf-life of the product, improve the application process, prevent degradation, or improve applicator safety (U.S. Environmental Protection Agency, 2021b). Pesticide formulations can be liquid (often in a concentrated form that is mixed with water before applying), dry formulations (granules or powders), or aerosols (frequently dispersing fine droplets in an enclosed area). Farmers also consider cost and the application method (e.g., equipment required) when selecting a specific product formulation.

Research studies examining health effects from pesticide exposure, typically focus only on the active ingredient and not the pesticide formulations. A systematic review of in vivo and in vitro studies compared the toxicity of both the active ingredient and the pesticide product formulation

(Nagy et al., 2020). The majority of the product formulations (24 out of 36) had increased toxicity compared to the active ingredient alone. In most cases the toxicity was attributed to the adjuvants contained in the formulations. For example, several studies examined various pyrethroid-based compounds (e.g., deltamethrin) and found that the product formulations (i.e., Decis, K-Othrin) were more toxic compared to pure deltamethrin. (Abbassy and Mossa, 2012; Moretti et al., 1997).

3. Pesticide registration

A registration process is used to approve the sale and use of pesticide products in most countries around the world. This process utilizes scientific evidence to determine that there are no unacceptable risks to humans or to the environment and that the product is effective for its intended use. It determines when, where, how, and how much a pesticide can be applied (e.g., type of crop, formulation, weather conditions) and who can apply the pesticide (e.g., certified pesticide applicator). This information is then included on the label, which in many countries serves as a legal document (Food and Agriculture Organization of the United Nations, 2015). It is illegal to use a pesticide other than for the purpose and manner stated on the label (U.S. Environmental Protection Agency, 2021f). The label contains information about proper handling and use, the hazards and risks of the pesticide, and information about safety. For an applicator or farmer, this may be the only information they receive about the product they are using. A pesticide should not be purchased, used, or stored unless it has a proper label attached.

This regulatory process is the first step for reducing pesticide exposure. The label is required to contain a hazard statement and signal words indicating the acute toxicity of the pesticide product. The label also contains precautionary statements that describes methods to prevent exposure, what to do if exposed, and safe storage and disposal practices. Recognizing that as scientific knowledge advances new hazards may be identified, pesticides will periodically go through a re-registration process to ensure that they continue to meet regulatory standards. This re-registration process has led to bans on pesticides that were previously approved (Damalas and Eleftherohorinos, 2011). However, these pesticides may still be available and sold to developing countries which may have less strict regulations. They may also be sold illegally.

4. Pesticide use

Since 1990, worldwide pesticide use has approximately doubled (Fig. 1). Herbicides are applied most frequently (50%), followed by insecticides (30%) and fungicides (18%) (Sharma et al., 2019). China, the United States (US), and Brazil use the greatest amount of pesticides, but when geographic area is taken into consideration, other small countries (e.g., the Maldives, Trinidad, Tobago, Costa Rica), apply more Kg per hectare than these larger countries (Worldometer, 2021). As populations grow and markets expand, pesticide use will continue to increase to improve crop production and storage. Crop storage is particularly important because it is estimated that up to 40% of the food produced each year is lost to pests (Abhilash and Singh, 2009; Savary et al., 2019), leading to more application of pesticides. Moreover, climate change (e.g., warmer temperatures allow pests to survive throughout the year and they are more likely to develop resistance to pesticides; (Ma et al., 2021)), labor shortages, and pest outbreaks (e.g., locust infestations, coffee leaf rust) also increase the use of pesticides (Food and Agriculture Organization of the United Nations, 2015). This high demand for pesticides, particularly in developing countries, has led to the sale of counterfeit pesticide products and the use of illegal or unlabeled pesticides (Nampeera et al., 2019).

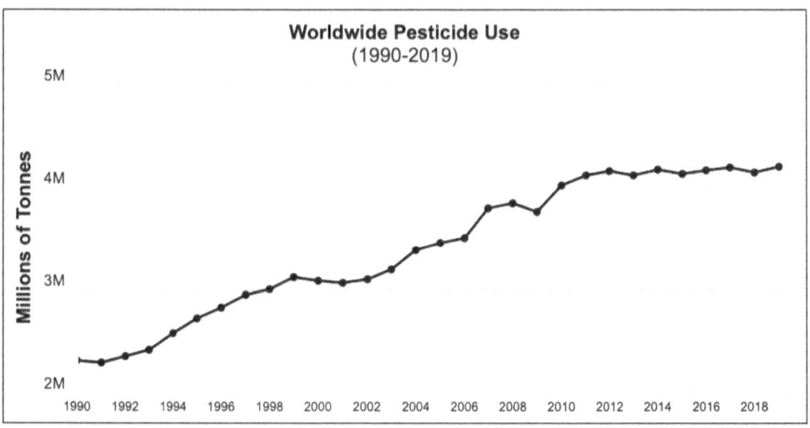

Fig. 1 Changes in worldwide pesticide use from 1990 to 2019 (FAO visualize http://www.fao.org/faostat/en/#data/RP/visualize). This includes data on pesticides (insecticides, herbicides, fungicides, plant growth regulators, and rodenticides) used in or sold to the agricultural sector.

Pesticide use may have a greater impact in developing countries. These countries have fewer regulations, lack safety equipment, including personal protective equipment (PPE), and have limited knowledge of hazards and safety practices. Labels on pesticide products may not be available in local languages and low literacy levels lead to a reliance on pesticide salespersons or dealers as the primary source of information. In addition to farmers, family and community members are also likely to be exposed due to the use of pesticide products in and around residential areas. Pesticide residues are higher in residential areas located near agricultural fields due to pesticide drift and take-home exposures. Pregnant women and young children working and living in agricultural communities are at greater risk for adverse health effects. The extensive use of pesticides on fields leads to contamination of water and soil. The improper disposal of pesticide containers may lead to their reuse, resulting in another source of unwanted exposure to pesticides.

5. Toxicity

The toxicity of a pesticide is determined based on the ability of the product to cause injury (i.e., health disorders) following exposure. Acute toxicity is the amount required to cause injury following a single (typically short duration) exposure. In this situation, effects typically appear soon after exposure. Chronic toxicity occurs when an individual is exposed to repeated small doses of a product over time and in this case, effects are typically delayed. Animal studies are used to assess toxicity following oral, dermal, or inhalation exposure to assess risks to human health. Specific outcomes that are examined include reproductive toxicity (e.g., effects on fertility, reproductive organ changes), carcinogenicity (e.g., initiating or promoting tumor development), mutagenicity (e.g., heritable lesions leading to altered phenotypes), developmental toxicity (e.g., mortality, structural or functional abnormalities, or growth retardation in the fetus), and neurotoxicity (e.g., behavior, function and motor activity deficits) (National Research Council, 1993). This information is used to categorize products based on toxicity and to identify appropriate precautions that must be implemented to reduce exposure when using the product (Hock, 2006).

As with all chemical compounds, the effects are dose related which takes into account both the toxicity of the product and the amount of exposure. The timing of exposure is also an important factor (e.g., during pregnancy or early childhood). Findings from animal studies are used to identify the dose that leads to harmful effects, and these are then used to calculate the harmful dose for humans. Information about the sources of pesticide exposure is also assessed. For example, how often you are exposed, when the exposure occurs, and the amount of contact with the pesticide (e.g., exposure through diet and drinking water, residential exposure, occupational exposure). This information is combined to characterize overall risk. Because of variations between animals and humans and between person to person, uncertainty factors are included in the risk assessment. For example, in the US the pesticide law requires an extra 10-fold safety factor to be added to protect infants and children (U.S. Environmental Protection Agency, 2021a). Based on the acute toxicity information, pesticides are classified into different categories using the LD_{50} for the rat (mg/kg body weight) for both oral and dermal exposure. The World Health Organization (WHO) uses the Acute Toxicity Hazards Categories from the Global Harmonized System (GHS) of classification (World Health Organization, 2020). With this system, pesticides are categorized into five categories: *extremely hazardous* (oral $LD_{50} < 5$ mg/kg, dermal $LD_{50} < 50$), *highly hazardous* (oral LD_{50} 5–50 mg/kg, dermal LD_{50} 50–200 mg/kg), *moderately hazardous* (oral LD_{50} 50–2000 mg/kg, dermal LD_{50} 200–2000 mg/kg), *slightly hazardous* (oral LD_{50} over 2000 mg/kg, dermal LD_{50} over 2000 mg/kg), or *unlikely to present acute hazard* (5000 mg/kg or higher).

Although these categories provide general guidelines for acute toxicity, many factors can impact exposure and subsequent health effects. In addition to the dose, exposure can be modified by an individual's behaviors, diet, and genetics (Kimbrough, 1984). Co-exposures to pesticide mixtures or product formulations can also enhance toxicity. For example, studies have examined the association between paraoxonase-1 (PON1) and organophosphate insecticide exposure on developmental outcomes. The PON1 enzyme, that is encoded by the PON1 gene, assists with metabolizing and detoxifying organophosphate pesticides (Furlong et al., 2005). Enzyme activity levels can vary across individuals and are associated with PON1 genetic polymorphisms (single nucleotide polymorphisms or SNPS).

Animals with low PON1 activity are more sensitive to the toxic effects of some organophosphate pesticides. Several studies have reported associations between PON1 genetic polymorphisms and PON1 activity on neurobehavioral and birth outcomes among children (Harley et al., 2011; Naksen et al., 2015).

6. Established mechanisms of action for pesticide classes

Because of their mechanisms of action (Table 1), insecticides are considered to be more acutely toxic than other types of pesticides (e.g., herbicides). Many are designed to act upon the nervous system of the insect and may also impact the nervous system of humans. In contrast, herbicides target specific metabolic pathways in plants (e.g., photosynthesis, regulation of cell division, plant hormone action) and may have very little impact in humans (Lushchak et al., 2018). Because of this, herbicides have low to moderate acute toxicity. The one exception to this is paraquat, which is highly toxic when inhaled or ingested. Cases of human poisoning are frequently due to exposure to organophosphate insecticides or exposure to paraquat (Gunnell and Eddleston, 2003). This review will primarily focus on insecticide pesticides.

Table 1 Insecticide sites and established mechanisms of action targeting nerves and muscles.

Site	Mechanism	Insecticide class
Chloride channel regulation	Inhibition of gamma-aminobutyric acid (GABA) receptor	Organochlorines
	Activation of chloride channels	Avermectins
	Agonists of the GABA-gated channel	Bifenazate
Cholinesterase inhibition	Inhibition of cholinesterase leading to overstimulation of the nerve	Organophosphate
		Carbamate
Sodium channel modulators	Bind to sodium channels fixing them open	Pyrethrins
		Pyrethroids
Acetylcholine receptor stimulation	Bind to acetylcholine receptors which results in prolonged stimulation	Neonicotinoid

Adapted from Lushchak, V.I., Matviishyn, T.M., Husak, V.V., Storey, J.M., Storey, K.B., 2018. Pesticide toxicity: a mechanistic approach. EXCLI J. 17, 1101.

7. Insecticides

7.1 Organochlorines

Changes in the types of insecticides applied over time demonstrate how scientific evidence and policy have shaped current pest control practices. As mentioned above, DDT an organochlorine, was the first modern synthetic insecticide developed. Organochlorines disrupt the nervous system and lead to uncontrolled neuronal excitation (Tables 1 and 2); however, they are not considered acutely toxic. During World War II there was widespread use of DDT for treatment of lice and mosquitos to control typhus and malaria (Hayes, 1991). After World War II, DDT continued to be used for vector control and in 1964 DDT was registered for use in agriculture. Several organochlorine pesticides, including DDT, aldrin, chlordane, and heptachlor, are considered persistent environmental pollutants or POPs.

As DDT became more widely used, it was found that many insects developed resistance to DDT. These insecticides also persist in the environment, take a long time to break down, move throughout the environment, are easily stored in fat, and bioaccumulate throughout the food chain. In addition, the breakdown product for DDT, dichlorodiphenyldichloroethylene (DDE) has similar properties as DDT. The primary route of exposure is through ingestion of contaminated food (Hassall, 1990) with the highest concentrations found in meat and dairy products. Due to the slow degradation of the parent compound, organochlorines will accumulate in adipose tissue and breast milk. Because of these concerns, DDT was banned for agricultural use in the United States in 1972 (U.S. Environmental Protection Agency, 1990). Other countries also banned or restricted its use in agriculture during the 1970–1980s (e.g., Argentina, Australia, Canada, Japan, Lebanon, the European Union [EU]). Although it is still used for agricultural purposes in several countries, particularly developing counties (Jayaraj et al., 2016).

Recognizing the benefits of DDT for malaria control, the World Health Organization has supported the indoor use of DDT in African countries. This aligns with the Stockholm Convention which allows the use of DDT for vector control when other affordable alternatives are not available (U.S. Environmental Protection Agency, 2021d). However, the environmental persistence of DDT and the development of pest resistance led to decreased use of organochlorines and the increased use of organophosphate and carbamate products which have the benefit of not persisting in the environment (Van Steenwyk, 2005).

Table 2 Classes of pesticides, example products, symptoms, and the toxicological and developmental effects.

Pesticide	Example products	Toxicology	Developmental effects	Symptoms
Insecticides				
Organochlorines	DDT, Dieldrin, Endosulfan, Heptachlor, Lindane	Bioaccumulate in humans (adipose tissue, muscle, blood, breast milk), because they are lipid soluble and degrade slowly	Transfer across the placenta and are excreted in breast milk	Sensory disturbances (hypersensitivity to light, sound, and touch), increased frequency of spontaneous movements (tremor, myoclonic jerking), paresthesia in the face or extremities, hyperreflexia, headache, dizziness, vomiting, incoordination, confusion, and tremor
		Hormonally active		Severe poisonings can lead to convulsions, respiratory depression, coma
				Pesticides (and/or metabolites) can be measured in blood, adipose tissue, and breast milk
Organophosphates	Azinphos-methyl, Methamidophos, Methyl parathion, Acephate, Chlorpyrifos, Diazinon, Malathion, Phosmet, Profenofos	Exposure leads to acute and subacute toxicology	Strong evidence of developmental deficits associated with prenatal exposure; some evidence of postnatal exposure associated with developmental deficits	Muscle twitching, hypersecretion, miosis, respiratory depression, headache, nausea, diarrhea, vomiting, tachycardia/bradycardia, bronchospasm, loss of consciousness
		Act through the phosphorylation of the acetylcholinesterase enzyme at nerve endings (irreversible cholinesterase inhibition) and results in muscarinic, nicotinic, and CNS effects		Depressed RBC AChE and/or butyrylcholinesterase
				Severe symptoms include loss of consciousness, incontinence, seizures, and respiratory depression. Primary cause of death is respiratory failure.
				Symptoms associated with cholinergic crisis are often identified with the acronyms SLUDGE and DUMBELS (see text for a description)
				Metabolites can be measured in urine

N-Methyl carbamates	Aldicarb, Carbofuran, Carbaryl, Propoxur	Acts through the reversable carbamylation of acetylcholinesterase enzyme (cholinesterase inhibition) and results in muscarinic, nicotinic, and CNS effects	Children are more likely than adults to experience CNS symptoms	Malaise, muscle weakness, dizziness, sweating
		Acute effects tend to be of shorter duration than organophosphate poisonings because of reversible AChE inhibition and more rapid metabolism of carbamates	Evidence of developmental deficits associated with pre- and postnatal exposure	May include blurred vision, muscle twitching, incoordination, slurred speech, salivation, headache, abdominal pain, diarrhea, nausea, vomiting
				Severe symptoms include seizures, coma, hypotonicity, hypertension, cardio/respiratory depression
				Metabolites can be measured in urine
				Blood cholinesterase levels not a reliable indicator
Pyrethroids	Allethrin, Cypermethrin, Fenvalerate, Permethrin	Low systemic toxicology through inhalation or dermal absorption	Prenatal exposure may lead to neurotoxicity due to inability of fetus to metabolize insecticide compound	Dizziness, headache, fatigue, vomiting, diarrhea
				Paresthesia with skin contact (typically will decrease within 24 h)
Type 1: usually does not contain a cyano group (e.g., permethrin)				Type I: fine tremor, reflex hyperexcitability
		Sites of action include: sodium and chloride channels, GABA, nicotinic acetylcholine, peripheral benzodiazepine receptors		Type II: severe salivation, hyperexcitability, choreoathetosis (usually more severe symptoms than Type I)
Type II: contain a cyano group (e.g., cypermethrin, deltamethrin)				Higher exposures may lead to pulmonary edema, seizures, coma
				Metabolites can be measured in urine

Continued

Table 2 Classes of pesticides, example products, symptoms, and the toxicological and developmental effects.—cont'd

Pesticide	Example products	Toxicology	Developmental effects	Symptoms
Neonicotinoids	Acetamiprid, Clothianidin, Imidacloprid, Thiamethoxam	Reported clinical toxicity in humans is rare	Limited information on developmental effects	Similar to acute nicotine poisoning
		Act by displacing acetylcholine from the nicotinic acetylcholine receptors (nAChRs)		Disorientation, agitation, confusion, headache, dizziness, drowsiness, weakness, tremor, unconsciousness
		Greater impact on insect nAChRs than human nAChRs		
		Most are excreted intact through the kidney		GI effects may occur due to solvent in the formulation (vomiting, sore throat, diarrhea, nausea, abdominal pain)
Herbicides				
Dipyridyls	Paraquat, Diquat	Paraquat has life threatening effects ("one sip can kill")	Limited information on developmental effects	Ingestion can lead to burning pain in the mouth and GI track, pulmonary edema, pancreatitis, renal and CNS effects (nervousness, irritability, diminished reflexes)
		Impacts GI track and other major organs (kidney, lung, liver, heart)		
		Pesticide compound can be measured in blood or urine		Dermal exposure of paraquat can led to contact dermatitis and long-term exposure can lead to blistering, abrasion, erythema, and fingernail changes
		Exposure associated with development of Parkinson's disease		

Chlorylphenoxy herbicides	2,4-Dichlorophenoxyacetic (2,4-D)	Typically see skin and eye irritation and GI symptoms	Limited information on developmental effects	Skin and eye irritation, GI symptoms
				Can inhibit human taste receptors for sweets
		Found in hundreds of products. Check label to confirm what is actually present		Most poisoning cases resulted in minimal signs of poisoning with mild GI symptoms
		Pesticide compound can be measured in blood or urine		Some patients present with vomiting, diarrhea, headache, confusion, aggressive behavior, peculiar odor on breath, elevated body temperature, muscle weakness, peripheral neuropathy, loss of reflexes
Phosphonate herbicides	Glyphosate	Most commonly used herbicide in the US with many reported poisonings	Limited information on developmental effects	GI symptoms, mouth/throat pain, nausea, vomiting, diarrhea
		Low risk for human toxicity (mechanism of action targets plant specific enzymes)		
		Actual toxicity likely come from adjuvants (surfactants, stabilizers, penetrants) added to pesticide formulation		More severe signs may be present in cases with high oral exposure
		Products may contain additional active ingredients – always read the label		Pesticide compound can be measured in plasma

Continued

Table 2 Classes of pesticides, example products, symptoms, and the toxicological and developmental effects.—cont'd

Pesticide	Example products	Toxicology	Developmental effects	Symptoms
Fungicides				
Fungicides	Hexachlorobenzene, Azoxystrobin, Thiram, Maneb, Captan, Triadimefon, Copper Acetate, Methyl mercury	Although they vary in toxicity, most are unlikely to cause acute toxicity	Limited information on developmental effects	May cause eye, skin, and respiratory irritation and some may cause GI symptoms
		These compounds are likely to cause acute toxicity: organomercury compounds, triazoles, dithiocarbamates, copper compounds		Some (e.g., methyl mercury, organotin compounds) may cause neurological symptoms such as tremor, headache, fatigue, emotional lability, difficulty thinking, incoordination, slurred speech
Fumigants				
Fumigants	Methyl bromide, Hydrogen cyanide, Naphthalene	Active ingredient is released as a gas or vapor	Limited information on developmental effects	Many are irritants
		Easily absorbed in the pulmonary membranes, GI tract, and skin		Can lead to central nervous system effects, pulmonary edema, and can impact the liver or kidney
		Some can easily penetrate through rubber and neoprene PPE		May cause convulsions and coma, particularly in children

Adapted from Roberts, J.R., Reigart, J.R., 2013. EPA Recognition and Management of Pesticide Poisonings, sixth ed.

7.2 Organophosphate (OP) insecticides

Originally developed in Germany, organophosphate (OP) compounds were commonly applied worldwide. For example, in 2000, approximately 70% of all insecticides used in the United States were OP insecticides, although their use has since declined, (e.g., 33% in 2012) (Atwood and Paisley-Jones, 2017). Because of the low cost of these chemicals, compared to newer insecticides, OP insecticides are still commonly used in most developing countries. In spite of this broad use, they are considered the most toxic pesticide. OPs use the same mechanism of action, cholinesterase inhibition, to kill insects as potent nerve agents used as chemical warfare agents.

OPs have been associated with acute and chronic health effects (Table 2). As they began to be used in agriculture, there were reports of memory impairment, anxiety, confusion, fatigue, irritability, and other long-term effects among previously poisoned worker populations (Ames et al., 1995; Richter et al., 1992; Rodnitzky et al., 1975; Rosentock et al., 1990; Steenland et al., 1994, 2000). Other studies examined occupationally exposed groups, with no history of poisoning, and identified neurobehavioral effects associated with exposure (Ismail et al., 2012; Meyer-Baron et al., 2015; Muñoz-Quezada et al., 2016; Perry et al., 2020; Rohlman et al., 2011; Takahashi and Hashizume, 2014). These health effects have led to bans and restrictions on use of organophosphate insecticides.

OP insecticides interfere with the transmission of nerve impulses across a synapse by inhibiting acetylcholinesterase. The acetylcholinesterase enzyme (AChE) is responsible for controlling the transmission of nerve impulses across the synapse by breaking down the neurotransmitter acetylcholine. Inhibition of AChE leads to an excess accumulation of acetylcholine which results in continuous firing of the nerve impulse. Exposure to organophosphate insecticides can result in muscarinic, nicotinic, and central nervous system (CNS) effects. Classic signs of acute OP exposure include twitching of the limbs, muscle contraction, secretion, and respiratory paralysis. Brain growth and development can also be affected (González-Alzaga et al., 2014; Muñoz-Quezada et al., 2013; Rohlman et al., 2019; Sagiv et al., 2018). OP compounds typically breakdown through enzymatic hydrolysis in the liver. However, this rate can vary based on the type of OP insecticide, and a slow rate can lead to temporary storage of the OP compound in fatty tissues (e.g., methyl parathion, diazinon; (Garcia-Repetto et al., 1995, Roberts and Aaron, 2007)).

There can be both immediate and delayed effects following exposure to OP pesticides. Symptoms occurring within minutes or hours following acute exposure are known as the cholinergic syndrome and are typically associated with >60–70% AChE inhibition (Abdollahi and Karami-Mohajeri, 2012; Richardson et al., 2019). Symptoms associated with this cholinergic crisis are often identified by the acronyms SLUDGE (Salivation, Lacrimation, Urination, Defecation, Gastric cramps, Emesis) or DUMBELS (Diarrhea, Urination, Miosis/Muscle weakness, Bronchospasm/Bradycardia, Emesis, Lacrimation, Salivation).

Approximately 20% of individuals will experience delayed symptoms following the acute cholinergic crisis. This is known as the intermediate syndrome (Richardson et al., 2019). This intermediate syndrome occurs 24–96 h after exposure, when acute symptoms have started to resolve. At this time, exposed individuals may begin to experience musculoskeletal weakness, including respiratory, neck, and proximal limb muscles, depressed deep tendon reflexes, and weakness in the cranial nerves (Abdollahi and Karami-Mohajeri, 2012). These symptoms typically resolve within 1 to 2 weeks. During this time, there may be elevated OP metabolite levels in the urine and depressed cholinesterase levels (Agency For Toxic Substances and Disease Registry, 2007).

Organophosphate-induced delayed neuropathy (OPIDN) occurs 1 to 5 weeks after acute cholinergic toxicity and may persist for several years, although some symptoms may be permanent. Initial symptoms include cramping, typically in the legs, unsteady balance, paresthesia and numbness, changes in gait, and can progress to paralysis. Only certain OPs, those that inhibit the neuropathy target esterase (NTE; (Johnson, 1990)), are associated with OPIDN (e.g., azinphos-methyl, chlorpyrifos, diazinon, dimethoate, malathion, methamidophos, methyl parathion, phosmet; (Lotti and Moretto, 2005)). More stringent regulations and testing requirements for NTE in organophosphate compounds has led to a reduction in cases (Richardson et al., 2020). OPIDN is typically associated with very high exposures to OP insecticides (e.g., suicide attempts). Although studies have examined low level, long-term exposure to OP insecticides and peripheral nervous system symptoms, an association has not been reported (Lotti, 2002).

Exposed individuals may also experience long-term, persistent CNS effects (e.g., cognitive and motor deficits, anxiety, depression, confusion insomnia, irritability). These symptoms have been attributed to both an acute, symptomatic exposure (e.g., poisoning) (Rosentock et al., 1990; Savage et al., 1988; Steenland et al., 1994) or repeated, asymptomatic exposures (Abou-Donia, 2003). There have been numerous studies examining the neurological impacts of chronic occupational exposures and although many of these studies have methodological limitations, systematic reviews and meta-analyses have identified associations between exposure to OP insecticides and impaired neurobehavioral functioning (i.e., executive function, working memory, visuospatial ability; (Ismail et al., 2012; Muñoz-Quezada et al., 2016; Ross et al., 2013). Much of this research has focused on exposure to chlorpyrifos, the most widely used OP pesticide worldwide (Burke et al., 2017). Similarly, studies have shown an association between pesticide exposure and Parkinson's disease (PD) (Yan et al., 2018), recognizing that the toxic effects of certain pesticides (e.g., organochlorines, OPs, carbamates) damage aspects of the nervous system that contribute to the development of PD (e.g., neuronal cell loss, oxidative stress, mitochondrial dysfunction) (Alavanja et al., 1996). However, with a few exceptions (e.g., rotenone and paraquat), the significant relationships are between reports of overall pesticide use or cumulative pesticide use, and not exposure to specific pesticide classes. Findings from the Agricultural Health Study, a prospective cohort study following pesticide applicators in the U.S.), have reported associations between specific pesticides and risk of PD (Shrestha et al., 2020).

7.2.1 Developmental neurotoxicity

Because of the widespread use of organophosphate insecticides in residential and agricultural settings prior to 2000 when chlorpyrifos and other OP insecticides were banned for residential use in the United States; (U.S. Environmental Protection Agency, 2021c), many studies have also examined developmental neurotoxicity. These studies have examined pre- and postnatal exposure among cohorts of children who had opportunities for exposure due to residential pest control, living in close proximity to agricultural fields, or living with agricultural workers (Muñoz-Quezada et al., 2013; Reiss et al., 2015; Sapbamrer and Hongsibsong, 2019). Other studies have looked at adolescents occupationally exposed to pesticides (Eckerman et al., 2007; Ismail et al., 2017b; Rasoul et al., 2008; Rohlman et al., 2007, 2016). Exposure was assessed through maternal and child biomarkers (cord blood, urinary organophosphate metabolites including the dialkyl phosphates, and the chlorpyrifos specific metabolite 3,5,5-trichloropyridinol or TCPy). These studies have examined birth outcomes, developmental milestones, mental and psychomotor development, child behaviors, and clinical disorders among infants, toddlers and preschool children, school-age children, and adolescents working as pesticide applicators. The methodological quality of these studies was generally high and included birth cohorts, longitudinal designs, as well as cross-sectional studies. Across all age groups, prenatal exposure to organophosphates was associated with cognitive, psychomotor and behavioral outcomes. Fewer studies examined postnatal exposure, however, the majority of these studies also found that exposure to OP insecticides was associated with adverse neurodevelopmental outcomes (González-Alzaga et al., 2014; Muñoz-Quezada et al., 2013; Sapbamrer and Hongsibsong, 2019). Limited research has examined adolescent exposure to pesticides. This population may be at greater risk due to their engagement in agricultural work, including applying pesticides, and the developmental changes that occur during puberty. More detail is described in the case study below.

7.2.2 Neurotoxic mechanisms

Findings from human and animal studies, which show associations between OP exposures and neurobehavioral outcomes (Meyer-Baron et al., 2015; Rohlman et al., 2011), neurodevelopmental impacts (Burke et al., 2017), and neurodegenerative diseases (Rohlman et al., 2019; Sagiv et al., 2018), that occur in the absence of cholinergic crisis, have called into question whether these effects are due different mechanism from AChE inhibition (Tsai and Lein, 2021). The primary criticism is the lack of a dose-response effect between AChE inhibition and neurobehavioral deficits (Meyer-Baron et al., 2015). Although studies have reported an association between biomarkers of exposure (e.g., urinary metabolite levels) and other indirect measures of exposure (e.g., job title), there is limited evidence of an association between AChE inhibition and neurobehavioral outcomes (Voorhees et al., 2017). The lack of association between cholinesterase inhibition and neurological outcomes in this and other human and animal studies has led to research examining other potential mechanisms of neurotoxicity, including oxidative stress and neuroinflammation (Costa, 2018; Guignet and Lein, 2019; Tsai and Lein, 2021). Other non-cholinergic mechanisms associated with organophosphate neurotoxicity include disruption of neural cell development, neurotransmitter systems (e.g., serotonin and the formation/activity of synapses in various brain regions (Aldridge et al., 2005; Barone et al., 2000; Judge et al., 2016; Qiao et al., 2003).

7.3 Carbamate insecticides

Similar to organophosphate insecticides, carbamate insecticides (derived from carbamic acid, whereas organophosphate pesticides are derived from phosphoric acid) are acetylcholinesterase inhibitors, leading to accumulation of acetylcholine in the brain and skeletal muscles and overstimulation of muscarinic and nicotinic receptors (Tables 1 and 2) (Gupta et al., 2018). Carbamate insecticides can lead to similar muscarinic, nicotinic and CNS effects as organophosphate insecticides. However, the cholinesterase inhibition for carbamate insecticides is reversible and will lead to faster recovery of enzyme activity (potentially within hours). This shortens the duration of the carbamate poisoning and lowers toxicity; however, it can limit the ability of using blood cholinesterase activity as a measure of exposure (Roberts and Reigart, 2013).

7.4 Pyrethrin and pyrethroid insecticides

Pyrethrins are natural insecticides derived from the chrysanthemum flower. They are primarily used for indoor pest control as sprays or foggers (e.g., bug bombs) and are thought to have low mammalian toxicity (Roberts and Reigart, 2013). Pyrethroid insecticides are the synthetic version of pyrethrins and are designed to be more stable. They do not break down as quickly in air or sunlight. Pyrethroid insecticides are used for vector control (e.g., West Nile virus, malaria), control of bed bugs, treatment for scabies or head lice, and in agricultural applications. As more restriction have been placed on the organophosphate insecticides, the use of pyrethroid insecticides has increased (Williams et al., 2008), accounting for up to 25% of insecticides applied worldwide (Shafer et al., 2005; Soderlund, 2012). Three compounds, permethrin, deltamethrin, and cypermethrin are the most commonly used (Chrustek et al., 2018). Although they are readily absorbed through the GI track, more limited dermal absorption and inhalation exposure generally leads to low systemic toxicity (Roberts and Reigart, 2013).

Similar to organochlorines, pyrethroids disrupt the nervous system and lead to uncontrolled neuronal excitation (Costa et al., 2008). Pyrethroid insecticides target the voltage-gated sodium channel, slowing their inactivation and leading to increased activation of the channel (Tables 1 and 2). Compared to other insecticides (i.e., organochlorines, organophosphate, and carbamate insecticides), pyrethroids are considered relatively safe (Soderlund, 2012). Animal studies and findings from clinical overdoses have identified two types of pyrethroids, Type I (e.g., permethrin) and Type II (e.g., cypermethrin, fenvalerate). Type I pyrethroids typically do not contain a cyano group, while Type II pyrethroids always contain a cyano group and high exposure to these pyrethroids is associated with more severe symptoms. Similarly, there are two types of neurotoxicity associated with exposure to pyrethroids, the T syndrome and the CS syndrome. The T syndrome, typically associated with Type I pyrethroids, includes fine tremor and reflex hyperexcitability (Ray and Forshaw, 2000). The CS syndrome, typically associated with Type II pyrethroids, results in salivation, hyperexcitability, choreoathetosis, and seizures. Dermal exposure typically results in paresthesia (a tingling or burning sensation) in the area that came in contact with the insecticide, which typically resolves within 24 h (Costa, 2015).

There are limited studies examining neurological effects associated with low levels of pyrethroid exposure. A recent review of the literature (Lucero and Muñoz-Quezada, 2021) identified 12 studies examining exposure among farmers, pesticide applicators, and children of agricultural workers. Eight of the twelve studies reported neurological symptoms (Huang et al., 2016; Kori et al., 2018; Lu, 2009; Motsoeneng and Dalvie, 2015; Shrestha et al., 2018) or cognitive and motor behavioral deficits (De Joode et al., 2016; Hansen et al., 2017; Ismail et al., 2018). There is a need for additional research utilizing robust methodologies, standardized assessment tools, and biomarkers of exposure to more fully understand the impact of exposure. However, based on the increased use of pyrethroids in agricultural and non-agricultural settings, these preliminary findings indicate a need to adopt methods to prevent exposure.

7.5 Neonicotinoid insecticides

Neonicotinoids are a newer class of insecticides that have a similar structure to nicotine. They are agonists of nicotinic acetylcholine receptors and are more toxic to insects than to humans (Tables 1 and 2) (Anadón et al., 2020; Roberts and Reigart, 2013). Developed in the 1980s, neonicotinoids were first registered for use in agriculture in the 1990s. Their use has been increasing and they are now the most widely used insecticide in the world, replacing organophosphate and carbamate insecticides (Jeschke et al., 2011). The three most common neonicotinoids, making up over 80% of all sales in 2012, are imidacloprid, thiamethoxam, and clothianidin (Bass et al., 2015).

Neonicotinoids are primarily used as seed treatments (prior to planting, seeds are coated with the insecticide) for a variety of crops. For some crops (e.g., corn, soybeans) it is difficult to find seeds that have not been treated with neonicotinoids. They are systemic insecticides and are absorbed throughout the plant which leads to the distribution of the pesticide in all parts of the plant (e.g., leaves, roots, stems, flowers, pollen; (Anadón et al., 2020)). They are considered safer and effective alternatives to the organophosphate and carbamate pesticides (Craddock et al., 2019). However, they have been linked with various environmental outcomes, including risks to pollinators and aquatic insects (Hladik et al., 2018). There are also concerns about human exposure through diet; one study in the US found detectable neonicotinoids in fruits (72%) and vegetables (45%); (Chen et al., 2014). There is limited research examining human health effects, but reviews have found some evidence of associations between exposure and neurotoxic outcomes (Cimino et al., 2017; Thompson et al., 2020). Two studies have reported neurological effects (e.g., tremor, fatigue, memory impairment) following acute poisoning (Marfo et al., 2015; Taira et al., 2011). A few studies have also reported associations between adverse birth outcomes and maternal exposures during pregnancy (Thompson et al., 2020). However, additional research is needed to better understand the impact of neonicotinoid exposure on neurotoxic outcomes.

8. Herbicides

Designed to eradicate unwanted plants, most herbicides utilize mechanisms that are unique to plants and are considered to have less acute toxicity than insecticides (Costa et al., 2008). Herbicides are the most frequently applied pesticides (Food and Agriculture Organization of the United Nations, 2021). While selective herbicides target specific plant types (e.g., glyphosate), non-selective herbicides kill all plants they come in contact with and have been used for defoliation (e.g., agent orange; 50/50 mix of 2,4-D [2, 4-dichlorophenoxyacetic acid] and 2,4,5-T [2,4,5-trichlorophenoxyacetic acid], with 2,3,7,8-tetrachlorodibenzo-p-dioxin (TCDD) as an unwanted contaminant). The first synthetic herbicide (2,4-D) was developed during World War II. Herbicides are applied at different times during the growing cycle to control weeds: prior to planting, before seedlings appear above the ground (preemergence), and after plants appear above the ground (postemergence). Similar to insecticides, the environmental impact and persistence of herbicides are a concern. Moreover, the high use of herbicides to control weeds has led to herbicide-resistant weeds. This has led to the development of genetically modified crops, including herbicide resistant transgenic plants (Schütte et al., 2017). The most common are glyphosate resistant crops (soybeans, maize, cotton, and canola) which were developed in the 1990s (Dill, 2005). In 1996 there were 1.66 million hectares planted with genetically modified crops, by 2013 there were over 168 million hectares (Dill, 2005).

With the exception of Agent Orange and paraquat, acute health effects associated with exposure to herbicides include dermatitis, eye irritation and mild GI Symptoms (Roberts and Reigart, 2013). Chronic outcomes including cancer and other health effects have been associated with Agent Orange (National Academies of Sciences, Engineering, and Medicine, 2018). Epidemiological studies have reported an association between Parkinson's Disease and exposure to paraquat (Berry et al., 2010; Moretto and Colosio, 2011; Tangamornsuksan et al., 2019; Vaccari et al., 2019).

9. Fungicides

Fungicides are used to protect plants from fungi (rusts, mildew, and blights) and to control mildew and mold in other settings. In agriculture they are designed to kill spores on the surface of the plant and are used to protect seed grain and fruit, and grain and vegetable crops during storage and shipment (Roberts and Reigart, 2013). The most well-known case of fungicide poisoning occurred in the 1970s when seed grain treated with a methylmercury fungicide was used to make flour which was then used to make bread exposing community members (Bakir et al., 1973). High exposure to methylmercury can damage the nervous system and lead to paresthesia, ataxia and other sensory deficits (e.g., vision, hearing). There are many different types of fungicide products with different health outcomes. The most neurotoxic products are organomercury compounds (methyl mercury), triazoles (triadimefon), copper compounds (copper acetate), and dithiocarbamates (maneb, mancozeb). Many of the most toxic compounds are no longer registered for use, although they may still be available in certain countries. Most registered products have low toxicity for humans and mammals and are applied in ways that are unlikely to lead to human exposure (e.g., wettable powders or granules).

10. Fumigants

In agriculture, fumigants are injected into the soil prior to planting or used in structures to protect stored crops. When injected into the soil they kill insects, fungi, weeds, bacteria, and nematodes. Often the injected area is covered in tarps to prevent dispersion into the air and restricted-entry intervals are in place to keep workers out of the treated area. Fumigation is also used to protect the harvested crops from pest infestations or decay in grain bins and other storage facilities. They are designed to release the active ingredient in the form of a gas or vapor which can easily be inhaled. Many products are restricted use pesticides because of their high toxicity to humans. Several fumigants (methyl bromide, ethylene oxide, carbon disulfide) have been associated with central nervous system effects (Keifer and Firestone, 2007; Roberts and Reigart, 2013).

11. Routes of exposure

Individuals are exposed to pesticides through inhalation, ingestion, or through contact with skin or eyes. In addition to local effects, once the pesticide enters the body it can be absorbed into the blood stream (e.g., through the intestinal or alveolar epithelium) and diffused throughout the body, including the brain. Ultrafine particles may also be transferred directly into the brain via the olfactory epithelium (Helwany and Bordoni, 2020). Exposure is also impacted by whether the product is in a liquid, solid, or gas formulation. Liquid or gas products can enter the body through all three routes. However, powders, dust, and granules are less likely to enter the body via dermal or inhalation routes. For farmers or pesticide applicators, the primary route of exposure is dermal or absorption through the skin (Costa et al., 2008; Damalas and Koutroubas, 2016). Although dermal exposure is the primary route of exposure, oral exposure or ingestion is more likely to be harmful, particularly in the case of intentional pesticide ingestion (suicidal intention) or accidental ingestion (e.g., pesticide placed in an unlabeled beverage container; Fig. 2). Oral exposures typically occur

Figure 2. Case Study: Paraquat poisoning.
Paraquat is an herbicide that is highly toxic and exposure to even a small amount can lead to death. Upon entering the body, typically through ingestion, paraquat causes toxic reactions in the gastrointestinal track, kidney, liver, heart, and other organs (Roberts and Reigart, 2013). Prolonged dermal exposure can lead to contact dermatitis. In the US, paraquat is only registered for agricultural applications and is classified as a restricted use pesticide which requires individuals to be certified in order to purchase or apply it. However, a review of poisoning cases in the US found a number of deaths associated with accidental ingestion of paraquat from a beverage container (Fortenberry, Beckman, Schwartz et al., 2016; Geller, Kezirian, Bangar et al., 2009). In order to prevent inwanted exposure, paraquat should be stored in the original container with the product label and should never be transferred to a beverage container. It should also be kept in a secure location away from children and unauthorized individuals (U.S. Environmental Protection Agency, *Paraquat Dichloride: One Sip Can Kill*).

Fig. 2 Case Study: Paraquat poisoning (Fortenberry et al., 2016; Geller et al., 2009; Roberts and Reigart, 2013; U.S. Environmental Protection Agency, 2021g).

when hands are not washed before eating or smoking, when pesticide containers are reused for drinking water, or from pesticide residues on food. Dermal exposure typically occurs during mixing or applying pesticides or by walking through recently treated fields, causing pesticides to come into contact with the skin or contaminate clothing. Pesticides can also be inhaled during aerial application of pesticides, through pesticide drift, or when powders or granules generate dust (Damalas and Eleftherohorinos, 2011).

12. Pathways of exposure

Agricultural workers, particularly applicators, typically have the highest exposures due to mixing and loading pesticides, applying pesticides, or when cleaning or maintaining the equipment. The use of concentrates, the type of equipment (backpack sprayer vs a closed cab sprayer), weather conditions (wind, humidity), walking through wet foliage, use of PPE, or entering recently treated fields can all impact exposure. Moreover, risks are increased when there are few regulations, easy access to illegal pesticides, and a lack of training, awareness of the potential hazards to health and the environment, and resources (e.g., high cost and/or lack of availability of PPE) (Abhilash and Singh, 2009). Low literacy levels can also increase risk (i.e., unable to read the pesticide label). Application practices that increase risk include poorly maintained equipment, mixing together inappropriate combinations of products, overuse of pesticides, and poor storage or disposal of pesticide products and containers. Personal behaviors such as smoking or eating while applying (or before washing hands), working long hours, hygiene practices, or lack of PPE also increase risk.

Non-applicators, including family and community members can be exposed through pesticide drift, take-home exposure, contact with contaminate soil or water, and re-use of pesticide containers. Take-home or para-occupational exposure occurs when pesticide residues are transferred from the workplace to the home by contaminated clothes, shoes, skin, or vehicles. This puts family members, particularly children, at risk of exposure. Studies examining biomarkers of exposure and/or environmental samples have found that families of farmworkers experience greater exposures than non-farmworker families (Table 3; Butler-Dawson et al., 2018; Hyland and Laribi, 2017; Lambert et al., 2005; López-Gálvez et al., 2019; Lu et al., 2000; McCauley et al., 2001).

Table 3 Factors associated with elevated levels of pesticide exposure in residential and biological samples.

Environment	Occupation	Work/home safety practices
Proximity of home to agricultural field	Parents working in agriculture	Use of PPE at work
Number of adults working in agriculture living in the home	Work tasks (applying pesticides, harvesting, pruning)	Home hygiene (removing shoes, washing clothes separately, changing out of work clothes)
Living in an agricultural community	Type of crops or season	Residential use of pesticides

13. Measuring exposure to pesticides

In order to examine the neurotoxic effects of pesticide exposure there is a need to have reliable and accurate methods to assess exposure to pesticides. Exposure assessment methods include both direct and indirect methods. Direct measures of exposure include biomonitoring (e.g., urinary metabolites, hair samples) or personal sampling (e.g., air samplers to assess inhalation exposure, skin patches to assess dermal exposure). Indirect measures typically utilize questionnaires to collect self-reported exposure and job history information and are the most commonly used (Ohlander et al., 2020). Environmental exposures such as home dust samples, can also be used to assess potential opportunities for exposure (Lu et al., 2000). These methods vary in their ability to correctly classify exposure. For example, indirect measures may be subject to recall bias, particularly when asked about exposure over the lifetime, and job title may not accurately reflect an individual's exposure to pesticides – not all farmers apply pesticides. Job exposure matrices that take into account the type of activities, crops, active ingredients, and use of PPE and hygiene behaviors may be more effective in measuring exposure (Carles et al., 2017). Although they provide more accurate measures of exposure, direct methods often are not feasible to use due to their high cost and the specialized equipment needed to collect, store, and analyze biological samples. Moreover, the timing of sample collection is important to accurately assess exposure.

Human studies examining health effects associated with pesticide exposure use a variety of direct and indirect measures of exposure. The variability and accuracy of these measures are commonly identified as a reason why research reviews do not show consistent findings across studies. Coupled with the fact that pesticide applicators are often exposed to several pesticide product formulations across the season or even at a single time point, it is not surprising that human studies report different outcomes. Exposure can occur through both environmental and occupational pathways, therefore, a biomarker that captures all routes of exposure (e.g., urinary biomarker such as DAPs or dialkylphosphate metabolites of organophosphate insecticides) may be more effective in capturing exposure than job category. This has led to initiatives to examine environmental exposures that are more representative of real-life exposures (National Institute of Environmental Health Sciences, 2021).

14. Biomarkers

There are three major types of biomarkers: biomarkers of exposure, effect, and susceptibility. A biomarker of exposure provides a direct measure of the contaminant or other proportionally related variable (metabolite of contaminant) in the body. Biological monitoring measures the body burden of a pollutant but not the source. For persistent pesticides, the substance measured may be either the contaminant itself (e.g., serum levels of DDT) or a persistent metabolite (e.g., serum levels of DDE). For non-persistent pesticides such as organophosphates or pyrethroids, biomarkers of exposure may include urinary levels of a metabolite specific to a given pesticide or a non-specific metabolite which is produced by several pesticides in a given class. Repeated measurements may be necessary to accurately classify exposure. Specifically, most OP insecticides are excreted from the body within days and a single spot urine sample may not reflect exposure (Barr, 2008).

Biomarkers of effect are indicators of a change in biologic function (e.g., biochemical, physiological, or behavioral changes) in response to a chemical exposure. One example of a biomarker of effect is blood cholinesterase (AChE, BChE) activity, which can become depressed following exposure to organophosphate and carbamate pesticides. Measuring changes in blood cholinesterase activity is useful in biomonitoring the relative health of pesticide applicators (Krenz et al., 2015).

Biomarkers of susceptibility are indicators of the natural characteristics of an organism that make it more susceptible to the effects of an exposure to a chemical. For example, a biomarker of susceptibility may be a functional polymorphism in a detoxifying or bioactivating enzyme, which may make an individual more susceptible or resistant to the adverse effects of exposure to a pesticide (e.g., PON1, GSTT1; (Costa et al., 2003, Goldman et al., 2012)). Epigenetic studies have also found that exposure to pesticides can influence DNA methylation (e.g., DDT, arsenic), histone modifications (e.g., paraquat, propoxur), and microRNA expression (e.g., fipronil, dichlorvos, triazophos) (Collotta et al., 2013).

The metabolism of chlorpyrifos has been extensively studied, highlighting the importance of the relative balance between bioactivation and detoxification reactions as a key factor regulating the susceptibility and relative safety in individuals (Buratti et al., 2003; Foxenberg et al., 2007; Sams et al., 2004). In humans, cytochrome P450 enzymes (CYPs) detoxify chlorpyrifos to 3,5,6-trichloro-2-pyridinol (TCPy) and dialkyl phosphates and bioactivate CPF to chlorpyrifos-oxon (CPF-O), which is the metabolite responsible for the inhibition of β-esterases, including acetylcholinesterase (AChE), butyrylcholinesterase (BChE) and carboxylesterase (CE; Fig. 3; Bushnell and Moser, 2006; Choi et al., 2006; Costa, 2006; Ross and Crow, 2007). A specific urinary biomarker of exposure to chlorpyrifos is TCPy, while dialkyl phosphates are a non-specific biomarker of exposure to chlorpyrifos and other organophosphate pesticides. Inhibition of blood AChE and BChE activities are sensitive biomarkers of effect for organophosphate pesticides.

In contrast to the majority of organophosphate pesticides, which require bioactivation to their oxon metabolite, the parent form of profenofos, a thiophosphate organophosphate, is a potent inhibitor of AChE (biomarker of effect) (Das et al., 2006; Nillos et al., 2007). CYP-mediated metabolism of profenofos results in oxidative detoxification to 4-bromo 2-chlorophenol (BCP), a specific urinary biomarker of exposure (Fig. 3; Abass et al., 2007; Wing et al., 1983). Animal studies suggest profenofos has similar toxicological activity and potency as chlorpyrifos (Das et al., 2006). With the exception of limited case reports on profenofos poisoning (Eddleston et al., 2009; Gotoh et al., 2000), and exposure assessment in Egyptian applicators (Dadson et al., 2013; Singleton et al., 2015), there are no human studies assessing exposure and health effects of profenofos. In 2009, the World Health Organization, highlighted the critical need for epidemiological data on profenofos. The lack of human data is particularly noteworthy since reports of pesticide residues on produce samples from Egypt, found that profenofos had the highest residue levels and was the most commonly-found organophosphate pesticide (World Health Organization, 2009).

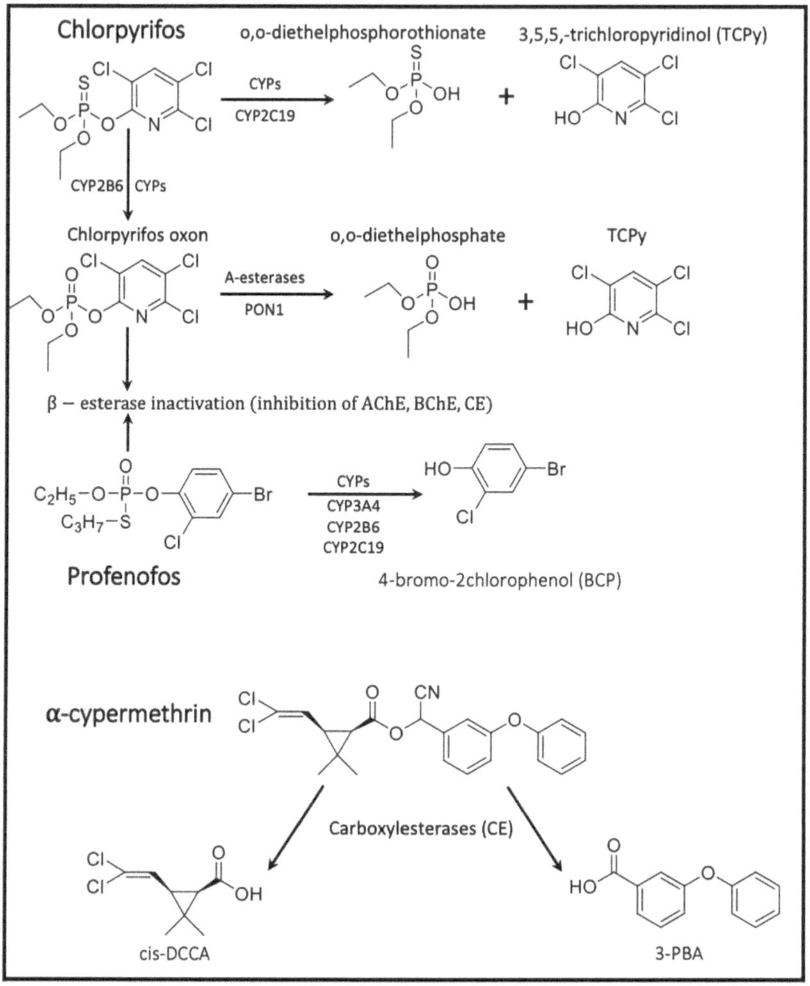

Fig. 3 Biomarkers of exposure to organophosphate (chlorpyrifos and profenofos) and pyrethroid (alpha-cypermethrin) insecticides.

In contrast to organophosphates, pyrethroids exert their toxic effects by prolonging the opening time of voltage-sensitive sodium channels (Soderlund et al., 2002). Alpha-cypermethrin (αCM) and lambda-cyhalothrin (λCH) are type II PYRs (containing a cyano group). Pyrethroids are active in their parent form, with metabolism by carboxylesterase and CYPs detoxifying the pesticide to 3-phenoxybenzoic acid (3-PBA, a non-specific urinary biomarker of exposure to pyrethroids), and cis-3-(2,2-dichlorovinyl)-2, 2-dimethylcyclopropane carboxylic acid (cis-DCCA/trans-DCCA, specific urinary biomarker of exposure to αCM) and lambda cyhalothrin acid, (λ-CHA, specific urinary biomarker of exposure to λ-CH), which serve as urinary biomarkers of exposure (Fig. 3). The most commonly reported symptom of pyrethroid exposure is paresthesia mostly limited to local effects of the exposed skin areas (biomarker of effect), which is temporary and reversible usually within hours after exposure (Ray and Forshaw, 2000; Vijverberg and Van Den Bercken, 1990; Wilks, 2000).

15. Controlling exposure to pesticides

The hierarchy of controls (Occupational Safety and Health Administration, 2021) is often used as a framework to evaluate and prioritize methods to control exposures. It provides guidance on the controls to reduce exposure and subsequent health effects and is organized from the most effects controls (elimination, substitution, engineering) to less effective control methods (administrative, PPE) (Table 4). A combination of control methods is often the most effective to reduce exposure.

15.1 Elimination and substitution

The most effective control methods are designed to eliminate exposure entirely. For example, policies that ban or restrict the use of hazardous pesticides will be the most effective. The pesticide registration process is used to identify health and environmental hazards associated with the use of the product. When pesticides are registered, they are classified into categories (e.g., highly hazardous, low risk) and guidance on safe handling practices is provided on the label. Registration is an ongoing process and considers new scientific evidence that has become available since the pesticide was first registered. For example, the US EPA requires review of all registered pesticides every 15 years to ensure they meet current scientific and regulatory standards (U.S. Environmental Protection Agency, 2021h). This review can lead to new restrictions on use, additional safety precautions, or the cancellation of a pesticide registration.

Another example of the effectiveness of controls that eliminate pesticide exposure are methods implemented to prevent suicide. One out of five deaths by suicide are due to self-poisoning by pesticides (World Health Organization, 2019) and the majority of these deaths occur in low-and-middle income countries where there are fewer pesticide regulations and access to highly toxic pesticides. Countries that have banned the most toxic

Table 4 Examples of methods to control exposure to pesticides organized by the Hierarchy of controls.

Controls	Examples
Elimination	Regulations or policies that ban or eliminate hazardous products
	Integrated pest management (IPM) strategies to eliminate or reduce the use of pesticides
	Eliminate pest breeding places
	Eliminate synthetic pesticides
Substitution	Use guidelines to select less toxic pesticides
	Use of biopesticides instead of chemical pesticides
Engineering controls	Use buffer zones between fields and areas of human activity
	Use precision application methods to reduce off target pesticide application (e.g., drift reduction technology)
	Ventilation systems in greenhouses
	Use an enclosed cab tractor for application
	Use a spill table when mixing pesticides
	Maintain and calibrate equipment
	Crop rotation and cover crops
Administrative controls	Restricted entry intervals
	Notify workers and neighboring properties of scheduled pesticide applications
	Use signage and flags to indicate chemically treated fields
	Limit daily maximum application time for applicators
	Hygiene behaviors (clean clothing and opportunities for washing or bathing)
	Consider weather conditions before applying
	Monitor and scout fields for pests to determine if application is needed
	Store pesticide in their original labeled container
	Written policies
	Training for pesticide applicators and handlers (e.g., EPA Worker Protection Standard)
	Monitoring programs (e.g., cholinesterase monitoring)
Personal protective equipment (PPE)	Wear protective clothing (e.g., chemically resistant gloves, boots, long pants, respirators, goggles)
	Maintain and clean PPE regularly

products (e.g., parathion, endosulfan, paraquat) have seen a reduction in deaths (Bonvoisin et al., 2020; Gunnell et al., 2017; Knipe et al., 2017).

Integrated pest management (IPM) strategies utilize both elimination and substitution to reduce exposure. These methods have led to reductions in the amount of pesticides applied on both large and small scale agricultural operations (Jepson et al., 2020). IPM utilizes knowledge about the pest environment and life cycle to mitigate pests. The goal is to use the least toxic methods prior to applying pesticides, particularly toxic pesticides (U.S. Environmental Protection Agency, 2021e). The process includes monitoring for pests and setting a threshold for when to actively control for pests (e.g., apply pesticides), managing the environment to prevent pests (e.g., crop rotation, selecting pest resistant plant varieties), and using less risky control methods (e.g., pheromones to prevent mating, mechanical controls) before utilizing more hazardous methods (e.g., broadcast spraying). A key part of this approach is to select pesticides that have fewer risks to human health and the environment. Jepson et al. (2020) developed guidelines that contain risk ratings for over 600 pesticides currently registered for use in the EU and in the US. The guideline groups these compounds into three categories, highly hazardous pesticides, high risk, and lower risk (Table 5). IPM techniques allow farmers, agricultural educators, and regulators to make informed decisions about the selection of products to control the pests (Prasanna et al., 2018).

15.2 Engineering controls

Engineering controls include designing or modifying equipment or the environment to reduce exposure. Adjusting the nozzle placement and height can reduce off target pesticide application (Hanna and Schaefer, 2008), using mixing tables designed to contain spills and reduce splash (Fig. 4; Pacific Northwest Agricultural Safety And Health Center, 2021), applying with enclosed cabs, or utilizing closed handling systems or direct injection systems are example engineering controls. With a closed handing system, the operator has no direct contact with the pesticide as it is loaded into the tank on the applicator without requiring human contact (e.g., water-soluble bags that rapidly dissolve in the spray tank). Direct injection systems are designed to reduce contact with the pesticide by directly injecting the chemical into the main sprayer line where it is then mixed with the water and applied to the crop. Newer technology uses unmanned vehicles to apply pesticides to the field while driven by an operator outside of the field (American Society Of Agricultural And Biological Engineers, 2019).

Table 5 A pesticide classification system based on acute and chronic health risks and environmental risks.

Pesticide classification	Description	Examples
Highly hazardous pesticides	Demonstrated high levels of acute or chronic health effects or adverse environmental effects	Aldicarb (carbamate)
		Atrazine (herbicide)
		Azinphos-methyl (organophosphate)
		Clothianidin (neonicotinoid)
		DDT (organochlorine)
		Imidacloprid (neonicotinoid)
		Methyl bromide (fumigant)
		Paraquat (herbicide)
High risk: Pesticides requiring risk mitigation	Require trained applicators, occupational risk mitigation (PPE), and/or environmental or bystander risk mitigation (restricted-entry Intervals, closed-mixing, closed cabs)	Acephate (organophosphate)
		Captan (fungicide)
		Chlorpyrifos (organophosphate)
		Cypermethrin (pyrethroid)
		Maneb (fungicide)
		Profenofos (organophosphate)
		Thiodicarb (carbamate)
Low risk pesticides	Require occupational health risk mitigation	2,4-D, (herbicide)
		Dicamba (herbicide)
		Hexaconazole (fungicide)
	OR	Picloram (herbicide)
		OR
		Azaconazole (fungicide)
	Only require basic PPE (long-sleeved shirt, pants, socks, shoes, and chemically resistant gloves)	*Bacillus thuringiensis* (biological)
		Carboxin (fungicide)
		Dichlorpop (herbicide)

Adapted from Jepson, P.C., Murray, K., Bach, O., Bonilla, M.A., Neumeister, L., 2020. Selection of pesticides to reduce human and environmental health risks: a global guideline and minimum pesticides list. Lancet Planet. Health 4, e56–e63.

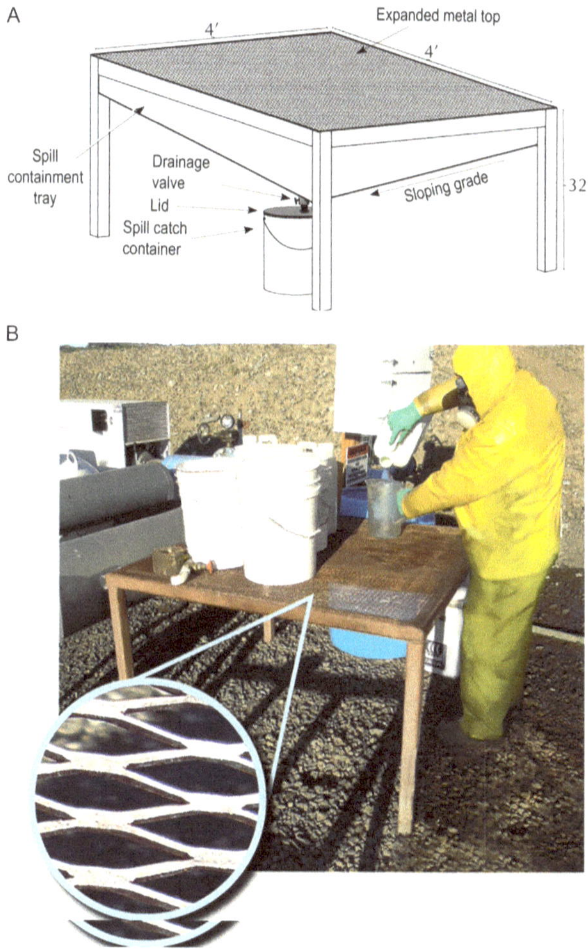

Fig. 4 (A, B) Examples of mixing tables designed to contain pesticide spills and splashes while mixing pesticides (Pacific Northwest Agricultural Safety and Health Center). https://deohs.washington.edu/pnash/practical_solutions.

15.3 Administrative controls

Administrative controls include training, policies, or work practices that reduce contact with pesticides. Most countries require pesticide applicators to be certified to apply the most harmful chemicals. Pesticide labels contain information about when and how to apply pesticides and often include information about restricted-entry intervals or how long workers with no PPE should stay out of recently sprayed fields. The Worker Protection Standard in the US requires employers to provide training to pesticide workers. The training includes safe handling practices, posting requirements for fields that have been sprayed, hygiene practices, and medical treatment. Restricted-entry intervals are used to indicate when it is safe to enter a field that was recently sprayed (e.g., use of signage to indicate recently sprayed fields, buffer zones). Workplace policies may require workers to change clothes after applying. PPE is described on the pesticide label and ranges from long sleeves and long pants, goggles, chemically resistant gloves, respirators, or chemically resistant clothing.

The types of crops and pests, field sizes, resources, and availability of equipment may impact how pesticides are applied and what controls are needed. For example, a vegetable farmer in Nigeria may have smaller fields and use different methods of application than a farmer in the US with 500 acres of corn. Access to safety equipment or PPE may also vary. In these cases, it is important to understand the application methods to identify feasible solutions to prevent exposure. The following case study describes a research project used to assess adolescent exposure to pesticides and the development of an intervention to reduce exposure.

16. Case study: Pesticide exposure among adolescents in Egypt

Organophosphate and pyrethroid insecticides are commonly applied throughout the world. Because of their widespread use, organophosphate and pyrethroid insecticides are ubiquitous in the environment and have been detected in air, water, soil, and in human tissues (Barr et al., 2011; Clune et al., 2012; Schnoor, 1992). Human and animal studies examining exposure to organophosphate insecticides, primarily chlorpyrifos, have consistently identified neurotoxicity as the primary endpoint of concern. Although banned from most indoor applications in the US, organophosphate insecticides are still extensively used in agricultural applications in the US and worldwide (Donley, 2019).

16.1 Adolescent pesticide exposure

Adolescents in Egypt are exposed to pesticides through non-occupational (e.g., diet, residential application) and para-occupational (e.g., take-home) pathways. In addition, male adolescents may mix and apply pesticides on family farms as private applicators or be hired as seasonal workers for the Ministry of Agriculture to apply pesticides to the cotton crop, often working for repeated seasons. Chlorpyrifos is the primary pesticide used for cotton production. During the application season it is applied daily by Ministry of Agriculture workers for up to four weeks. Research indicates that adolescents applying pesticides have higher exposures compared to adults and engage in behaviors that increase their risk (e.g., wearing shorts instead of pants, wearing wet clothing, and washing less frequently) (Kearney et al., 2015). Although females do not apply pesticides, they may work in the fields and come in contact with pesticide contaminated clothing (Abdel Rasoul et al., 2019). Residential pesticide use in Egypt is also high and includes both organophosphate and pyrethroid insecticides.

Adolescence is a time of rapid development. In addition to the hormonal and physiological changes associated with puberty, there are also significant developmental changes in the brain, primarily the prefrontal cortex (Crews et al., 2007; Spear, 2010; Steinberg, 2008). These changes are associated with behavioral changes, including increases in novelty seeking and risk-taking behavior, emotional reactivity, changes in information processing speed and tasks of executive function (e.g., response inhibition, working memory, and attention). The adolescent brain also has heightened synaptic plasticity indicating more malleability in response to environmental exposures (Selemon, 2013). Additionally, vulnerability is impacted by the ability to metabolize pesticides and other xenobiotics, which can vary across ages (Connors et al., 2008; Eskenazi et al., 2010; Kalia, 2008) and sex (Trentini et al., 2019). The impact of environmental exposures can vary across developmental periods (Acheson et al., 1998; Dwyer et al., 2009; Selemon, 2013) and consequences of prolonged exposure can last into adulthood (Zeid et al., 2018).

16.2 Low-level exposure to organophosphate insecticides

While the impact of acute high exposures (i.e., poisoning) to organophosphate insecticides is well understood, the impact of low-level exposures, particularly on neurobehavioral functioning, is not as well understood. Even less information is available examining the impact of cumulative exposure. Despite several reviews supporting an association between low-level organophosphate exposure and deficits in neurobehavioral performance (González-Alzaga et al., 2014; Jurewicz and Hanke, 2008; Meyer-Baron et al., 2015; Muñoz-Quezada et al., 2013), other reviews have emphasized inconsistent findings and limited exposure assessments as reasons for the continued uncertainty (Burns et al., 2013; Li et al., 2012). Moreover, most studies have examined only a single time point. Those that have addressed multiple time points have primarily focused only on pre- and post-application comparisons (Bazylewicz-Walczak et al., 1999; Daniell et al., 1992; Nguyen et al., 2015), while few studies have examined exposure across multiple years (Baldi et al., 2011; Berent et al., 2014; Roldan-Tapia et al., 2005, 2006). Furthermore, these previous studies were conducted with adult workers and not adolescents. Animal studies have indicated that chlorpyrifos exposure, at levels below cholinesterase inhibition, is associated with long-term impacts on brain structure and function that are likely irreversible (Rauh and Margolis, 2016; Slotkin, 2004). Examining the impact of low-level pesticide exposure is necessary to determine and quantify risk.

16.3 Examining changes over time

Cohorts of male adolescent pesticide applicators and non-applicators in Egypt were recruited to characterize changes in biomarkers and neurobehavioral outcomes across the application season (Rohlman et al., 2014, 2016). Test sessions included the collection of biological samples (i.e., urine, blood, saliva), questionnaires, and neurobehavioral testing. In addition, field observations were used to observe application practices to understand pathways of exposure. While the organophosphate chlorpyrifos is the primary insecticide applied to the cotton crop in Egypt, applied daily for up to 4 weeks, this is generally followed by two additional application cycles of the organophosphate profenofos and the pyrethroids, alpha-cypermethrin and lambda-cyhalothrin.

16.4 Study outcomes

Changes in biomarkers and neurobehavioral outcomes were characterized across the application season. Biomarkers of effect, e.g., BChE activity, showed a pattern of inhibition and recovery associated with chlorpyrifos application that is repeated across the years (Fig. 5). Biomarker data also indicate that non-applicators received environmental exposures to chlorpyrifos, profenofos, and the pyrethroids, alpha-cypermethrin and lambda-cyhalothrin (Crane et al., 2013; Himmel et al., 2016; Singleton et al., 2014). Questionnaire data identified several reasons for the elevated metabolite levels among non-applicators, including working as private applicators or applying pesticides on family farms, residential pesticide use, and living in close proximity to agricultural fields. Urinary TCPy levels were negatively associated with BChE activity. Importantly, adolescents showed BChE

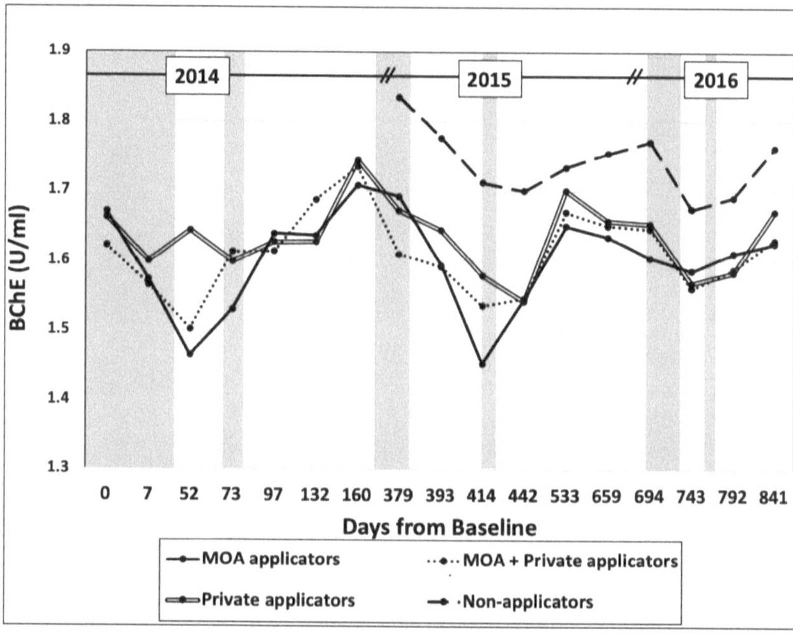

Fig. 5 Changes in BChE activity from 2014 to 2016. Shaded areas indicate chlorpyrifos application.

inhibition at much lower urinary levels of TCPy (28.9 μg/g creatinine; (Ismail et al., 2017a) than adults (>110 μg/g creatinine; (Farahat et al., 2011; Garabrant et al., 2009).

Increased symptoms (Ismail et al., 2017b; Khan et al., 2014), reduced lung function (Callahan et al., 2014) and neurobehavioral deficits (Eadeh et al., 2021; Ismail et al., 2017a; Rohlman et al., 2014, 2016) were identified in applicators compared to non-applicators. A dose response gradient between Attention Deficit Hyperactivity Disorder (ADHD) symptoms and urinary TCPy was identified (Rohlman et al., 2019). Our overarching hypothesis was that the inherent plasticity of the developing adolescent brain would allow recovery from selected neurobehavioral deficits associated with short- and long-term exposures to pesticides, however, neurobehavioral deficits among applicators *increased* during the application season and *persisted* after application ended (Fig. 6; Rohlman et al., 2016). Moreover, these sustained deficits were present in the absence of red blood cell acetyl cholinesterase (AChE) inhibition.

A pilot study found that adolescent females living with a pesticide applicator were also exposed to organophosphate insecticides (Abdel

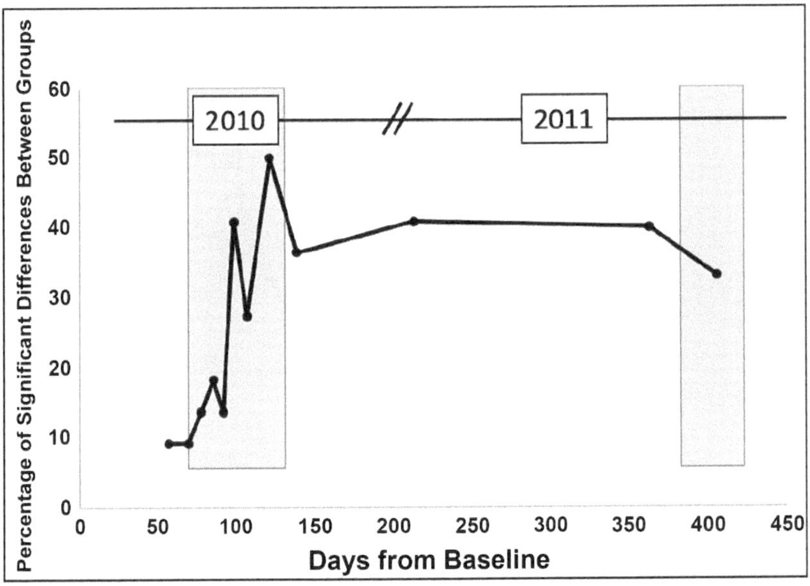

Fig. 6 Neurobehavioral differences between applicators and non-applicators increased during application and continue into the next year. Shaded areas indicate chlorpyrifos application.

Rasoul et al., 2019). These exposures were associated with neurobehavioral deficits, irregular menstrual cycles, and increased intermenstrual bleeding. Although females typically do not apply pesticides in Egypt, questionnaire data indicated that females are primarily responsible for household laundry and may have exposure while washing work clothes.

16.5 Intervention to reduce exposure

Recognizing the need to reduce exposures, an intervention was developed in partnership with the Ministry of Agriculture. An observational checklist identified behaviors during pesticide application, hygiene practices, and minimal use of PPE that increased contact with pesticides and therefore, exposure (e.g., mixing with hands, contact with pesticides during loading and applying, reentry into sprayed fields; Fig. 7). Furthermore, increased urinary TCPy levels were associated with increased time applying, and lower peak TCPy levels were associated with bathing immediately after work and using a stick to mix pesticides (instead of hands) (Callahan et al., 2017). The intervention focused on three low cost behavioral and hygiene methods: using a stick (instead of hands) when mixing pesticides,

Fig. 7 Adolescent pesticide applicators mixing (a and b) and applying (c) pesticides.

staying out of fields where pesticides were recently applied, and bathing after application and wearing clean clothes. These behaviors were selected because they were in the control of the adolescents and did not require access to PPE or controls not available to the adolescents.

Focus groups were held with staff from the Ministry of Agriculture, adolescents, and parents to present study findings and discuss *feasible methods* to reduce exposure. It was found that an in-person training delivered by researchers or agricultural officials was preferred. Although the MOA provides limited PPE and has adopted application procedures to reduce exposure (e.g., point nozzle downward, maintain distance between applicators, consider wind direction when applying), applicators receive no formal training. Most workers reported bathing after work, however, there was variation in how frequently work clothes were washed (e.g., daily, monthly). An educational intervention, utilizing a behavioral change theory, was administered to applicators (Rohlman et al., 2020). Findings showed an immediate increase in knowledge and awareness of the risks that was sustained over time (8 months after the training). Following the training there was an increase in the proportion of applicators who viewed pesticides as being a long-term health risk. At 8-month follow-up applicators were observed wearing googles, shoes, and masks more frequently. This is an example of a low-cost intervention that can be applied in other countries with similar safety culture surrounding pesticide application.

Organophosphorus and pyrethroid insecticides are among the most commonly used pesticides worldwide, with documented neurotoxic effects. Although adolescents around the world are frequently employed as applicators, few studies have examined the impact of exposure among this vulnerable population. The adolescent developing brain may be uniquely sensitive to the neurotoxic effects of these agents. We identified neurobehavioral deficits and increased symptoms of ADHD associated with pesticide exposure among Egyptian male adolescents. Furthermore, elevated urinary levels of organophosphate and pyrethroid metabolites were found in non-applicator males. These data suggest that adolescents living in agricultural communities in Egypt are receiving substantial environmental exposure to pesticides.

17. Conclusions

Pesticides are important tools for our food system and for public health. Designed to eradicate pests, they can also adversely impact human health. In spite of this, there are numerous examples throughout history that demonstrate the need for pesticides (e.g., plagues of locusts eating crops, the spread of diseases such as malaria and the bubonic plague through insects and rodents, the potato blight in Ireland that led to widespread emigration from the country (Matthews, 2018)). The development of manufactured synthetic pesticides led to a shift in food production, including the increased use of pesticides. Genetically modified crops have also contributed to increased pesticide use. Although these changes have led to improved crop yields and better storage methods, there have also led to adverse impacts on both the environment and human health, particularly among the most vulnerable members of society. There is a need to continue to develop new methods to control pests that minimize human and environmental impact. Understanding the history of pesticide use, the types of products and their mechanisms, and the pathways of exposure are key to understanding how to prevent exposure to pesticides and the subsequent neurotoxic effects. The challenge is to find ways to utilize these tools both safely and effectively.

Acknowledgments

The authors would like to acknowledge the Ministry of Agriculture and the adolescents who participated in the research study. We would also like to acknowledge the research team at Menoufia University who were responsible for data collection, quality control checks, and assisted with data analysis, Mahmoud Ismail, Mohammed Fou'ad, Tameem Abou Eleinein, Ahmed Nassef, Mohamed Nowishy, Mohamed Hella, Tarek Khater, Mohamed Abdel-Hamid, Mabrook Abdel-Halim, Mahmoud Nassef, Mohamed Abdelgawad, Ahmed Gamal Eldin, and Mohamed Elbayoumy. We would also like to thank Megan TePoel for her assistance with the technical formatting of the chapter. This work was supported from funding by the National Institute for Environmental Health Science and the Fogarty Institute (R01 ES022163 and R21 ES017223).

Conflict of interest

Oregon Health and Science University and Dr. Rohlman have a significant financial interest in Northwest Education Training and Assessment, LLC, a company that may have a commercial interest in the results of this research and technology. This potential conflict of interest was reviewed by the University of Iowa and Oregon Health and Science University and an approved Conflict of Interest in Research management plan was implemented.

References

Abass, K., Reponen, P., Jalonen, J., Pelkonen, O., 2007. In vitro metabolism and interaction of profenofos by human, mouse and rat liver preparations. Pest. Biochem. Physiol. 87, 238–247.

Abbassy, M., Mossa, A., 2012. Haemato-biochemical effects of formulated and technical cypermethrin and deltamethrin insecticides in male rats. J. Pharmacol. Toxicol. 7, 312–321.

Abdel Rasoul, G., Salem, E., Elbadry, A., Hendy, O., Rohlman, D., Abdel-Latif, A., 2019. Neurobehavioral and menstrual disorders among adolescent females environmentally exposed to pesticides, Menoufia Governate. Egypt. J. Occup. Med. 43, 331–343.

Abdollahi, M., Karami-Mohajeri, S., 2012. A comprehensive review on experimental and clinical findings in intermediate syndrome caused by organophosphate poisoning. Toxicol. Appl. Pharmacol. 258, 309–314.

Abhilash, P., Singh, N., 2009. Pesticide use and application: an Indian scenario. J. Hazard. Mater. 165, 1–12.

Abou-Donia, M.B., 2003. Organophosphorus ester-induced chronic neurotoxicity. Arch. Environ. Health Int. J. 58, 484–497.

Acheson, S.K., Stein, R.M., Swartzwelder, H.S., 1998. Impairment of semantic and figural memory by acute ethanol: age-dependent effects. Alcohol. Clin. Exp. Res. 22, 1437–1442.

Agency For Toxic Substances and Disease Registry, 2007. Case Studies in Environmental Medicine: Cholinesterase Inhibitors: Including Insecticides and Chemical Warfare Nerve Agents. U.S. Department of Health and Human Services.

Aktar, M.W., Sengupta, D., Chowdhury, A., 2009. Impact of pesticides use in agriculture: their benefits and hazards. Interdiscip. Toxicol. 2, 1.

Alavanja, M.C., Sandler, D.P., McMaster, S.B., Zahm, S.H., McDonnell, C.J., Lynch, C.F., Pennybacker, M., Rothman, N., Dosemeci, M., Bond, A.E., 1996. The agricultural health study. Environ. Health Perspect. 104, 362–369.

Aldridge, J.E., Levin, E.D., Seidler, F.J., Slotkin, T.A., 2005. Developmental exposure of rats to chlorpyrifos leads to behavioral alterations in adulthood, involving serotonergic mechanisms and resembling animal models of depression. Environ. Health Perspect. 113, 527–531.

American Society Of Agricultural And Biological Engineers, 2019. The Global Unmanned Spray System (Guss). Engineering and Technology for a Sustainable World.
Ames, R.G., Steenland, K., Jenkins, B., Chrislip, D., Russo, J., 1995. Chronic neurologic sequelae to cholinesterase inhibition among agricultural pesticide applicators. Arch. Environ. Health Int. J. 50, 440–444.
Anadón, A., Ares, I., Martínez, M., Martínez-Larrañaga, M.-R., Martínez, M.-A., 2020. Neurotoxicity of neonicotinoids. Adv. Neurotoxicol.
Atwood, D., Paisley-Jones, C., 2017. Pesticides Industry Sales and Usage: 2008–2012 Market Estimates. US Environmental Protection Agency, Washington, DC, p. 20460.
Bakir, F., Damluji, S.F., Amin-Zaki, L., Murtadha, M., Khalidi, A., Al-Rawi, N., Tikriti, S., Dhahir, H., Clarkson, T., Smith, J., 1973. Methylmercury poisoning in Iraq. Science 181, 230–241.
Baldi, I., Gruber, A., Rondeau, V., Lebailly, P., Brochard, P., Fabrigoule, C., 2011. Neurobehavioral effects of long-term exposure to pesticides: results from the 4-year follow-up of the Phytoner study. Occup. Environ. Med. 68, 108–115.
Barone Jr., S., Das, K.P., Lassiter, T.L., White, L.D., 2000. Vulnerable processes of nervous system development: a review of markers and methods. Neurotoxicology 21, 15–36.
Barr, D.B., 2008. Biomonitoring of exposure to pesticides. J. Chem. Health Saf. 15, 20–29.
Barr, D.B., Wong, L.Y., Bravo, R., Weerasekera, G., Odetokun, M., Restrepo, P., Kim, D.G., Fernandez, C., Whitehead Jr., R.D., Perez, J., Gallegos, M., Williams, B.L., Needham, L.L., 2011. Urinary concentrations of dialkylphosphate metabolites of organophosphorus pesticides: National Health and Nutrition Examination Survey 1999-2004. Int. J. Environ. Res. Public Health 8, 3063–3098.
Bass, C., Denholm, I., Williamson, M.S., Nauen, R., 2015. The global status of insect resistance to neonicotinoid insecticides. Pest. Biochem. Physiol. 121, 78–87.
Bazylewicz-Walczak, B., Majczakowa, W., Szymczak, M., 1999. Behavioral effects of occupational exposure to organophosphorous pesticides in female greenhouse planting workers. Neurotoxicology 20, 819–826.
Berent, S., Giordani, B., Albers, J.W., Garabrant, D.H., Cohen, S.S., Garrison, R.P., Richardson, R.J., 2014. Effects of occupational exposure to chlorpyrifos on neuropsychological function: a prospective longitudinal study. Neurotoxicology 41, 44–53.
Berry, C., La Vecchia, C., Nicotera, P., 2010. Paraquat and Parkinson's disease. Cell Death Differ. 17, 1115–1125.
Bonvoisin, T., Utyasheva, L., Knipe, D., Gunnell, D., Eddleston, M., 2020. Suicide by pesticide poisoning in India: a review of pesticide regulations and their impact on suicide trends. BMC Public Health 20, 1–16.
Buratti, F.M., Volpe, M.T., Meneguz, A., Vittozzi, L., Testai, E., 2003. Cyp-specific bioactivation of four organophosphorothioate pesticides by human liver microsomes. Toxicol. Appl. Pharmacol. 186, 143–154.
Burke, R.D., Todd, S.W., Lumsden, E., Mullins, R.J., Mamczarz, J., Fawcett, W.P., Gullapalli, R.P., Randall, W.R., Pereira, E.F., Albuquerque, E.X., 2017. Developmental neurotoxicity of the organophosphorus insecticide chlorpyrifos: from clinical findings to preclinical models and potential mechanisms. J. Neurochem. 142, 162–177.
Burns, C.J., Mcintosh, L.J., Mink, P.J., Jurek, A.M., Li, A.A., 2013. Pesticide exposure and neurodevelopmental outcomes: review of the epidemiologic and animal studies. J. Toxicol. Environ. Health B Crit. Rev. 16, 127–283.
Bushnell, P.J., Moser, V.C., 2006. Behavioral toxicity of cholinesterase inhibitors. In: Gupta, R.C. (Ed.), Toxicology of Organophosphate and Carbamate Compounds. Elsevier, San Diego, CA.
Butler-Dawson, J., Galvin, K., Thorne, P.S., Rohlman, D.S., 2018. Organophosphorus pesticide residue levels in homes located near orchards. J. Occup. Environ. Hyg. 15, 847–856.

Callahan, C.L., Al-Batanony, M., Ismail, A.A., Abdel-Rasoul, G., Hendy, O., Olson, J.R., Rohlman, D.S., Bonner, M.R., 2014. Chlorpyrifos exposure and respiratory health among adolescent agricultural workers. Int. J. Environ. Res. Public Health 11, 13117–13129.

Callahan, C.L., Hamad, L.A., Olson, J.R., Ismail, A.A., Abdel-Rasoul, G., Hendy, O., Rohlman, D.S., Bonner, M.R., 2017. Longitudinal assessment of occupational determinants of chlorpyrifos exposure in adolescent pesticide workers in Egypt. Int. J. Hyg. Environ. Health 220 (8), 1356–1362.

Carles, C., Bouvier, G., Lebailly, P., Baldi, I., 2017. Use of job-exposure matrices to estimate occupational exposure to pesticides: a review. J. Expo. Sci. Environ. Epidemiol. 27, 125–140.

Chen, M., Tao, L., Mclean, J., Lu, C., 2014. Quantitative analysis of neonicotinoid insecticide residues in foods: implication for dietary exposures. J. Agric. Food Chem. 62, 6082–6090.

Choi, H., Jedrychowski, W., Spengler, J., Camann, D.E., Whyatt, R.M., Rauh, V., Tsai, W.Y., Perera, F.P., 2006. International studies of prenatal exposure to polycyclic aromatic hydrocarbons and fetal growth. Environ. Health Perspect. 114, 1744–1750.

Chrustek, A., Hołyńska-Iwan, I., Dziembowska, I., Bogusiewicz, J., Wróblewski, M., Cwynar, A., Olszewska-Słonina, D., 2018. Current research on the safety of pyrethroids used as insecticides. Medicina 54, 61.

Cimino, A.M., Boyles, A.L., Thayer, K.A., Perry, M.J., 2017. Effects of neonicotinoid pesticide exposure on human health: a systematic review. Environ. Health Perspect. 125, 155–162.

Clune, A.L., Ryan, P.B., Barr, D.B., 2012. Have regulatory efforts to reduce organophosphorus insecticide exposures been effective? Environ. Health Perspect. 120, 521–525.

Collotta, M., Bertazzi, P.A., Bollati, V., 2013. Epigenetics and pesticides. Toxicology 307, 35–41.

Connors, S.L., Levitt, P., Matthews, S.G., Slotkin, T.A., Johnston, M.V., Kinney, H.C., Johnson, W.G., Dailey, R.M., Zimmerman, A.W., 2008. Fetal mechanisms in neurodevelomental disorders. Pediatr. Neurol. 38, 163–176.

Cooper, J., Dobson, H., 2007. The benefits of pesticides to mankind and the environment. Crop Prot. 26, 1337–1348.

Cooper, J., Dobson, H., Maritime, C., 2007. Pesticides and Humanity: The Benefits of Using Pesticides. Natural Resources Institute, University of Greenwich, UK.

Costa, L.G., 2006. Current issues in organophosphate toxicology. Clin. Chim. Acta 366, 1–13.

Costa, L.G., 2015. The neurotoxicity of organochlorine and pyrethroid pesticides. Handb. Clin. Neurol. 131, 135–148.

Costa, L.G., 2018. Organophosphorus compounds at 80: some old and new issues. Toxicol. Sci. 162, 24–35.

Costa, L.G., Cole, T.B., Furlong, C.E., 2003. Polymorphisms of paraoxonase (Pon1) and their significance in clinical toxicology of organophosphates. J. Toxicol. Clin. Toxicol. 41, 37–45.

Costa, L.G., Giordano, G., Guizzetti, M., Vitalone, A., 2008. Neurotoxicity of pesticides: a brief review. Front. Biosci. 13, 1240–1249.

Craddock, H.A., Huang, D., Turner, P.C., Quirós-Alcalá, L., Payne-Sturges, D.C., 2019. Trends in neonicotinoid pesticide residues in food and water in the United States, 1999–2015. Environ. Health 18, 1–16.

Crane, A.L., Abdel Rasoul, G., Ismail, A.A., Hendy, O., Bonner, M.R., Lasarev, M.R., Al-Batanony, M., Singleton, S.T., Khan, K., Olson, J.R., Rohlman, D.S., 2013. Longitudinal assessment of chlorpyrifos exposure and effect biomarkers in adolescent Egyptian agricultural workers. J. Expo. Sci. Environ. Epidemiol. 23, 356–362.

Crews, F., He, J., Hodge, C., 2007. Adolescent cortical development: a critical period of vulnerability for addiction. Pharmacol. Biochem. Behav. 86, 189–199.
Dadson, O.A., Ellison, C.A., Singleton, S.T., Chi, L.H., McGarrigle, B.P., Lein, P.J., Farahat, F.M., Farahat, T., Olson, J.R., 2013. Metabolism of profenofos to 4-bromo-2-chlorophenol, a specific and sensitive exposure biomarker. Toxicology 306, 35–39.
Damalas, C.A., Eleftherohorinos, I.G., 2011. Pesticide exposure, safety issues, and risk assessment indicators. Int. J. Environ. Res. Public Health 8, 1402–1419.
Damalas, C.A., Koutroubas, S.D., 2016. Farmers' Exposure to Pesticides: Toxicity Types and Ways of Prevention. Multidisciplinary Digital Publishing Institute.
Daniell, W., Barnhart, S., Demers, P., Costa, L.G., Eaton, D.L., Miller, M., Rosenstock, L., 1992. Neuropsychological performance among agricultural pesticide applicators. Environ. Res. 59, 217–228.
Das, G.P., Jamil, K., Rahman, M.F., 2006. Effect of four organophosphorus compounds on human blood acetylcholinesterase: in vitro studies. Toxicol. Mech. Methods 16, 455–459.
De Joode, B.V.W., Mora, A.M., Lindh, C.H., Hernández-Bonilla, D., Córdoba, L., Wesseling, C., Hoppin, J.A., Mergler, D., 2016. Pesticide exposure and neurodevelopment in children aged 6–9 years from Talamanca, Costa Rica. Cortex 85, 137–150.
Dill, G.M., 2005. Glyphosate-resistant crops: history, status and future. Pest Manag. Sci. 61, 219–224.
Donley, N., 2019. The Usa lags behind other agricultural nations in banning harmful pesticides. Environ. Health 18, 44.
Dwyer, J.B., Mcquown, S.C., Leslie, F.M., 2009. The dynamic effects of nicotine on the developing brain. Pharmacol. Ther. 122, 125–139.
Eadeh, H.-M., Ismail, A.A., Rasoul, G.M.A., Hendy, O.M., Olson, J.R., Wang, K., Bonner, M.R., Rohlman, D.S., 2021. Evaluation of occupational pesticide exposure on Egyptian male adolescent cognitive and motor functioning. Environ. Res. 197, 111137.
Eckerman, D.A., Gimenes, L.S., De Souza, R.C., Galvão, P.R.L., Sarcinelli, P.N., Chrisman, J.R., 2007. Age related effects of pesticide exposure on neurobehavioral performance of adolescent farm workers in Brazil. Neurotoxicol. Teratol. 29, 164–175.
Eddleston, M., Worek, F., Eyer, P., Thiermann, H., Von Meyer, L., Jeganathan, K., Sheriff, M.H., Dawson, A.H., Buckley, N.A., 2009. Poisoning with the S-alkyl organophosphorus insecticides profenofos and prothiofos. QJM 102, 785–792.
Eskenazi, B., Huen, K., Marks, A., Harley, K.G., Bradman, A., Barr, D.B., Holland, N., 2010. Pon1 and neurodevelopment in children from the Chamacos study exposed to organophosphate pesticides in utero. Environ. Health Perspect. 118, 1775–1781.
Farahat, F.M., Ellison, C.A., Bonner, M.R., McGarrigle, B.P., Crane, A.L., Fenske, R.A., Lasarev, M.R., Rohlman, D.S., Anger, W.K., Lein, P.J., Olson, J.R., 2011. Biomarkers of chlorpyrifos exposure and effect in Egyptian cotton field workers. Environ. Health Perspect. 119, 801–806.
Food And Agriculture Organization Of The United Nations, 2015. International Code of Conduct on Pesticide Management: Guidelines on Good Labelling Practice for Pesticides.
Food and Agriculture Organization of the United Nations, 2021. Pesticides Use (Online]. Available: https://www.fao.org/faostat/en/#data/Rp [Accessed October 5, 2021).
Fortenberry, G.Z., Beckman, J., Schwartz, A., Prado, J.B., Graham, L.S., Higgins, S., Lackovic, M., Mulay, P., Bojes, H., Waltz, J., 2016. Magnitude and characteristics of acute paraquat-and diquat-related illnesses in the Us: 1998–2013. Environ. Res. 146, 191–199.

Foxenberg, R., McGarrigle, B., Knaak, J., Kransler, K., Kostyniak, P., Olson, J., 2007. Human hepatic cytochrome P450–Specific metabolism of parathion and chlorpyrifos. Drug Metab. Dispos. 35, 189–193.

Furlong, C.E., Cole, T.B., Jarvik, G.P., Pettan-Brewer, C., Geiss, G.K., Richter, R.J., Shih, D.M., Tward, A.D., Lusis, A.J., Costa, L.G., 2005. Role of paraoxonase (Pon1) status in pesticide sensitivity: genetic and temporal determinants. Neurotoxicology 26, 651–659.

Garabrant, D.H., Aylward, L.L., Berent, S., Chen, Q., Timchalk, C., Burns, C.J., Hays, S.M., Albers, J.W., 2009. Cholinesterase inhibition in chlorpyrifos workers: characterization of biomarkers of exposure and response in relation to urinary Tcpy. J. Expo. Sci. Environ. Epidemiol. 19, 634–642.

Garcia-Repetto, R., Martinez, D., Repetto, M., 1995. Coefficient of distribution of some organophosphorous pesticides in rat tissue. Vet. Hum. Toxicol. 37, 226–229.

Geller, R., Kezirian, R., Bangar, P., Strong, D., Carlson, T., 2009. Epidemiology of accidental poisoning caused by storage of non-food substances in food containers and unmarked bottles/containers. Clin. Toxicol., 705. Informa Healthcare 52 Vanderbilt Ave, New York, NY 10017 USA.

Goldman, S.M., Kamel, F., Ross, G.W., Bhudhikanok, G.S., Hoppin, J.A., Korell, M., Marras, C., Meng, C., Umbach, D.M., Kasten, M., Chade, A.R., Comyns, K., Richards, M.B., Sandler, D.P., Blair, A., Langston, J.W., Tanner, C.M., 2012. Genetic modification of the association of paraquat and Parkinson's disease. Mov. Disord. 27, 1652–1658.

González-Alzaga, B., Lacasaña, M., Aguilar-Garduño, C., Rodríguez-Barranco, M., Ballester, F., Rebagliato, M., Hernández, A., 2014. A systematic review of neurodevelopmental effects of prenatal and postnatal organophosphate pesticide exposure. Toxicol. Lett. 230, 104–121.

Gotoh, M., Sakata, M., Endo, T., Hayashi, H., Seno, H., Suzuki, O., 2000. Profenofos metabolites in human poisoning. Forensic Sci. Int. 116, 221–226.

Guignet, M., Lein, P.J., 2019. Neuroinflammation in organophosphate-induced neurotoxicity. Adv. Neurotoxicol. 3, 35–79.

Gunnell, D., Eddleston, M., 2003. Suicide by intentional ingestion of pesticides: a continuing tragedy in developing countries. Int. J. Epidemiol. 32, 902–909.

Gunnell, D., Knipe, D., Chang, S.-S., Pearson, M., Konradsen, F., Lee, W.J., Eddleston, M., 2017. Prevention of suicide with regulations aimed at restricting access to highly hazardous pesticides: a systematic review of the international evidence. Lancet Glob. Health 5, e1026–e1037.

Gupta, R.C., Sachana, M., Mukherjee, I.M., Doss, R.B., Malik, J.K., Milatovic, D., 2018. Chapter 37—Organophosphates and carbamates. In: Gupta, R.C. (Ed.), Veterinary Toxicology, third ed. Academic Press.

Hanna, H.M., Schaefer, K.J., 2008. Factors affecting pesticide drift.

Hansen, M.R.H., Jørs, E., Lander, F., Condarco, G., Debes, F., Tirado Bustillos, N., Schlünssen, V., 2017. Neurological deficits after long-term pyrethroid exposure. Environ. Health Insights 11. 1178630217700628.

Harley, K.G., Huen, K., Aguilar Schall, R., Holland, N.T., Bradman, A., Barr, D.B., Eskenazi, B., 2011. Association of organophosphate pesticide exposure and paraoxonase with birth outcome in Mexican-American women. PLoS One 6, e23923.

Hassall, K.A., 1990. Biochemistry and Uses of Pesticides. Macmillan Press Ltd.

Hayes Jr., W., 1991. Chapter 1, introduction; Section 1.4. 1.2 Benefits from disease control. In: Handbook of Pesticide Toxicology. vol. 1, pp. 10–13.

Helwany, M., Bordoni, B., 2020. Neuroanatomy, Cranial Nerve 1 (Olfactory).

Himmel, M., McGarrigle, B., Singh, R., Ismail, A., Abdel Rasoul, G., Hendy, O., Bonner, M.R., Rohlman, D., Aga, D., Olson, J., 2016. In: Pyrethroid Metabolism and Exposure Assessment in Egyptian Agricultural Workers. Society of Toxicology (Sot) 55th Annual Meeting, 2016 New Orleans, LA.

Hladik, M.L., Main, A.R., Goulson, D., 2018. Environmental Risks and Challenges Associated With Neonicotinoid Insecticides. ACS Publications.

Hock, W., 2006. Toxicity of Pesticides [Online]. Available: https://extension.psu.edu/toxicity-of-pesticides [Accessed September 5, 2021].

Huang, X., Zhang, C., Hu, R., Li, Y., Yin, Y., Chen, Z., Cai, J., Cui, F., 2016. Association between occupational exposures to pesticides with heterogeneous chemical structures and farmer health in China. Sci. Rep. 6, 1–7.

Hyland, C., Laribi, O., 2017. Review of take-home pesticide exposure pathway in children living in agricultural areas. Environ. Res. 156, 559–570.

Ismail, A.A., Bodner, T., Rohlman, D., 2012. Neurobehavioral performance among agricultural workers and pesticide applicators: a meta-analytic study. Occup. Environ. Med. 69, 457–464.

Ismail, A., Wang, K., Olson, J., Bonner, M.R., Hendy, O., Abdel Rasoul, G., Rohlman, D., 2017a. The impact of repeated organophosphorus pesticide exposure on biomarkers and neurobehavioral outcomes among adolescents. J. Toxicol. Environ. Health A 80 (10–12), 542–555.

Ismail, A.A., Bonner, M.R., Hendy, O., Abdel Rasoul, G., Wang, K., Olson, J.R., Rohlman, D.S., 2017b. Comparison of neurological health outcomes between two adolescent cohorts exposed to pesticides in Egypt. PLoS One 12, e0172696.

Ismail, A.A., Almalki, M., Agag, A., Solan, Y.M., Bani, I.A., 2018. Pesticide application and khat chewing as predictors of the neurological health outcomes among pesticide applicators in a vector control Unit, Saudi Arabia. Int. J. Occup. Environ. Med. 9, 32.

Jayaraj, R., Megha, P., Sreedev, P., 2016. Organochlorine pesticides, their toxic effects on living organisms and their fate in the environment. Interdiscip. Toxicol. 9, 90.

Jepson, P.C., Murray, K., Bach, O., Bonilla, M.A., Neumeister, L., 2020. Selection of pesticides to reduce human and environmental health risks: a global guideline and minimum pesticides list. Lancet Planet. Health 4, e56–e63.

Jeschke, P., Nauen, R., Schindler, M., Elbert, A., 2011. Overview of the status and global strategy for neonicotinoids. J. Agric. Food Chem. 59, 2897–2908.

Johnson, M.K., 1990. Organophosphates and delayed neuropathy—is NTE alive and well? Toxicol. Appl. Pharmacol. 102, 385–399.

Judge, S.J., Savy, C.Y., Campbell, M., Dodds, R., Gomes, L.K., Laws, G., Watson, A., Blain, P.G., Morris, C.M., Gartside, S.E., 2016. Mechanism for the acute effects of organophosphate pesticides on the adult 5-Ht system. Chem. Biol. Interact. 245, 82–89.

Jurewicz, J., Hanke, W., 2008. Prenatal and childhood exposure to pesticides and neurobehavioral development: review of epidemiological studies. Int. J. Occup. Med. Environ. Health 21, 121–132.

Kalia, M., 2008. Brain development: anatomy, connectivity, adaptive plasticity, and toxicity. Metabolism 57 (Suppl. 2), S2–S5.

Kearney, G.D., Rodriguez, G., Quandt, S.A., Arcury, J.T., Arcury, T.A., 2015. Work safety climate, safety behaviors, and occupational injuries of youth farmworkers in North Carolina. Am. J. Public Health 105, 1336–1343.

Keifer, M.C., Firestone, J., 2007. Neurotoxicity of pesticides. J. Agromed. 12, 17–25.

Khan, K., Ismail, A.A., Abdel Rasoul, G., Bonner, M.R., Lasarev, M.R., Hendy, O., Al-Batanony, M., Crane, A.L., Singleton, S.T., Olson, J.R., Rohlman, D.S., 2014. Longitudinal assessment of chlorpyrifos exposure and self-reported neurological symptoms in adolescent pesticide applicators. BMJ Open 4, e004177.

Kimbrough, R.D., 1984. Relationship between dose and health effects. Clin. Lab. Med. 4, 507–519.
Knipe, D.W., Gunnell, D., Eddleston, M., 2017. Preventing deaths from pesticide self-poisoning—learning from Sri Lanka's success. Lancet Glob. Health 5, e651–e652.
Kori, R.K., Singh, M.K., Jain, A.K., Yadav, R.S., 2018. Neurochemical and behavioral dysfunctions in pesticide exposed farm workers: a clinical outcome. Ind. J. Clin. Biochem. 33, 372–381.
Krenz, J.E., Hofmann, J.N., Smith, T.R., Cunningham, R.N., Fenske, R.A., Simpson, C.D., Keifer, M., 2015. Determinants of butyrylcholinesterase inhibition among agricultural pesticide handlers in Washington State: an update. Ann. Occup. Hyg. 59, 25–40.
Lambert, W.E., Lasarev, M., Muniz, J., Scherer, J., Rothlein, J., Santana, J., McCauley, L., 2005. Variation in organophosphate pesticide metabolites in urine of chidlren living in agricultural communities. Environ. Health Perspect. 113, 504–508.
Li, A.A., Lowe, K.A., Mcintosh, L.J., Mink, P.J., 2012. Evaluation of epidemiology and animal data for risk assessment: chlorpyrifos developmental neurobehavioral outcomes. J. Toxicol. Environ. Health B Crit. Rev. 15, 109–184.
López-Gálvez, N., Wagoner, R., Quirós-Alcalá, L., Ornelas Van Horne, Y., Furlong, M., Avila, E.G., Beamer, P., 2019. Systematic literature review of the take-home route of pesticide exposure via biomonitoring and environmental monitoring. Int. J. Environ. Res. Public Health 16, 2177.
Lotti, M., 2002. Low-level exposures to organophosphorus esters and peripheral nerve function. Muscle Nerve 25, 492–504.
Lotti, M., Moretto, A., 2005. Organophosphate-induced delayed polyneuropathy. Toxicol. Rev. 24, 37–49.
Lu, J.L., 2009. Comparison of pesticide exposure and physical examination, neurological assessment, and laboratory findings between full-time and part-time vegetable farmers in the Philippines. Environ. Health Prevent. Med. 14, 345–352.
Lu, C., Fenske, R.A., Simcox, N.J., Kalman, D., 2000. Pesticide exposure of children in an agricultural community: evidence of household proximity to farmland and take home exposure pathways. Environ. Res. 84, 290–302.
Lucero, B., Muñoz-Quezada, M.T., 2021. Neurobehavioral, neuromotor, and neurocognitive effects in agricultural workers and their children exposed to pyrethroid pesticides: a review. Front. Hum. Neurosci. 369.
Lushchak, V.I., Matviishyn, T.M., Husak, V.V., Storey, J.M., Storey, K.B., 2018. Pesticide toxicity: a mechanistic approach. EXCLI J. 17, 1101.
Ma, C.-S., Zhang, W., Peng, Y., Zhao, F., Chang, X.-Q., Xing, K., Zhu, L., Ma, G., Yang, H.-P., Rudolf, V.H.W., 2021. Climate warming promotes pesticide resistance through expanding overwintering range of a global pest. Nat. Commun. 12, 5351.
Marfo, J.T., Fujioka, K., Ikenaka, Y., Nakayama, S.M., Mizukawa, H., Aoyama, Y., Ishizuka, M., Taira, K., 2015. Relationship between urinary N-desmethyl-acetamiprid and typical symptoms including neurological findings: a prevalence case-control study. PLoS One 10, e0142172.
Matthews, G.A., 2018. A History of Pesticides. CABI.
McCauley, L.A., Beltran, M., Phillips, J., Lasarev, M., Sticker, D., 2001. The Oregon migrant farmworker community: an evolving model for participatory research. Environ. Health Perspect. 109, 449–455.
Meyer-Baron, M., Knapp, G., Schaper, M., Van Thriel, C., 2015. Meta-analysis on occupational exposure to pesticides—neurobehavioral impact and dose-response relationships. Environ. Res. 136, 234–245.

Moretti, M., Villarini, M., Scassellati-Sforzolini, G., Pasquini, R., Monarca, S., 1997. Applicability of aspecific noninvasive methods for biomonitoring of occupational exposure to deltamethrin: preliminary study using an animal model. Arch. Environ. Contam. Toxicol. 33, 323–328.

Moretto, A., Colosio, C., 2011. Biochemical and toxicological evidence of neurological effects of pesticides: the example of Parkinson's disease. Neurotoxicology 32, 383–391.

Motsoeneng, P.M., Dalvie, M.A., 2015. Relationship between urinary pesticide residue levels and neurotoxic symptoms among women on farms in the Western Cape, South Africa. Int. J. Environ. Res. Public Health 12, 6281–6299.

Muñoz-Quezada, M.T., Lucero, B.A., Barr, D.B., Steenland, K., Levy, K., Ryan, P.B., Iglesias, V., Alvarado, S., Concha, C., Rojas, E., 2013. Neurodevelopmental effects in children associated with exposure to organophosphate pesticides: a systematic review. Neurotoxicology 39, 158–168.

Muñoz-Quezada, M.T., Lucero, B.A., Iglesias, V.P., Muñoz, M.P., Cornejo, C.A., Achu, E., Baumert, B., Hanchey, A., Concha, C., Brito, A.M., 2016. Chronic exposure to organophosphate (Op) pesticides and neuropsychological functioning in farm workers: a review. Int. J. Occup. Environ. Health 22, 68–79.

Nagy, K., Duca, R.C., Lovas, S., Creta, M., Scheepers, P.T., Godderis, L., Ádám, B., 2020. Systematic review of comparative studies assessing the toxicity of pesticide active ingredients and their product formulations. Environ. Res. 181, 108926.

Naksen, W., Prapamontol, T., Mangklabruks, A., Chantara, S., Thavornyutikarn, P., Srinual, N., Panuwet, P., Ryan, P.B., Riederer, A.M., Barr, D.B., 2015. Associations of maternal organophosphate pesticide exposure and Pon1 activity with birth outcomes in Sawasdee birth cohort, Thailand. Environ. Res. 142, 288–296.

Nampeera, E.L., Nonnecke, G.R., Blodgett, S.L., Tusiime, S.M., Masinde, D.M., Wesonga, J.M., Murungi, L.K., Baidu-Forson, J.J., Abukutsa-Onyango, M.O., 2019. Farmers' knowledge and practices in the management of insect pests of leafy amaranth in Kenya. J. Integ. Pest Manage. 10, 31.

National Academies of Sciences, Engineering & Medicine, 2018. Veterans and Agent Orange: Update 11.

National Institute of Environmental Health Sciences, 2021. Mixtures [Online]. Available: https://www.niehs.nih.gov/research/supported/exposure/mixtures/index.cfm [Accessed September 12, 2021].

National Research Council, 1993. Pesticides in the Diets of Infants and Children.

Nguyen, H.T., Quandt, S.A., Summers, P., Morgan, T.M., Chen, H., Walker, F.O., Howard, T.D., Galvan, L., Arcury, T.A., 2015. Learning ability as a function of practice: does it apply to farmworkers? J. Occup. Environ. Med. 57, 676–681.

Nillos, M.G., Rodriguez-Fuentes, G., Gan, J., Schlenk, D., 2007. Enantioselective acetylcholinesterase inhibition of the organophosphorous insecticides profenofos, fonofos, and crotoxyphos. Environ. Toxicol. Chem. 26, 1949–1954.

Occupational Safety and Health Administration, 2021. Recommended Practices for Safety and Health Programs [Online]. Available: https://www.osha.gov/safety-management [Accessed September 14, 2021].

Ohlander, J., Fuhrimann, S., Basinas, I., Cherrie, J.W., Galea, K.S., Povey, A.C., Van Tongeren, M., Harding, A.-H., Jones, K., Vermeulen, R., 2020. Systematic review of methods used to assess exposure to pesticides in occupational epidemiology studies, 1993–2017. Occup. Environ. Med. 77, 357–367.

Pacific Northwest Agricultural Safety And Health Center, 2021. Practical Solutions for Pesticide Safety [Online]. Available: https://deohs.washington.edu/pnash/practical_solutions [Accessed September 28, 2021].

Perry, J., Cotton, J., Rahman, M.A., Brumby, S.A., 2020. Organophosphate exposure and the chronic effects on farmers: a narrative review. Rural Remote Health 20, 4508.

Prasanna, B., Huesing, J., Eddy, R., Peschke, V., 2018. Fall Armyworm in Africa: A Guide for Integrated Pest Management. CIMMYT Publications.
Qiao, D., Seidler, F.J., Tate, C.A., Cousins, M.M., Slotkin, T.A., 2003. Fetal chlorpyrifos exposure: adverse effects on brain cell development and cholinergic biomarkers emerge postnatally and continue into adolescence and adulthood. Environ. Health Perspect. 111, 536–544.
Rasoul, G.M.A., Abou Salem, M.E., Mechael, A.A., Hendy, O.M., Rohlman, D.S., Ismail, A.A., 2008. Effects of occupational pesticide exposure on children applying pesticides. Neurotoxicology 29, 833–838.
Rauh, V.A., Margolis, A.E., 2016. Research review: environmental exposures, neurodevelopment, and child mental health—new paradigms for the study of brain and behavioral effects. J. Child Psychol. Psychiatry 57, 775–793.
Ray, D.E., Forshaw, P.J., 2000. Pyrethroid insecticides: poisoning syndromes, synergies, and therapy. J. Toxicol. Clin. Toxicol. 38, 95–101.
Reiss, R., Chang, E.T., Richardson, R.J., Goodman, M., 2015. A review of epidemiologic studies of low-level exposures to organophosphorus insecticides in non-occupational populations. Crit. Rev. Toxicol. 45, 531–641.
Richardson, J.R., Fitsanakis, V., Westerink, R.H., Kanthasamy, A.G., 2019. Neurotoxicity of pesticides. Acta Neuropathol. 138, 343–362.
Richardson, R.J., Fink, J.K., Glynn, P., Hufnagel, R.B., Makhaeva, G.F., Wijeyesakere, S.J., 2020. Neuropathy target esterase (Nte/Pnpla6) and organophosphorus compound-induced delayed neurotoxicity (Opidn). Adv. Neurotoxicol. Elsevier.
Richter, E.D., Chuwers, P., Levy, Y., Gordon, M., Grauer, F., Marzouk, J., Levy, S., Barron, S., Gruener, N., 1992. Health effects from exposure to organophosphate pesticides in workers and residents in Israel. Isr. J. Med. Sci. 28, 584–598.
Roberts, D.M., Aaron, C.K., 2007. Management of acute organophosphorus pesticide poisoning. BMJ 334, 629–634.
Roberts, J.R., Reigart, J.R., 2013. Recognition and Management of Pesticide Poisonings.
Rodnitzky, R.L., Levin, H.S., Mick, D.L., 1975. Occupational exposure to organophosphate pesticides: a neurobehavioral study. Arch. Environ. Health Int. J. 30, 98–103.
Rohlman, D.S., Lasarev, M., Anger, W.K., Scherer, J., Stupfel, J., McCauley, L., 2007. Neurobehavioral performance of adult and adolescent agricultural workers. Neurotoxicology 28, 374–380.
Rohlman, D.S., Anger, W.K., Lein, P.J., 2011. Correlating neurobehavioral performance with biomarkers of organophosphorous pesticide exposure. Neurotoxicology 32, 268–276.
Rohlman, D.S., Ismail, A.A., Abdel-Rasoul, G., Lasarev, M., Hendy, O., Olson, J.R., 2014. Characterizing exposures and neurobehavioral performance in Egyptian adolescent pesticide applicators. Metab. Brain Dis. 29, 845–855.
Rohlman, D.S., Ismail, A.A., Rasoul, G.A., Bonner, M.R., Hendy, O., Mara, K., Wang, K., Olson, J.R., 2016. A 10-month prospective study of organophosphorus pesticide exposure and neurobehavioral performance among adolescents in Egypt. Cortex 74, 383–395.
Rohlman, D.S., Ismail, A., Bonner, M.R., Rasoul, G.A., Hendy, O., Dickey, L.O., Wang, K., Olson, J.R., 2019. Occupational pesticide exposure and symptoms of attention deficit hyperactivity disorder in adolescent pesticide applicators in Egypt. Neurotoxicology 74, 1–6.
Rohlman, D.S., Davis, J.W., Ismail, A., Abdel Rasoul, G.M., Hendy, O., Olson, J.R., Bonner, M.R., 2020. Risk perception and behavior in Egyptian adolescent pesticide applicators: an intervention study. BMC Public Health 20, 1–10.
Roldan-Tapia, L., Parron, T., Sanchez-Santed, F., 2005. Neuropsychological effects of long-term exposure to organophosphate pesticides. Neurotoxicol. Teratol. 27, 259–266.

Roldan-Tapia, L., Nieto-Escamez, F.A., Del Aguila, E.M., Laynez, F., Parron, T., Sanchez-Santed, F., 2006. Neuropsychological sequelae from acute poisoning and long-term exposure to carbamate and organophosphate pesticides. Neurotoxicol. Teratol. 28, 694–703.

Rosentock, L., Keifer, M., Daniell, W., Macconnell, R., Claypoole, K., 1990. Chronic central nervous system effects of acute organophosphate pesticide intoxication. Lancet 338, 223–227.

Ross, M.K., Crow, J.A., 2007. Human carboxylesterases and their role in xenobiotic and endobiotic metabolism. J. Biochem. Mol. Toxicol. 21, 187–196.

Ross, S.M., McManus, I., Harrison, V., Mason, O., 2013. Neurobehavioral problems following low-level exposure to organophosphate pesticides: a systematic and meta-analytic review. Crit. Rev. Toxicol. 43, 21–44.

Sagiv, S.K., Harris, M.H., Gunier, R.B., Kogut, K.R., Harley, K.G., Deardorff, J., Bradman, A., Holland, N., Eskenazi, B., 2018. Prenatal organophosphate pesticide exposure and traits related to autism spectrum disorders in a population living in proximity to agriculture. Environ. Health Perspect. 126, 047012.

Sams, C., Cocker, J., Lennard, M.S., 2004. Biotransformation of chlorpyrifos and diazinon by human liver microsomes and recombinant human cytochrome P450s (Cyp). Xenobiotica 34, 861–873.

Sapbamrer, R., Hongsibsong, S., 2019. Effects of prenatal and postnatal exposure to organophosphate pesticides on child neurodevelopment in different age groups: a systematic review. Environ. Sci. Pollut. Res. 26, 18267–18290.

Savage, E.P., Keefe, T.J., Mounce, L.M., Heaton, R.K., Lewis, J.A., Burcar, P.J., 1988. Chronic neurological sequelae of acute organophosphate pesticide poisoning. Arch. Environ. Health Int. J. 43, 38–45.

Savary, S., Willocquet, L., Pethybridge, S.J., Esker, P., McRoberts, N., Nelson, A., 2019. The global burden of pathogens and pests on major food crops. Nat. Ecol. Evol. 3, 430–439.

Schnoor, J., 1992. Fate of Pesticides and Chemicals in the Environment. Wiley, New York.

Schütte, G., Eckerstorfer, M., Rastelli, V., Reichenbecher, W., Restrepo-Vassalli, S., Ruohonen-Lehto, M., Saucy, A.-G.W., Mertens, M., 2017. Herbicide resistance and biodiversity: agronomic and environmental aspects of genetically modified herbicide-resistant plants. Environ. Sci. Eur. 29, 1–12.

Selemon, L.D., 2013. A role for synaptic plasticity in the adolescent development of executive function. Transl. Psychiatry 3, e238.

Shafer, T.J., Meyer, D.A., Crofton, K.M., 2005. Developmental neurotoxicity of pyrethroid insecticides: critical review and future research needs. Environ. Health Perspect. 113, 123–136.

Sharma, A., Kumar, V., Shahzad, B., Tanveer, M., Sidhu, G., Handa, N., Kohli, S., Yadav, P., Bali, A., Parihar, R., 2019. Worldwide pesticide usage and its impacts on ecosystem. Sn Appl. Sci. 1, 1446.

Shrestha, S., Kamel, F., Umbach, D.M., Fan, Z., Freeman, L.E.B., Koutros, S., Alavanja, M., Blair, A., Sandler, D.P., Chen, H., 2018. Factors associated with dream enacting behaviors among Us farmers. Parkinsonism Relat. Disord. 57, 9–15.

Shrestha, S., Parks, C.G., Umbach, D.M., Richards-Barber, M., Hofmann, J.N., Chen, H., Blair, A., Beane Freeman, L.E., Sandler, D.P., 2020. Pesticide use and incident Parkinson's disease in a cohort of farmers and their spouses. Environ. Res. 191, 110186.

Singleton, S.T., Lein, P.J., Farahat, F.M., Farahat, T., Bonner, M.R., Knaak, J.B., Olson, J.R., 2014. Characterization of alpha-cypermethrin exposure in Egyptian agricultural workers. Int. J. Hyg. Environ. Health 217, 538–545.

Singleton, S.T., Lein, P.J., Dadson, O.A., McGarrigle, B.P., Farahat, F.M., Farahat, T., Bonner, M.R., Fenske, R.A., Galvin, K., Lasarev, M.R., Anger, W.K., Rohlman, D.S., Olson, J.R., 2015. Longitudinal assessment of occupational exposures to the organophosphorous insecticides chlorpyrifos and profenofos in Egyptian cotton field workers. Int. J. Hyg. Environ. Health 218, 203–211.

Slotkin, T.A., 2004. Guidelines for developmental neurotoxicity and their impact on organophosphate pesticides: a personal view from an academic perspective. Neurotoxicology 25, 631–640.

Soderlund, D.M., 2012. Molecular mechanisms of pyrethroid insecticide neurotoxicity: recent advances. Arch. Toxicol. 86, 165–181.

Soderlund, D.M., Clark, J.M., Sheets, L.P., Mullin, L.S., Piccirillo, V.J., Sargent, D., Stevens, J.T., Weiner, M.L., 2002. Mechanisms of pyrethroid neurotoxicity: implications for cumulative risk assessment. Toxicology 171, 3–59.

Spear, L.P., 2010. The Behavioral Neuroscience of Adolesence. W.W. Norton & Company, New York.

Steenland, K., Jenkins, B., Ames, R.G., O'malley, M., Chrislip, D., Russo, J., 1994. Chronic neurological sequelae to organophosphate pesticide poisoning. Am. J. Public Health 84, 731–736.

Steenland, K., Dick, R.B., Howell, R.J., Chrislip, D.W., Hines, C.J., Reid, T.M., Lehman, E., Laber, P., Krieg Jr., E.F., Knott, C., 2000. Neurologic function among termiticide applicators exposed to chlorpyrifos. Environ. Health Perspect. 108, 293–300.

Steinberg, L., 2008. A social neuroscience perspective on adolescent risk-taking. Dev. Rev. 28, 78–106.

Taira, K., Aoyama, Y., Kawakami, T., Kamata, M., Aoi, T., 2011. Detection of chloropyridinyl neonicotinoid insecticide metabolite 6-chloronicotinic acid in the urine: six cases with subacute nicotinic symptoms. Chudoku Kenkyu 24, 222–230.

Takahashi, N., Hashizume, M., 2014. A systematic review of the influence of occupational organophosphate pesticides exposure on neurological impairment. BMJ Open 4, e004798.

Tangamornsuksan, W., Lohitnavy, O., Sruamsiri, R., Chaiyakunapruk, N., Norman Scholfield, C., Reisfeld, B., Lohitnavy, M., 2019. Paraquat exposure and Parkinson's disease: a systematic review and meta-analysis. Arch. Environ. Occup. Health 74, 225–238.

Thompson, D.A., Lehmler, H.-J., Kolpin, D.W., Hladik, M.L., Vargo, J.D., Schilling, K.E., Lefevre, G.H., Peeples, T.L., Poch, M.C., Laduca, L.E., 2020. A critical review on the potential impacts of neonicotinoid insecticide use: current knowledge of environmental fate, toxicity, and implications for human health. Environ. Sci. Processes Impacts 22, 1315–1346.

Trentini, A., Bellini, T., Bonaccorsi, G., Cavicchio, C., Hanau, S., Passaro, A., Cervellati, C., 2019. Sex difference: an important issue to consider in epidemiological and clinical studies dealing with serum paraoxonase-1. J. Clin. Biochem. Nutr. 64, 250–256.

Tsai, Y.-H., Lein, P.J., 2021. Mechanisms of Organophosphate Neurotoxicity. Curr. Opin. Toxicol.

U.S. Environmental Protection Agency, 1990. Suspended, Cancelled, and Restricted Pesticides. Office of Pesticides and Toxic Substances, Washington, DC.

U.S. Environmental Protection Agency, 2021a. Assessing Human Health Risk from Pesticides. [Online]. Available: https://www.epa.gov/pesticide-science-and-assessing-pesticide-risks/assessing-human-health-risk-pesticides (Accessed 21 September 2021).

U.S. Environmental Protection Agency, 2021b. Basic Information About Pesticide Ingredients [Online]. Available: https://www.epa.gov/ingredients-used-pesticide-products/basic-information-about-pesticide-ingredients [Accessed September 21, 2021].

U.S. Environmental Protection Agency, 2021c. Chlorpyrifos [Online]. Available: https://www.epa.gov/ingredients-used-pesticide-products/chlorpyrifos [Accessed September 21, 2021].

U.S. Environmental Protection Agency, 2021d. DDT—A Brief History and Status [Online]. Available: https://www.epa.gov/ingredients-used-pesticide-products/ddt-brief-history-and-status [Accessed September 21, 2021].

U.S. Environmental Protection Agency, 2021e. Integrated Pest Management (IPM) Principles [Online]. Available: https://www.epa.gov/safepestcontrol/integrated-pest-management-ipm-principles [Accessed September 21, 2021].

U.S. Environmental Protection Agency, 2021f. Introduction to Pesticide Labels [Online]. Available: https://www.epa.gov/pesticide-labels/introduction-pesticide-labels [Accessed September 21, 2021].

U.S. Environmental Protection Agency, 2021g. Paraquat Dichloride: One Sip Can Kill [Online]. Available: https://www.epa.gov/pesticide-worker-safety/paraquat-dichloride-one-sip-can-kill#poisoning [Accessed September 21, 2021].

U.S. Environmental Protection Agency, 2021h. Registration Review Process.

U.S. Environmental Protection Agency, 2021i. What is a Pesticide? [Online]. Available: https://www.epa.gov/minimum-risk-pesticides/what-pesticide [Accessed September 21, 2021].

Vaccari, C., El Dib, R., Gomaa, H., Lopes, L.C., De Camargo, J.L., 2019. Paraquat and Parkinson's disease: a systematic review and meta-analysis of observational studies. J. Toxicol. Environ. Health Part B 22, 172–202.

Van Steenwyk, R., 2005. Food Quality Protection Act launches search for pest management alternatives. Calif. Agric. 59, 7–10.

Vijverberg, H.P., Van Den Bercken, J., 1990. Neurotoxicological effects and the mode of action of pyrethroid insecticides. Crit. Rev. Toxicol. 21, 105–126.

Voorhees, J.R., Rohlman, D.S., Lein, P.J., Pieper, A.A., 2017. Neurotoxicity in preclinical models of occupational exposure to organophosphorus compounds. Front. Neurosci. 10, 590.

Wilks, M.F., 2000. Pyrethroid-induced paresthesia—a central or local toxic effect? J. Toxicol. Clin. Toxicol. 38, 103–105.

Williams, M.K., Rundle, A., Holmes, D., Reyes, M., Hoepner, L.A., Barr, D.B., Camann, D.E., Perera, F.P., Whyatt, R.M., 2008. Changes in pest infestation levels, self-reported pesticide use, and permethrin exposure during pregnancy after the 2000–2001 US Environmental Protection Agency restriction of organophosphates. Environ. Health Perspect. 116, 1681–1688.

Wing, K.D., Glickman, A.H., Casida, J.E., 1983. Oxidative bioactivation of S-alkyl phosphorothiolate pesticides: stereospecificity of profenofos insecticide activation. Science 219, 63–65.

World Health Organization, 2009. Pesticide Residues in Food: 2007, Toxicological Evaluations. Sponsored Jointly by Food and Agricultural Organization of the United Nations (FAO) and WHO, With the Support of the International Programme on Chemical Safety/Joint Meeting of the FAO Panel of Experts on Pesticide Residues in Food and the Environment and WHO Core Assessment Group. 18–27 September 2007 Geneva, Switzerland.

World Health Organization, 2019. Preventing Suicide: A Resource for Pesticide Registrars and Regulators.

World Health Organization, 2020. The WHO Recommended Classification of Pesticides by Hazard and Guidelines to Classification 2019. World Health Organization.

Worldometer, 2021. Pesticide Use by Country [Online]. Available: https://www.worldometers.info/food-agriculture/pesticides-by-country/ [Accessed].
Yan, D., Zhang, Y., Liu, L., Shi, N., Yan, H., 2018. Pesticide exposure and risk of Parkinson's disease: Dose-response meta-analysis of observational studies. Regul. Toxicol. Pharmacol. 96, 57–63.
Zeid, D., Kutlu, M.G., Gould, T.J., 2018. Differential effects of nicotine exposure on the hippocampus across lifespan. Curr. Neuropharmacol. 16, 388–402.

CHAPTER SEVEN

Diagnosis and management of occupational and environmental carbon monoxide neurotoxicity

Jonathan Rutchik[a,b,*,†] **and Marcia H. Ratner**[c,†]
[a]Division of Occupational & Environmental Medicine, University of California San Francisco, San Francisco, CA, United States
[b]Neurology, Environmental and Occupational Medicine Associates, Mill Valley, CA, United States
[c]Department of Pharmacology and Experimental Therapeutics, Boston University School of Medicine, Boston, MA, United States
*Corresponding author: e-mail address: jsrutch@neoma.com

Contents

1. Introduction

Carbon monoxide (CO) is a naturally occurring ubiquitous odorless non-irritating gas. It is a common occupational and environmental hazard

† Dr. Rutchik and Dr. Ratner contributed equally to this work.

Advances in Neurotoxicology, Volume 7
ISSN 2468-7480
https://doi.org/10.1016/bs.ant.2022.04.002

formed by the incomplete combustion of hydrocarbons including gasoline and propane. Despite significant increased regulation and improvements in industrial hygiene practices CO poisoning continues to be a significant cause of morbidity and mortality in the workplace (Mattiuzzi and Lippi, 2020; Downs et al., 2021). The percentage of patients who die from CO poisoning increases in parallel with aging (Mattiuzzi and Lippi, 2020).

Patients are referred to an occupational and environmental medicine (OEM) provider when exposure to CO is suspected as a cause of their symptoms. Because exposure to CO is often associated with neurological sequelae, OEM providers need to be familiar with the diagnosis and treatment of neurological effects of CO poisoning. This chapter serves as a guide to the effective diagnosis and management of patients who present to OEM providers with neurological signs and symptoms attributed to CO exposures. The chapter begins by reviewing common occupational and environmental sources of CO exposures and the industrial hygiene practices and exposure regulations designed to reduce exposure. The neurotoxic mechanisms of action implicated in CO poisoning are reviewed in terms of how these relate to clinical manifestations. Best practices for reducing risk for CO exposure and the approach to documenting CO exposures when these do occur are reviewed. The diagnosis and effective management of neurological sequelae of CO poisoning as well as a discussion of disability along with the prognostic value of carboxyhemoglobin and findings from neuroimaging studies and neuropsychological assessments are explored in detail.

2. Occupational and environmental sources of CO exposure

The effective diagnosis and management of patients with neurological symptoms suspected of being related to CO exposure depends on recognizing the circumstances associated with an increased risk for exposure to this neurotoxic gas. Although the sources of exposure to CO implicated in neurotoxicity are well-recognized today, this was not always the case. The connection between coal mining and risk for neurotoxic effects of occupational exposure to CO was first recognized in the late nineteenth century by John Scott Haldane (Haldane, 1895). Haldane subsequently performed experiments on himself which allowed him to experience firsthand the symptoms associated with occupational type exposures to CO at various concentrations and durations during different types of behaviors

(Haldane, 1895). Haldane noted that his symptoms of exposure were "brought on or much aggravated by exertion" and, that even moderate exertion such as occurs in many occupational settings gives rise to "a loss or impairment of the senses." He also observed that the time required for ambient air containing CO to produce symptoms of poisoning depends on the time required for the blood to reach a certain degree of saturation with a much longer duration of CO exposure being required at lower concentrations than at higher concentrations. Haldane noted the onset of "distinctly appreciable symptoms at rest occurring when blood saturation reaches about a third" (Haldane, 1895).

The early research of Haldane calls our attention to the fact that work-related activities requiring physical exertion, such as manipulating heavy equipment, increase the risk for adverse effects of CO exposure (Hauck and Neuberger, 1984; Rutchik et al., 2021). The risk for CO poisoning is also increased by using small gasoline powered tools such as pressure washers and concrete saws in poorly ventilated areas. The risk for adverse effects of CO exposure is also increased by not using an appropriate respirator (Rutchik et al., 2021).

Despite improvements in industrial hygiene policies and practices since Haldane did his early experiments, miners continue to be at risk for CO poisoning (Markey and Zumwalt, 2001; Zhou et al., 2018). Other occupations associated with a well-recognized risk for CO exposure include automotive mechanic, Zamboni driver, welder, concrete saw operator, firefighter and truck driver (Antonsson et al., 2013; Banjoko et al., 2007; Cox et al., 2019; Rutchik et al., 2021; Semmens et al., 2021; Weaver et al., 2019). In addition to the risk for occupational CO poisoning resulting from running motor vehicles and power equipment inside inadequately ventilated locations, occupational exposure to CO can also occur during the operation of motor vehicles with faulty exhaust systems (Fisher-Hubbard et al., 2018; Jovanović et al., 1999). Contaminated air supplies are also a source of CO exposure that can result in toxicity (Holt and Weaver, 2012). With more employees working from home, this once "non-occupational" source of exposure has become an occupational source as well.

While many occupational situations create an obvious risk for exposure to CO, others are less obvious. For example, the home office where many employees now work remotely for extended periods of time has become a non-traditional source of occupational as well as environmental exposures to CO. Common sources of CO exposure in the home include water heaters, boilers, and heaters (Annane et al., 2001). The risk for occupational CO poisoning associated with working from home is greater during the winter months and during power outages when portable gasoline-power generators are more likely to be used in poorly ventilated areas such as a garage attached to the house (CDC, 2006; Worsham et al., 2022). Therefore, it is prudent for employees who work remotely to install a CO detector in their home workspace (Johnson-Arbor et al., 2012; Yoon et al., 1998).

The Occupational Safety and Health Administration (OSHA) permissible exposure limit for CO as an 8-hour time-weighted average is 50 parts per million (ppm) in ambient air (Table 1). According to OSHA maritime workers must be removed from the source of CO exposure when ambient air concentration exceeds 100 ppm. The OSHA peak ambient air CO concentration for workers engaged in loading and unloading of cargo is 200 ppm. The Occupational Safety and Health Act of 1970 applies to work performed in an employee's home (US Dept of Labor, 2000). However, it should be noted that OSHA does not conduct inspections of employees' home offices and OSHA will not hold employers liable for employees' home offices.

Methylene chloride which is metabolized to CO is also well-recognized as a source of CO exposure and therefore, the risk for CO poisoning occurs in any industrial or environmental situation where methylene chloride is present (Rodkey, 1977). Inhalation of methylene chloride vapor is associated with an increase in percent COHb (Stewart et al., 1972). Exposure to methylene chloride vapor concentrations of 1000 ppm is associated with signs and symptoms of central nervous system (CNS) depression (Stewart et al., 1972). The effects associated with simultaneous exposure to this common paint stripper and CO derived from other sources are additive and thereby create a potential occupational hazard that employers and/or employees may not recognize (Banjoko et al., 2007). Fatal occupational and non-occupational exposures to methylene chloride have been reported among furniture and bathtub refinishers (CDC, 2012; Hoang et al., 2021;

Table 1 Occupational and environmental exposure limits for carbon monoxide.

	OSHA (ppm)	NIOSH (ppm)	ACGIH (ppm)
PEL/REL	50	35	25
STEL		200	
Ceiling			400
IDLH		1200	

Raphael et al., 2002; Rosenman, 2021; Shusterman et al., 1990; Stewart and Hake, 1976). Since 2019, retailers in the United States have been prohibited from selling products that contained methylene chloride to consumers, but commercial use by independent contractors and employees is still allowed (Rosenman, 2021). The OSHA PEL for methylene chloride is 25 ppm as an eight-hour time-weighted average with a short -term exposure limit of 125 ppm.

The risks for CO poisoning due to exposure to either CO or methylene chloride is reduced by the use of an appropriate OSHA approved respirator or self-contained breathing apparatus with an oxygen supply. It is important to determine if the exposed worker was wearing a respirator and if so, if the respirator was the appropriate type for the specific exposure circumstances (Rutchik et al., 2021). In addition to wearing a respirator, exposure to hazardous concentrations of CO is reduced by the use of exposure monitors (Lin et al., 2018; Thomas et al., 2018; Yoon et al., 1998).

3. Mechanisms of CO neurotoxicity

The effects of CO on biological functions follow an inverted U shaped curve. The affinity of CO for hemoglobin is 250 times greater than that of oxygen. Because CO binds to hemoglobin with a greater affinity than oxygen, the percent carboxyhemoglobin (COHb) increases with increased respiration rate brought on by physical exertion. As the percent COHb in blood increases, the body attempts to compensate for a lower oxygen level by increasing the respiration rate further (Ilano and Raffin, 1990). The high oxygen demand of the brain makes it particularly vulnerable to the neurotoxic effects of CO. A common misconception is that the neurotoxicity of CO is entirely dependent on binding of CO to hemoglobin and, that the percent COHb accurately predicts severity of neurotoxicity. In fact, the mechanisms of neurotoxicity implicated in CO poisoning include tissue hypoxia, arterial hypotension, disruption of mitochondrial respiration mediated by inhibition of cytochrome c oxidase, formation of free radicals and inflammation (Alonso et al., 2003). It is important to recognize that CO inhibits cellular respiration by binding to cytochrome *C* oxidase (Alonso et al., 2003; Miró et al., 1998). Furthermore, studies indicate that oxidative stress and increased lipid peroxidation continue to play a role in the loss of functional mitochondria and depletion of glutathione stores in non-fatal CO poisoning after the percent COHb in blood has normalized (Savolainen et al., 1980). These additional mechanisms of action account for why the signs and symptoms associated with non-fatal CO poisoning cannot all be explained by percent COHb (Hampson and Hauff, 2008). While there is currently no reliable method to predict the clinical prognosis following CO poisoning, determination of mitochondrial respiration using intact cells obtained from patients exposed to CO has been proposed as a potential biomarker (Jang et al., 2017).

Non-fatal CO poisoning is often associated with residual parkinsonism and encephalopathy. Displacement of oxygen from hemoglobin by CO leads to tissue hypoxia and disruption of adenosine triphosphate (ATP) dependent processes including reuptake of excitatory amino acids such as glutamate. In the absence of ATP, the excitatory amino acid transporter (EAAT) ceases to function. The initial clinical manifestations of CO poisoning are attributable to hypoxia and primarily mediated by reversible disruption of ATP-dependent neuronal functions. At autopsy the brain shows evidence of necrosis and apoptosis following CO poisoning (Uemura et al., 2001, 2003). The neurological symptoms which occur in close chronological relationship to CO exposure are primarily due to necrosis of neurons while those which emerge over the course of the ensuing 72 h or more following a period of "*pseudo recovery*" are attributable to a combination of necrosis mediated by reactive oxygen species, apoptosis and demyelination (Piantadosi et al., 1997; Sohn et al., 2000; Uemura et al., 2003). Disruption of EAAT function allows glutamate released from presynaptic neurons to accumulate in the extracellular space where it activates post-synaptic glutamate receptors and promotes calcium influx. Unchecked calcium influx eventually leads to calcium overload and induction of cell death by necrosis. The unchecked influx of calcium also induces the apoptotic cascade in neurons that survive the initial insult (Ankarcrona et al., 1995). The subsequent death of these neurons which occurs over the course of the ensuing 72 h produces the delayed neurological sequelae commonly associated with CO poisoning. CO poisoning in a preclinical rat model has also been associated with adduct formation between myelin basic protein and malonylaldehyde resulting in an immunological cascade. Although the immunological responses seen in CO poisoning differs from those seen in experimental allergic encephalomyelitis which is commonly used as a model of MS, the lymphocytes from CO-poisoned rats do nevertheless exhibit an auto-reactive proliferative response to myelin basic protein which has been associated with memory deficits in rats (Thom et al., 2004).

4. Documenting CO exposure

Whenever possible, the actual ambient air and breathing zone concentrations of CO or methylene chloride should be determined via onsite air sampling by a certified industrial hygienist. When direct measurement of exposure levels is not possible, a certified industrial hygienist should be consulted to model the ambient air and breathing zone CO concentrations which can then be related to observed and reported symptoms of exposure (Rodricks, 2011). When direct onsite sampling is not possible company records, gas company reports, and/or insurance claims should be reviewed to obtain additional documentation of the exposure circumstances (Devine et al., 2002). Exposure modelling serves as a surrogate to air sampling and should be used to estimate the CO concentrations whenever it is not possible to document the exposure by other methods. For occupational exposures these data should be interpreted in relation to the OSHA permissible exposure limits. A Time-Exposure Symptoms Line is a useful tool for documenting the onset of symptoms in relationship to CO exposure (Feldman et al., 1999; Ratner et al., 2020).

The percent COHb in blood should be determined in all suspected cases of occupational and environmental CO poisoning even if the exposed person reports experiencing only mild symptoms such as headache and dizziness. It is not uncommon for a substantial period of time to have elapsed between removal of the patient from the source of CO exposure and measurement of his or her percent COHb (Devine et al., 2002). Because of this delay, the percent COHb may not correlate well with observed signs and symptoms of CO exposure (Hampson and Hauff, 2008). When the elapsed duration of time is known, the patient's percent COHb at the time of exposure can be calculated from the half-life of COHb which is 74 ± 25 min in patients breathing 100% oxygen at atmospheric pressure (Rutchik et al., 2021; Weaver et al., 2000). The half-life of COHb is not influenced by gender or age (Weaver et al., 2000). The air concentration of CO in the patient's breathing zone at the time of exposure is modeled mathematically using the equation $[\text{CO ppm}] = \%\ \text{COHb}/KT$ where K is a constant equal to 0.048 during light work and T equals time in hours (Beckett, 2007; Rutchik et al., 2021) (Table 2).

Table 2 Relationship between percent COHb, time and ambient air CO concentrations.

	Ambient air carbon monoxide concentration (ppm)				
COHb[a]	0.5 (h)	1 (h)	2 (h)	3 (h)	4 (h)
0	0	0	0	0	0
10	416	208	104	69	52
20	833	417	208	139	104
30	1250	625	313	208	156
40	1666	833	417	278	208
50	2083	1042	521	347	260
60	2500	1250	625	417	313
70	2916	1458	729	486	365
80	3333	1667	833	556	417
90	3750	1875	938	625	469
100	4166	2083	1042	694	521

[a]Percent COHb in adult non-smoker.

5. Diagnosis and management of CO poisoning

5.1 Diagnosis and management of acute CO poisoning

OEM providers often are called upon to diagnose and manage patients who survive acute occupational CO poisoning. Because exposed persons who are not removed from the source of CO are at risk for loss of consciousness and death, persons reporting headache, fatigue, nausea or dizziness should immediately be removed from the source of exposure and their exposure circumstances documented (Antonsson et al., 2013; Neufeld et al., 1981; Sohn et al., 2000). Removal of non-ambulatory patients from the source of CO exposure should be carried out as quickly as possible without jeopardizing the safety of other workers, family members, or first responders. Persons who have been unconscious for extended periods due to acute CO poisoning are also at risk for rhabdomyolysis and acute renal failure with associated uremic encephalopathy (Jha et al., 1994; Kim et al., 2019). Compression type peripheral neuropathy has also been associated with sustained periods of unconsciousness due to acute CO poisoning. (Rahmani et al., 2013).

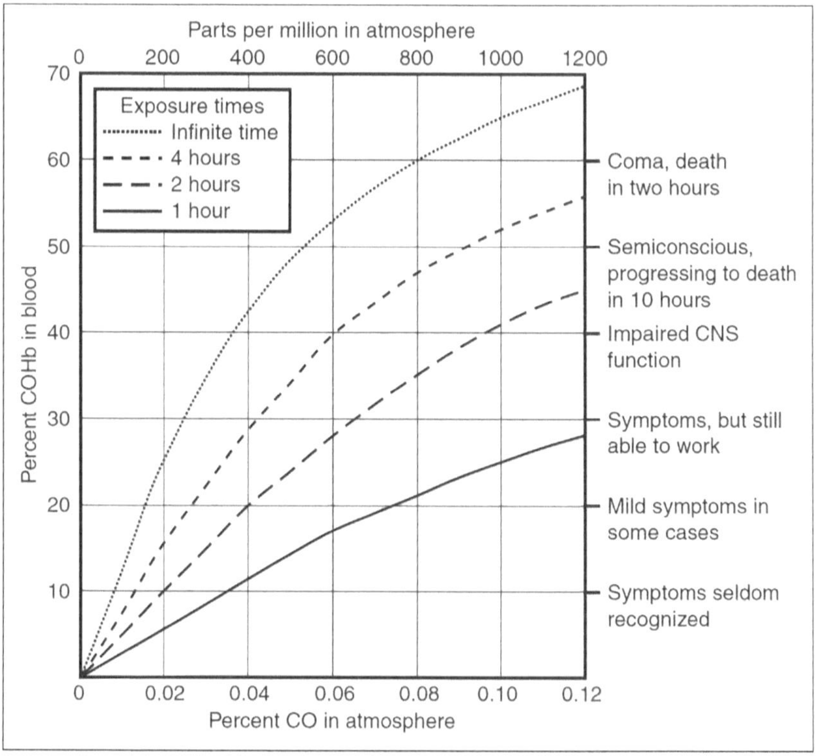

Fig. 1 Percent COHb in blood vs atmospheric CO concentrations and corresponding symptoms observed at 1 h, 2 h, 4 h, and infinite time. These are the expected symptoms observed when the percentage of COHb in blood is rapidly raised and maintained at that level in healthy adult non-smokers. *With permission from Forbes, W.H., 1970. Carbon monoxide uptake via the lungs. Ann. N. Y. Acad. Sci. 174 (1), 72–75.*

Percent COHb remains the gold standard for diagnosis of CO poisoning (Fig. 1). A blood sample to document the percent COHb should be obtained as soon as possible after removal of the patient from the source of CO exposure. Mortality risk is greatest when COHb is 40% or higher. However, a person with a 25% COHb level can also succumb to CO poisoning (Nelson, 1998; National Research Council (US) Committee on Acute Exposure Guideline Levels, 2010). The relationship between oxygen administration and the timing of percent COHb determination should be taken into consideration when interpreting this laboratory result (Ratner and Jabre, 2006; Rutchik et al., 2021). Cardiovascular disease increases the risk of an adverse outcome of CO exposure (Nelson, 1998). Arterial Blood gases are not significantly influenced by saturation of hemoglobin with CO and thus, are considered to be of limited clinical value in the diagnosis of CO poisoning (Ilano and Raffin, 1990). On the contrary, because metabolic derangement is associated with CO poisoning serum pH, electrolytes, lactate levels as well as the anion and osmol gaps should be determined and used to guide clinical management (Kim et al., 2018; Ratner and Jabre, 2006; Simonsen et al., 2021).

High-flow 100% oxygen should be administered via a non-rebreather mask to the CO poisoned patient until their COHb is <5% (Buboltz and Robins, 2021; Weaver, 2009). Unconscious patients should be intubated and placed on 100% oxygen. Hyperbaric oxygen therapy should only be considered as a treatment option for unconscious patients with a COHb >25% (Weaver, 2009; Weaver et al., 2002). It is important to recognize that hyperbaric oxygen treatment also has risks which include hyperoxic seizures and barotrauma (Weaver et al., 2002). Although hyperbaric oxygen therapy does not appear to decrease risk for mortality, treated patients do appear to have a lower risk for cognitive problems and impaired ability to perform activities of daily living on the Barthel index (i.e., feeding, grooming and dressing, bowel and bladder control, transfers and stair climbing) (Mahoney and Barthel, 1965; Nakajima et al., 2020; Weaver et al., 2002). Some patients with chronic symptoms that an OEM provider has attributed to CO poisoning may seek treatment with hyperbaric oxygen based on anecdotal evidence of a therapeutic benefit (Pappalardo et al., 2019). At this time further studies are needed to provide evidence-based guidance on the risks and benefits of hyperbaric oxygen for the treatment of residual sequelae of CO poisoning (Pappalardo et al., 2019). The Food and Drug Administration (FDA) has however recently approved the use of ClearMate, a medical device which works employs passive isocapnic hyperventilation to enhance the elimination CO by delivering a mixture of oxygen and carbon dioxide to the blood which causes the patient to breathe faster (FDA, 2019). Clinical studies indicated that patients treated with this device do not experience any device-related complications.

Seizures associated with acute CO poisoning are typically a self-limiting acute clinical manifestation of poisoning (Durnin, 1987; Neufeld et al., 1981). Seizure should be managed symptomatically to prevent status epilepticus (Abdulaziz et al., 2012). The initial electroencephalogram (EEG) of the CO poisoned patient may reveal epileptiform activity. This abnormal neural activity typically resolves with time and recovery of function (Neufeld et al., 1981). Seizures may also occur as a consequence of hyperbaric oxygen therapy (Heyboer et al., 2014; Lee et al., 2021;

Marziali et al., 2016). Although hyperbaric-oxygen therapy can cause hyperoxic seizures, treatment of patients with acute, symptomatic CO poisoning with hyperbaric-oxygen may reduce the risk for cognitive sequelae (Weaver et al., 2002).

5.1.1 Neurobehavioral assessments for acute cognitive deficits due to CO poisoning

The results of an initial bedside neurological and cognitive function assessments are of great value to understanding both the severity of the acute neurological and neurobehavioral changes induced by CO. These results of these bedside assessments can be related to residual neurological and behavioral sequelae which have been shown to correlate in severity with the duration of unconsciousness due to CO poisoning (Parkinson et al., 2002). The acute changes in cognitive function of patients presenting with acute CO poisoning is effectively assessed using objective measures of cognitive function such as the Carbon Monoxide Neuropsychological Screening Battery (CONSB) or the Mini Mental Status Examination (Amitai et al., 1998; Kao and Nañagas, 2006; Ku et al., 2010; Messier and Myers, 1991). As always, neuropsychological testing must include validity testing of the patient's effort and engagement (Greher and Wodushek, 2017).

5.2 Diagnosis and management of chronic CO poisoning

A small portion of patients may present with signs and symptoms that are not readily attributable to CO poisoning especially during winter months. Symptoms seen in these patients are often nonspecific and flu-like emerged over days or weeks during chronic exposure to low concentrations of CO (Heckerling, 1987; Heckerling et al., 1990). The diagnosis of these patients is complicated further by their percent COHb which may not be remarkably elevated on admission (Heckerling, 1987; Heckerling et al., 1990). Patients admitted to the emergency department or other urgent care settings with flu-like symptoms should be questioned about working in poorly ventilated spaces and whether exposure to CO may have occurred. If there are any lingering doubts or concerns about their risk for CO exposure determination of their percent COHb is warranted especially during the winter months (Fig. 1).

5.3 Diagnosis and management of residual neurological sequelae of CO poisoning

Although CO poisoning is often fatal, patients who survive the acute reversible toxic effects of CO poisoning often have residual neurological complications which compel them to seek medical attention (Huang et al., 2019). Non-fatal CO poisoning is often associated with parkinsonism and toxic encephalopathy which emerge days, weeks or even months after CO exposure (Borras et al., 2009; Choi, 1983; Gordon, 1965; Hopkins and Woon, 2006; Kesler et al., 2001; Rutchik et al., 2021; Sohn et al., 2000). Frequently observed symptoms and signs of neurological dysfunction include cognitive deficits, gait disturbances, mutism, masked face, tremor, increased muscle tone, and retropulsion (Choi, 1983; Rutchik et al., 2021). The delayed neurological sequelae of CO poisoning are more likely to be observed in older adults (Choi, 2001). The diagnosis of neurological symptoms in patients with a well-documented history of CO exposure is typically made based on the history, clinical manifestations, laboratory data, and neuroimaging findings. Even mild CO poisoning which does not result in a loss of consciousness can be associated with residual neurological symptoms. Annane et al. (2001) noted that patients with a history of non-fire-related, domestic, CO poisoning who presented with dizziness before hospital admission or headaches upon admission have an increased risk of minor persistent neurological symptoms. Patients who experience acute difficulty concentrating or confusion but no loss of consciousness in association with exposure to CO are at greater risk for residual neurological sequelae (Ely et al., 1995). Many but not all of these patients show recovery of function to the point where they are eventually able to resume their former occupations (Ely et al., 1995). Positive histories of prior head injury, exposures to other neurotoxic chemicals, advanced age, and medical comorbidities such as hypertension and vascular disease as well as neurological disorders that involve myelin and the extrapyramidal system such as MS and PD are anticipated to be associated with a poorer prognosis (Huang et al., 2014; Olson and Smollin, 2008; Weaver et al., 2007).

Although symptoms such as tremor associated with bilateral pallidal lesions are frequently reported as residual effects of CO poisoning, bilateral demyelination of the cerebral white matter has been shown to also play a major role in delayed neurological sequelae (Nah et al., 2020; Sohn et al., 2000). The periventricular white matter and centrum semiovale are particularly vulnerable to CO poisoning induced demyelination (Fujiwara et al., 2012; Guo et al., 2014; Parkinson et al., 2002). Although the association between CO poisoning and exacerbation of multiple sclerosis (MS) has not been elucidated, because CO poisoning is associated with a delayed post hypoxic leukoencephalopathy and with a neuroinflammatory response the adverse effects of CO can interact with neuropathological process implicated in MS (Thom et al., 2004).

Factors associated with COP	Item	Factors coincidental, contradictory, or c/w multiple DXS
Symptom onset post exposure, LOC, cardiac abnormality	History	Preexisting symptoms
Best estimate consistent with exposure	Risk assessment/ industrial hygiene	None or inconsistent documentation of exposure
COHb elevated; history of smoking, hemolytic anemia	Laboratory testing	No COHb testing
Mental status, movement disorder, exam findings, cardiac abnormality	Examination	No exam findings
Cognitive complaints, cognitive deficits	Neuropsychological testing	Invalid, inconsistent or c/w alternative diagnosis
C/W COP: white matter, ± pallidal lesions, ± hippocampal atrophy	Neuroimaging	Normal or abnormalities c/w alternative diagnosis
Stability, nonprogression or improvement	Progression	Progression

Abbreviations: COP, carbon monoxide poisoning; LOC, loss of consciousness; Dx, diagnosis.

The clinical presentation of parkinsonism associated with CO poisoning is complex and variable. Symptoms can emerge immediately due to cell death via necrosis. Parkinsonian symptoms may not be present on initial neurological examination but may emerge after a period of pseudo recovery. Tremor is not always seen in subjects with CO-induced parkinsonism and, when present it does not always correlate well with other parkinsonian symptoms (Sohn et al., 2000; Sun et al., 2018). Resting and intention type tremors are the type of tremor most frequently associated with CO poisoning (Choi, 2002; Rutchik et al., 2021). Holmes type tremor which includes resting as well as postural and intentional components has also been reported (Rutchik et al., 2021). When tremor is present in patients with a history of CO poisoning neurodegenerative parkinsonism needs to be included in the differential diagnosis. The lesions due to CO poisoning involve damage to the extrapyramidal system which can exacerbate the symptoms of PD. An increased risk for Parkinson's disease (PD) has also been reported among patients with a history of CO poisoning (Lai et al., 2015).

Damage to the cerebellum although rare has also been associated with CO poisoning (Fan et al., 2009; Hamcan et al., 2013; Mascalchi et al., 1996; Xu et al., 2020). When present the tremor associated with cerebellar dysfunction is typically of the wing-beating type. Cerebellar tremor may be accompanied by other cerebellar signs and symptoms including nystagmus, problems with balance, gait ataxia, and titubation (Holt and Weaver, 2012; Savoldi et al., 1973; Yang and Jeon, 2011).

5.3.1 Neurobehavioral assessments for residual cognitive deficits due to CO poisoning

Formal neuropsychological assessment should be performed to confirm the patient's complaints of cognitive deficits due to CO poisoning. The functional domains assessed should include attention and executive function, memory, psychomotor function, visuospatial ability, language and verbal skills, and affect and personality. In addition to detecting performance deficits due to CO poisoning neuropsychological testing is also sensitive to performance deficits due to developmental disorders and neurodegenerative diseases such as Alzheimer's which are associated with specific cognitive deficits such as verbal intelligence quotient (IQ) lower than performance IQ which is often seen in patients with developmental disabilities and anomia which is a common finding in patients with Alzheimer's disease. Therefore, all cognitive findings observed both at the bedside and at subsequent examinations should be interpreted in the context of the developmental, academic, occupational, medical and psychiatric histories of the patient (Devine et al., 2002; Ratner and Jabre, 2017). Because patients with CO poisoning present with a variety of cognitive deficits across multiple cognitive domains a battery of neuropsychological tests sensitive for detecting cognitive changes associated with toxic encephalopathy is recommended (Devine et al., 2002; Ratner and Jabre, 2017; Ratner et al., 2002; Watt et al., 2018). Dynamic cognitive process such as attention and executive function, learning and memory function, psychomotor function, and visuospatial information processing are particularly vulnerable to CO poisoning. Cognitive deficits reported to be associated with CO poisoning include memory impairments, deficits of attention and executive function, and slowing of processing speed (Hopkins and Woon, 2006; Pages et al., 2014; Parkinson et al., 2002; Porter et al., 2002; Sohn et al., 2000; Tamura et al., 2021; Watt et al., 2018). Tests of attention and task switching such as Trails A and B have been shown to be sensitive to cognitive deficits due to CO poisoning (Amitai et al., 1998; Devine et al., 2002;

Parkinson et al., 2002). The severity and nature of the observed cognitive deficits is often but not always reflected in the findings on neuroimaging studies (see Section 6). For example, hyperintensities in the centrum semi-ovale have been associated with more severe cognitive impairments than are lesions of the periventricular white matter (Parkinson et al., 2002). Hippocampal volume lose due to CO poisoning is reflected in performance deficits on tests of memory function (Tamura et al., 2021). Tests of visuo-spatial and visuomotor function may also reveal performance deficits (Amitai et al., 1998). Cognitive deficits may be present even in the apparent absence of neuroimaging findings (Devine et al., 2002; Kesler et al., 2001).

Follow-up testing is useful for directing treatment and vocational retraining if necessary (Ratner and Jabre, 2017; Ratner et al., 2002). Studies looking at the results of serial neuropsychological testing of CO poisoned- patients indicate that recovery of cognitive function occurs in some but not all patients (Sohn et al., 2000; Watt et al., 2018). Other studies suggest that CO poisoning is a risk factor for dementia (Wong et al., 2016). Because there is overlap between the performance deficits due to CO poisoning and those associated with developmental disorders, psychiatric conditions and age-related neurodegenerative diseases positive histories for these disorders should be considered in the differential diagnosis as well as in the development of a treatment plan.

Because CO exposure can be life threatening it is not uncommon for persons at risk for exposure to CO to also experience acute stress disorder and to even develop symptoms of PTSD following a documented exposure (Pages et al., 2014). It is important to rule out functional neurological disorders in persons at risk for exposure to CO when the exposure cannot be documented (Nordt et al., 2012). Functional neurological disorder should be considered when patients or large groups present with rapid onset of non-specific acute symptoms and normal physical examinations and laboratory findings (Nordt et al., 2012). Functional tremors show changes in frequency and amplitude, periods when the tremor is absent are also observed. In addition, tremor entrainment is seen in response to voluntary movements of the unaffected limb (Roper et al., 2013). The aforementioned findings are not observed in patients with a tremor entirely attributable to CO poisoning (Rutchik et al., 2021). Neuropsychological testing with invalid data may be suggestive of a functional syndrome, or intentional exaggeration and should be reviewed by experienced a neuropsychologist. Neuropsychological tests sensitive to effort such as the Test of Memory Malingering assist with detection of functional neurological disorders and symptom magnification (Greher and Wodushek, 2017; Greve et al., 2006). The evaluator must consider the motivations of the patient. Psychopharmacology in combination with cognitive behavioral therapy administered by qualified psychological counselor or psychiatrist may be necessary when a patient is resistant to treatment, experiences anxiety about returning to work, or shows a lack of motivation to facilitate their own recovery. Patients presenting with anxiety or PTSD symptoms who also have a suspected but undocumented history of CO exposure, and invalid neuropsychological testing with negative or non-specific imaging findings are difficult to diagnose. The patient's progress should be reassessed periodically to determine if the patient is showing functional improvements and performance consistency on behavioral testing.

5.3.2 Treatment of residual neurological sequelae

There are currently no FDA approved pharmaceuticals specifically approved the treatment of the residual neurological effects of CO poisoning. Therefore, when necessary residual symptoms of parkinsonism and cognitive dysfunction, and PTSD should be treated symptomatically. Depending on the clinical presentation and nature of the patient specific neuropathology the response to dopaminergic therapeutics in patients with CO poisoning may be good or poor (Klawans et al., 1982; Rutchik et al., 2021).

6. Neuroimaging

Neuroimaging and autopsy studies of CO poisoned patients often reveal abnormalities in the globus pallidus and white matter (Chen et al., 2015; Kinoshita et al., 2005; Porter et al., 2002; Sato et al., 2012; Tsai et al., 2017). These lesions which are often bilateral may also involve the corpus collosum (Chen et al., 2015; Porter et al., 2002; Tsai et al., 2017). It is important to recognize that white matter lesions are associated with non-demented aging (Burns et al., 2005). Therefore, the presence of white matter lesions alone should not be relied upon to diagnose CO poisoning in older adults. Lesions to the globus pallidus appear as areas of increased signal intensity on T2-weighted and FLAIR magnetic resonance images (MRI) (Lo et al., 2007). The T1-weighted images are typically hypointense but contrast-enhanced T1-weighted images may show patchy or peripheral enhancement in these same brain regions during the acute stages of CO poisoning (Lo et al., 2007). Cytotoxic edema associated with tissue necrosis influences the findings on an MRI performed early in the clinical course of CO poisoning. Diffusion-weighted MR imaging (DWI) permits detection of neuropathology with greater specificity than conventional MRI and it is increasingly used to assess for brain damage due to CO poisoning (Beppu, 2014; Kim et al., 2017; Kinoshita et al., 2005). Cytotoxic edema is also associated with areas of restricted diffusion which appear as areas of high-signal intensity on the DWIs of patients with acute CO poisoning (Jeon et al., 2018; Kim et al., 2017; Sert et al., 2021). Patients at increased risk for delayed neurological sequelae typically have lower initial Glasgow Coma Scale scores, longer durations of CO exposure, and abnormal findings on DWI studies performed on admission to the emergency department (Sert et al., 2021). It is important that OEM providers recognize neuroimaging markers do not always correlate well with the clinical findings (Choi, 2002; Devine et al., 2002; McKinney et al., 2009; Pavese et al., 1999; Rutchik et al., 2021).

Striatal dopaminergic hypofunction associated with CO poisoning can be measured with 6-[(18)F]fluoro-L-dopa Positron Emission Tomography (Rissanen et al., 2010). Single photon emission computed tomography scans using 99mTcTRODAT-1 as a dopamine transporter imaging agent indicate that signal strength in the caudate correlates with risk for tremor and bradykinesia (Sun et al., 2018). Yang et al. (2021) noted that the extent to which dopamine transporter density is reduced in the striatum following CO poisoning correlates with the severity of cognitive deficits. A reduction in the vesicular monoamine transporter type 2 signals in the caudate, amygdala, and hippocampus is seen on ^{18}F-FP-(+)-DTBZ PET scans of patients with CO poisoning (Hsiao et al., 2022). This finding is in contrast, to the low uptake seen in the putamen and substantia nigra of patients with PD reflecting the differences in the underlying neuropathology (Hsiao et al., 2022).

Recovery from neurological dysfunction associated with leukoencephalopathy and regression of neuroimaging biomarkers of CNS demyelination has been reported (Hansen et al., 2014; Law-Ye et al., 2018; Sohn et al., 2000). Magnetic resonance spectroscopy (MRS) metrics include the ratios of choline [Cho]/creatine [Cr], N-acetyl aspartate [NAA]/Cr, and Cho/NAA. These biomarkers are sensitive measure of neuronal health and viability following CO poisoning and are therefore useful in understanding recovery of neurological function. MRS studies of the white matter and cortex performed on admission following CO poisoning show an increased choline- (Cho) to creatine-phosphocreatine ratio and a decrease in the N-acetylaspartate (NAA) to Cr ratio. A reduction in the Cho/Cr ratio and an increase in the NAA/Cr ratio in the white matter may be seen at follow-up as compared with initial MRS (Kuroda et al., 2016; Sohn et al., 2000). The presence of an early lactate peak in the white matter on MRS is associated with a poorer long-term clinical prognosis (Kuroda et al., 2016). Reversal of initially decreased NAA/Cr ratio in cerebral gray matter is associated with recovery of function following CO poisoning (Hansen et al., 2014). Although MRS often provides additional prognostic data, these biomarkers of CO poisoning are not always sensitive to cognitive deficits (Prockop, 2005).

Neuroimaging is helpful for the diagnosis and estimating prognosis of patients with CO exposure and toxicity. However, neuroimaging findings alone without considering the medical and occupational and environmental exposure histories of the patient, as well as the neurological and neuropsychological test results should not be used to formulate a diagnosis.

7. Prognosis

Prognosis for patients with residual sequelae is difficult to assess due to factors such as motivation, access to medical care and provider creativity. While motivation cooperation may a characteristic of the neurological consequence of disease secondary to exposure based on imaging and consistent neuropsychological testing results, if may also be influences by PTSD and depression either unrelated or related to the exposure events. This must be considered by the evaluator. One should also consider factors such as denied treatment or delays due to the patients medical insurance or workers compensation laws in the state where the patient resides. Some medical systems also do not encourage repeat testing and reassessment when symptoms persist leaving the patient feeling abandoned, which is another factor to be considered. Repeat neuropsychological testing and imaging may be important to review for consistency and/or progression. Support for behavioral change with focus on sleep hygiene, stress management and meditation, diet and nutrition, exercise and time management are often necessary in patients with any chronic medical problem and often improves the overall outlook, functional abilities and potential for return to gainful employment.

8. Conclusions

The diagnosis and management of patients with CO neurotoxicity is facilitated by recognizing the complex mechanisms of action including glutamatergic excitotoxicity, apoptosis, oxidative stress, and neuroinflammation that are initiated during exposure and continue to contribute to neurological dysfunction after cessation of exposure thereby giving rise to the delayed onset of neurological symptoms. It is also important to recognize that there are many overlaps between the underlying neuropathology as well as the mechanisms of action implicated in neurodegenerative diseases and CO poisoning which can exacerbate preexisting neurological disorders. The limitations of percent COHb determined at the time of exposure with respect to predicting the onset and severity of delayed neurological sequelae of CO poisoning must also be considered. Likewise, OEM providers should be aware that neuroimaging findings do not always correlate well with clinical findings. While both of these diagnostic biomarkers provide useful supplemental data, the diagnosis of CO neurotoxicity remains a clinical one which should never be based on any single diagnostic metric. The prognosis for recovery of neurological function following CO poisoning is highly variable and depends on the nature of the lesions, motivation of the patient as well as the clinical acumen of the primary care provider. Patients should be offered creative treatment options that include pharmacotherapy but also behavioral therapy and a multidisciplinary support which often can lead to improvement in their activities of daily living. Serial neuropsychological assessments aid in understanding the nature of organic versus functional symptoms as well as the need for vocational retraining and work conditioning.

References

Abdulaziz, S., Dabbagh, O., Arabi, Y., Kojan, S., Hassan, I., 2012. Status epilepticus and cardiopulmonary arrest in a patient with carbon monoxide poisoning with full recovery after using a neuroprotective strategy: a case report. J. Med. Case Rep. 6, 421.

Alonso, J.R., Cardellach, F., López, S., Casademont, J., Miró, O., 2003. Carbon monoxide specifically inhibits cytochrome c oxidase of human mitochondrial respiratory chain. Pharmacol. Toxicol. 93 (3), 142–146.

Amitai, Y., Zlotogorski, Z., Golan-Katzav, V., Wexler, A., Gross, D., 1998. Neuropsychological impairment from acute low-level exposure to carbon monoxide. Arch. Neurol. 55 (6), 845–848.

Ankarcrona, M., Dypbukt, J.M., Bonfoco, E., Zhivotovsky, B., Orrenius, S., Lipton, S.A., Nicotera, P., 1995. Glutamate-induced neuronal death: a succession of necrosis or apoptosis depending on mitochondrial function. Neuron 15 (4), 961–973.

Annane, D., Chevret, S., Jars-Guincestre, C., Chillet, P., Elkharrat, D., Gajdos, P., Raphael, C., 2001. Prognostic factors in unintentional mild carbon monoxide poisoning. Intensive Care Med. 27 (11), 1776–1781.

Antonsson, A.B., Christensson, B., Berge, J., Sjögren, B., 2013. Fatal carbon monoxide intoxication after acetylene gas welding of pipes. Ann. Occup. Hyg. 57 (5), 662–666.

Banjoko, S.O., Sridhar Mynapelli, K.C., Ogunkola, I.O., Masheyi, O.O., 2007. Methylene chloride exposure and carboxyhemoglobin levels in cabinetmakers. Indian J. Occup. Environ. Med. 11 (2), 56–60.

Beckett, W.S., 2007. Chemical asphyxiants. In: Rom, W.N., Markowitz, S.B. (Eds.), Environment and Occupational Medicine, fourth ed. Wolters Klover/Lippincott Williams and Wilkins, pp. 562–569.

Beppu, T., 2014. The role of MR imaging in assessment of brain damage from carbon monoxide poisoning: a review of the literature. Am. J. Neuroradiol. 35 (4), 625–631.

Borras, L., Constant, E., De Timary, P., Huguelet, P., Khazaal, Y., 2009. Intoxication au monoxyde de carbone: quelles séquelles neuropsychiatriques? A propos d'un cas clinique et revue de la littérature [Long-term psychiatric consequences of carbon monoxide poisoning: a case report and literature review]. Rev. Med. Intern. 30 (1), 43–48.

Buboltz, J.B., Robins, M., 2021. Hyperbaric treatment of carbon monoxide toxicity. 2021 Apr 21. In: StatPearls [Internet]. StatPearls Publishing, Treasure Island, FL.

Burns, J.M., Church, J.A., Johnson, D.K., Xiong, C., Marcus, D., Fotenos, A.F., et al., 2005. White matter lesions are prevalent but differentially related with cognition in aging and early Alzheimer disease. Arch. Neurol. 62 (12), 1870–1876.

Centers for Disease Control and Prevention (CDC), 2006. Carbon monoxide poisonings after two major hurricanes—Alabama and Texas, August–October 2005. Morb. Mortal. Wkly. Rep. 55 (9), 236–239.

Centers for Disease Control and Prevention (CDC), 2012. Fatal exposure to methylene chloride among bathtub refinishers—United States, 2000–2011. Morb. Mortal Wkly. Rep. 61 (7), 119–122.

Chen, P.C., Chen, M.H., Chen, H.L., Lu, C.H., Chou, K.H., Wu, R.W., et al., 2015. Callosal damage and cognitive deficits in chronic carbon monoxide intoxication: a diffusion tensor imaging study. J. Neurol. Sci. 355 (1—2), 101–107.

Choi, I.S., 1983. Delayed neurologic sequelae in carbon monoxide intoxication. Arch. Neurol. 40 (7), 433–435.

Choi, I.S., 2001. Carbon monoxide poisoning: systemic manifestations and complications. J. Korean Med. Sci. 16 (3), 253–261.

Choi, I.S., 2002. Parkinsonism after carbon monoxide poisoning. Eur. Neurol. 48 (1), 30–33.

Cox, A., Sleeth, D., Handy, R., Alaves, V., 2019. Characterization of CO and NO_2 exposures of iceskating rink maintenance workers. J. Occup. Environ. Hyg. 16 (2), 101–108.

Devine, S.A., Kirkley, S.M., Palumbo, C.L., White, R.F., 2002. MRI and neuropsychological correlates of carbon monoxide exposure: a case report. Environ. Health Perspect. 110 (10), 1051–1055.

Downs, J.W., Wills, B.K., Cumpston, K.L., Rose, S.R., 2021. Descriptive epidemiology of clinically significant occupational poisonings, United States, 2008–2018. Clin. Toxicol. (Phila) 59 (12), 1259–1263.

Durnin, C., 1987. Carbon monoxide poisoning presenting with focal epileptiform seizures. Lancet 1 (8545), 1319.

Ely, E.W., Moorehead, B., Haponik, E.F., 1995. Warehouse workers' headache: emergency evaluation and management of 30 patients with carbon monoxide poisoning. Am. J. Med. 98 (2), 145–155.

Fan, H.C., Wang, A.C., Lo, C.P., Chang, K.P., Chen, S.J., 2009. Damage of cerebellar white matter due to carbon monoxide poisoning: a case report. Am. J. Emerg. Med. 27 (6). 757.e5-7.

Feldman, R.G., Ratner, M.H., Feldman, E.S., 1999. Approach to neurotoxicity tort cases. Neurol. Clin. Med. Leg. Issues Facing Neurol. 17 (2), 267–281.

Fisher-Hubbard, A.O., Appleford, C., Kesha, K., Schmidt, C.J., Gupta, A., 2018. Accidental carbon monoxide poisoning while driving: a case report with review of the literature. Am. J. Forensic Med. Pathol. 39 (3), 270–272.

Food and Drug Administration (FDA), 2019. FDA allows marketing of new device to help treat carbon monoxide poisoning. https://www.fda.gov/news-events/press-announcements/fda-allows-marketing-new-device-help-treat-carbon-monoxide-poisoning. Published March 14, 2019.

Fujiwara, S., Beppu, T., Nishimoto, H., Sanjo, K., Koeda, A., Mori, K., et al., 2012. Detecting damaged regions of cerebral white matter in the subacute phase after carbon monoxide poisoning using voxel-based analysis with diffusion tensor imaging. Neuroradiology 54 (7), 681–689.

Gordon, E.B., 1965. Carbon-monoxide encephalopathy. Br. Med. J. 1 (5444), 1232.

Greher, M.R., Wodushek, T.R., 2017. Performance validity testing in neuropsychology: scientific basis and clinical application—a brief review. J. Psychiatr. Pract. 23 (2), 134–140.

Greve, K.W., Bianchini, K.J., Black, F.W., Heinly, M.T., Love, J.M., Swift, D.A., Ciota, M., 2006. Classification accuracy of the Test of Memory Malingering in persons reporting exposure to environmental and industrial toxins: results of a known-groups analysis. Arch. Clin. Neuropsychol. 21 (5), 439–448.

Guo, J., Meng, J., Han, T., 2014. MRI-based comparison of brain damage between acute carbon monoxide poisoning and delayed encephalopathy after acute carbon monoxide poisoning. Zhonghua Lao Dong Wei Sheng Zhi Ye Bing Za Zhi. 32 (7), 533–536.

Haldane, J., 1895. The action of carbonic oxide on man. J. Physiol. 18 (5-6), 430–462.

Hamcan, S., Akgun, V., Yilmaz, O., Turan, A., 2013. Isolated cerebellar damage caused by carbon monoxide intoxication. BMJ Case Rep. 2013:bcr2013201647.

Hampson, N.B., Hauff, N.M., 2008. Carboxyhemoglobin levels in carbon monoxide poisoning: do they correlate with the clinical picture? Am. J. Emerg. Med. 26 (6), 665–669.

Hansen, M.B., Kondziella, D., Danielsen, E.R., Larsen, V.A., Jansen, E.C., Hyldegaard, O., 2014. Cerebral proton magnetic resonance spectroscopy demonstrates reversibility of N-acetylaspartate/creatine in gray matter after delayed encephalopathy due to carbon monoxide intoxication: a case report. J. Med. Case Rep. 8, 211.

Hauck, H., Neuberger, M., 1984. Carbon monoxide uptake and the resulting carboxyhemoglobin in man. Eur. J. Appl. Physiol. Occup. Physiol. 53 (2), 186–190.

Heckerling, P.S., 1987. Occult carbon monoxide poisoning: a cause of winter headache. Am. J. Emerg. Med. 5 (3), 201–204.

Heckerling, P.S., Leikin, J.B., Terzia, N.C.G., Maturen, A., 1990. Occult carbon monoxide poisoning in patients with neurologic illness. J. Toxicol. Clin. Toxicol. 28 (1), 29–44.

Heyboer 3rd, M., Jennings, S., Grant, W.D., Ojevwe, C., Byrne, J., Wojcik, S.M., 2014. Seizure incidence by treatment pressure in patients undergoing hyperbaric oxygen therapy. Undersea Hyperb. Med. 41 (5), 379–385.

Hoang, A., Fagan, K., Cannon, D.L., et al., 2021. Assessment of methylene chloride–related fatalities in the United States, 1980-2018. JAMA Intern. Med. 181 (6), 797–805.

Holt, J., Weaver, L.K., 2012. Carbon monoxide poisoning mimicking arterial gas embolism in a commercial diver. Undersea Hyperb. Med. 39 (2), 687–690.

Hopkins, R.O., Woon, F.L., 2006. Neuroimaging, cognitive, and neurobehavioral outcomes following carbon monoxide poisoning. Behav. Cogn. Neurosci. Rev. 5 (3), 141–155.

Hsiao, I.T., Chang, Y.T., Weng, Y.H., Hsu, S.W., Lin, K.J., Lu, C.S., et al., 2022. Comparisons of vesicular monoamine transporter type 2 signals in Parkinson's disease and parkinsonism secondary to carbon monoxide poisoning. Neurotoxicology 88, 178–186.

Huang, C.C., Chung, M.H., Weng, S.F., Chien, C.C., Lin, S.J., Lin, H.J., et al., 2014. Long-term prognosis of patients with carbon monoxide poisoning: a nationwide cohort study. PLoS One 9 (8), e105503.

Huang, C.C., Lee, J.C., Lin, K.C., Lin, H.J., Su, S.B., Hsu, C.C., Guo, H.R., 2019. Exposure duration and history of hypertension predicted neurological sequelae in patients with carbon monoxide poisoning. Epidemiology 30 (Suppl. 1), S76–S81.
Ilano, A.L., Raffin, T.A., 1990. Management of carbon monoxide poisoning. Chest 97 (1), 165–169.
Jang, D.H., Kelly, M., Hardy,.K., Lambert, D.S., Shofer, F.S., Eckmann, D.M., 2017. A preliminary study in the alterations of mitochondrial respiration in patients with carbon monoxide poisoning measured in blood cells. Clin. Toxicol. (Phila) 55 (6), 579–584.
Jeon, S.B., Sohn, C.H., Seo, D.W., Oh, B.J., Lim, K.S., Kang, D.W., et al., 2018. Acute brain lesions on magnetic resonance imaging and delayed neurological sequelae in carbon monoxide poisoning. JAMA Neurol. 75 (4), 436–443.
Jha, R., Kher, V., Kale, S.A., Jain, S.K., Arora, P., 1994. Carbon monoxide poisoning: an unusual cause of acute renal failure. Ren. Fail. 16 (6), 775–779.
Johnson-Arbor, K., Liebman, D.L., Carter, E.M., 2012. A survey of residential carbon monoxide detector utilization among Connecticut Emergency Department patients. Clin. Toxicol. (Phila) 50 (5), 384–389.
Jovanović, J., Jovanović, M., Dordević, D., 1999. Professional exposure of drivers to carbon monoxide as a possible risk factor for the occurrence of traffic accidents in the road traffic. Vojnosanit. Pregl. 56 (6), 587–592.
Kao, L.W., Nañagas, K.A., 2006. Toxicity associated with carbon monoxide. Clin. Lab. Med. 26 (1), 99–125.
Kesler, S.R., Hopkins, R.O., Weaver, L.K., Blatter, D.D., Edge-Booth, H., Bigler, E.D., 2001. Verbal memory deficits associated with fornix atrophy in carbon monoxide poisoning. J. Int. Neuropsychol. Soc. 7 (5), 640–646.
Kim, D.M., Lee, I.H., Park, J.Y., Hwang, S.B., Yoo, D.S., Song, C.J., 2017. Acute carbon monoxide poisoning: MR imaging findings with clinical correlation. Diagn. Interv. Imaging 98 (4), 299–306.
Kim, H., Choi, S., Park, E., Yoon, E., Min, Y., Lampotang, S., 2018. Serum markers and development of delayed neuropsychological sequelae after acute carbon monoxide poisoning: anion gap, lactate, osmolarity, S100B protein, and interleukin-6. Clin. Exp. Emerg. Med. 5 (3), 185–191.
Kim, S.G., Woo, J., Kang, G.W., 2019. A case report on the acute and late complications associated with carbon monoxide poisoning: acute kidney injury, rhabdomyolysis, and delayed leukoencephalopathy. Medicine 98 (19), e15551.
Kinoshita, T., Sugihara, S., Matsusue, E., Fujii, S., Ametani, M., Ogawa, T., 2005. Pallidoreticular damage in acute carbon monoxide poisoning: diffusion-weighted MR imaging findings. Am. J. Neuroradiol. 26 (7), 1845–1848.
Klawans, H.L., Stein, R.W., Tanner, C.M., Goetz, C.G., 1982. A pure parkinsonian syndrome following acute carbon monoxide intoxication. Arch. Neurol. 39 (5), 302–304.
Ku, H.L., Yang, K.C., Lee, Y.C., Lee, M.B., Chou, Y.H., 2010. Predictors of carbon monoxide poisoning-induced delayed neuropsychological sequelae. Gen. Hosp. Psychiatry 32 (3), 310–314.
Kuroda, H., Fujihara, K., Mugikura, S., Takahashi, S., Kushimoto, S., Aoki, M., 2016. Altered white matter metabolism in delayed neurologic sequelae after carbon monoxide poisoning: a proton magnetic resonance spectroscopic study. J. Neurol. Sci. 360, 161–169.
Lai, C.Y., Chou, M.C., Lin, C.L., Kao, C.H., 2015. Increased risk of Parkinson disease in patients with carbon monoxide intoxication: a population-based cohort study. Medicine (Baltimore) 94 (19), e869.
Law-Ye, B., Dodet, P., Hermann, B., Trunet, S., Dormont, D., Pyatigorskaya, N., Leclercq, D., 2018. Progressive white-matter demyelination in delayed CO poisoning encephalopathy. J. Neuroradiol. 45 (1), 59–62.

Lee, C.H., Choi, J.G., Lee, J.S., Lee, Y., Kim, H., Kim, Y.S., Cha, Y.S., 2021. Seizure during hyperbaric oxygen therapy: experience at a single academic hospital in Korea. Undersea Hyperb. Med. 48 (1), 43–51.

Lin, C., Xian, X., Qin, X., Wang, D., Tsow, F., Forzani, E., Tao, N., 2018. High performance colorimetric carbon monoxide sensor for continuous personal exposure monitoring. ACS Sens. 3 (2), 327–333.

Lo, C.P., Chen, S.Y., Lee, K.W., Chen, W.L., Chen, C.Y., Hsueh, C.J., Huang, G.S., 2007. Brain injury after acute carbon monoxide poisoning: early and late complications. Am. J. Roentgenol. 189 (4), W205–W211.

Mahoney, F.I., Barthel, D.W., 1965. Functional evaluation: The barthel index. Md. State Med. J. 14, 61–65.

Markey, M.A., Zumwalt, R.E., 2001. Fatal carbon monoxide poisoning after the detonation of explosives in an underground mine: a case report. Am. J. Forensic Med. Pathol. 22 (4), 387–390.

Marziali, S., Di Giuliano, F., Picchi, E., Natoli, S., Leonardis, C., Leonardis, F., et al., 2016. Non-convulsive status epilepticus in a patient with carbon-monoxide poisoning treated with hyperbaric oxygen therapy. Neuroradiol. J. 29 (6), 431–435.

Mascalchi, M., Petruzzi, P., Zampa, V., 1996. MRI of cerebellar white matter damage due to carbon monoxide poisoning: case report. Neuroradiology 38 (Suppl. 1), S73–S74.

Mattiuzzi, C., Lippi, G., 2020. Worldwide epidemiology of carbon monoxide poisoning. Hum. Exp. Toxicol. 39 (4), 387–392.

McKinney, A.M., Kieffer, S.A., Paylor, R.T., SantaCruz, K.S., Kendi, A., Lucato, L., 2009. Acute toxic leukoencephalopathy: potential for reversibility clinically and on MRI with diffusion-weighted and FLAIR imaging. Am. J. Roentgenol. 193 (1), 192–206.

Messier, L.D., Myers, R.A., 1991. A neuropsychological screening battery for emergency assessment of carbon-monoxide-poisoned patients. J. Clin. Psychol. 47 (5), 675–684.

Miró, O., Casademont, J., Barrientos, A., Urbano-Márquez, A., Cardellach, F., 1998. Mitochondrial cytochrome c oxidase inhibition during acute carbon monoxide poisoning. Pharmacol. Toxicol. 82 (4), 199–202.

Nah, S., Choi, S., Kim, H.B., Lee, J., Lee, S.U., Lee, Y.H., et al., 2020. Cerebral white matter lesions on diffusion-weighted images and delayed neurological sequelae after carbon monoxide poisoning: a prospective observational study. Diagnostics (Basel) 10 (9), 698.

Nakajima, M., Aso, S., Matsui, H., Fushimi, K., Yasunaga, H., 2020. Hyperbaric oxygen therapy and mortality from carbon monoxide poisoning: a nationwide observational study. Am. J. Emerg. Med. 38 (2), 225–230.

National Research Council (US) Committee on Acute Exposure Guideline Levels, 2010. Acute Exposure Guideline Levels for Selected Airborne Chemicals: Volume 8. National Academies Press (US), Washington, DC, p. 2. Carbon Monoxide Acute Exposure Guideline Levels.

Nelson, G.L., 1998. Carbon monoxide and fire toxicity: a review and analysis of recent work. Fire Technol. 34, 39–58.

Neufeld, M.Y., Swanson, J.W., Klass, D.W., 1981. Localized EEG abnormalities in acute carbon monoxide poisoning. Arch. Neurol. 38 (8), 524–527.

Nordt, S.P., Minns, A., Carstairs, S., Kreshak, A., Campbell, C., Tomaszweski, C., Hayden, S.R., Clark, R.F., Joshua, A., Ly, B.T., 2012. Mass sociogenic illness initially reported as carbon monoxide poisoning. J. Emerg. Med. 42 (2), 159–161.

Olson, K., Smollin, C., 2008. Carbon monoxide poisoning (acute). BMJ Clin. Evid. 2008, 2103.

Pages, B., Planton, M., Buys, S., Lemesle, B., Birmes, P., Barbeau, E.J., et al., 2014. Neuropsychological outcome after carbon monoxide exposure following a storm: a case-control study. BMC Neurol. 14, 153.

Pappalardo, I., Lorenzo, C., Arena, L., Testa, L., Giovanniello, A., Cava, M., Colla, L., Ghiazza, G., Morelli, N., Rota, E., 2019. Is "delayed" hyperbaric therapy effective for "delayed" encephalopathy after carbon monoxide intoxication? eNeurolog. Sci. 18, 100219.

Parkinson, R.B., Hopkins, R.O., Cleavinger, H.B., Weaver, L.K., Victoroff, J., Foley, J.F., Bigler, E.D., 2002. White matter hyperintensities and neuropsychological outcome following carbon monoxide poisoning. Neurology 58 (10), 1525–1532.

Pavese, N., Napolitano, A., De Iaco, G., Canapicchi, R., Collavoli, P.L., Lucetti, C., et al., 1999. Clinical outcome and magnetic resonance imaging of carbon monoxide intoxication. A long-term follow-up study. Ital. J. Neurol. Sci. 20 (3), 171–178.

Piantadosi, C., Zhang, J., Levin, E., Folz, R., Schmechel, D., 1997. Apoptosis and delayed neuronal damage after carbon monoxide poisoning in the rat. Exp Neurol. 147 (1), 103–114. https://doi.org/10.1006/exnr.1997.6584.

Porter, S.S., Hopkins, R.O., Weaver, L.K., Bigler, E.D., Blatter, D.D., 2002. Corpus callosum atrophy and neuropsychological outcome following carbon monoxide poisoning. Arch. Clin. Neuropsychol. 17 (2), 195–204.

Prockop, L.D., 2005. Carbon monoxide brain toxicity: clinical, magnetic resonance imaging, magnetic resonance spectroscopy, and neuropsychological effects in 9 people. J. Neuroimaging 15 (2), 144–149.

Rahmani, M., Belaidi, H., Benabdeljlil, M., Bouchhab, W., El Jazouli, N., Brini, E., et al., 2013. Bilateral brachial plexus injury following acute carbon monoxide poisoning. BMC Pharmacol. Toxicol. 14, 61.

Raphael, M., Nadiras, P., Flacke-Vordos, N., 2002. Acute methylene chloride intoxication—a case report on domestic poisoning. Eur. J. Emerg. Med. 9 (1), 57–59.

Ratner, M., Feldman, R., White, R., 2002. Neurobehavioral toxicology. In: Ramachandran, V. (Ed.), Encyclopedia of the Human Brain. 3. Elsevier, New York, pp. 423–439.

Ratner, M.H., Jabre, J., 2006. Treatment of neurotoxic effects of gases: carbon monoxide, hydrogen sulfide, and the nitrogen oxides. In: Noseworthy, J.H. (Ed.), Neurological Therapeutics: Principles and Practices. Chapter 148. Martin Dunitz, London, pp. 1701–1706.

Ratner, M.H., Jabre, J.F., 2017. Neurobehavioral Toxicology. In: Reference Module in Neuroscience and Biobehavioral Psychology. Elsevier.

Ratner, M.H., Ewing, W.M., Rutchik, J.S., 2020. Neurological effects of chronic occupational exposure to alcohol mists and vapors in a machinist. Toxicol. Commun. 4 (1), 43–48.

Rissanen, E., Paavilainen, T., Virta, J., Marttila, R., Rinne, J., Airas, L., 2010. Carbon monoxide poisoning-induced nigrostriatal dopaminergic dysfunction detected using positron emission tomography (PET). Neurotoxicology 31 (4), 403–407. https://doi.org/10.1016/j.neuro.2010.03.006.

Rodkey, F.L., 1977. Biological oxidation of [14C] methylene chloride to carbon monoxide and carbon dioxide by the rat. Toxicol. Appl. Pharmacol. 40 (1), 33–38.

Rodricks, J.V., 2011. Reference Guide on Exposure Science. National Research Council Reference Manual on Scientific Evidence, third ed. National Academies Press, Washington, DC, USA.

Roper, L.S., Saifee, T.A., Parees, I., Rickards, H., Edwards, M.J., 2013. How to use the entrainment test in the diagnosis of functional tremor. Pract. Neurol. 13 (6), 396–398.

Rosenman, K., 2021. Deaths from methylene chloride exposure: when chemicals used at home or work can kill. JAMA Intern. Med. 181 (6), 806–807.

Rutchik, J., Bowler, R.M., Ratner, M.H., 2021. A rare case of Holmes tremor in a worker with occupational carbon monoxide poisoning. Am. J. Ind. Med. 64 (5), 435–449.

Sato, H., Tanaka, T., Kasai, K., Tanaka, N., 2012. An autopsy case of acute carbon monoxide poisoning after a long-term vegetative state. Am. J. Forensic Med. Pathol. 33 (4), 341–343.

Savolainen, H., Kurppa, K., Tenhunen, R., Kivistö, H., 1980. Biochemical effects of carbon monoxide poisoning in rat brain with special reference to blood carboxyhemoglobin and cerebral cytochrome oxidase activity. Neurosci. Lett. 19 (3), 319–323.
Savoldi, F., Mazzella, G.L., Bo, P., Massarini, M., 1973. Atassia cerebellare da intossicazione acuta da monossido di carbonio [Cerebellar ataxia due to acute carbon monoxide poisoning]. Riv. Neurobiol. 19 (3–4), 197–206.
Semmens, E.O., Leary, C.S., West, M.R., Noonan, C.W., Navarro, K.M., Domitrovich, J.W., 2021. Carbon monoxide exposures in wildland firefighters in the United States and targets for exposure reduction. J. Expo. Sci. Environ. Epidemiol. 31 (5), 923–929.
Sert, E.T., Kokulu, K., Mutlu, H., 2021. Clinical predictors of delayed neurological sequelae in charcoal-burning carbon monoxide poisoning. Am. J. Emerg. Med. 48, 12–17.
Shusterman, D., Quinlan, P., Lowengart, R., Cone, J., 1990. Methylene chloride intoxication in a furniture refinisher. A comparison of exposure estimates utilizing workplace air sampling and blood carboxyhemoglobin measurements. J. Occup. Med. 32 (5), 451–454.
Simonsen, C., Magnusdottir, S.O., Andreasen, J.J., Wimmer, R., Rasmussen, B.S., Kjaergaard, B., Maltesen, R.G., 2021. Metabolic changes during carbon monoxide poisoning: an experimental study. J. Cell Mol. Med. 25 (11), 5191–5201.
Sohn, Y.H., Jeong, Y., Kim, H.S., Im, J.H., Kim, J., 2000. The brain lesion responsible for parkinsonism after carbon monoxide poisoning. Arch. Neurol. 57 (8), 1214–1218.
Stewart, R.D., Hake, C.L., 1976. Paint-remover hazard. JAMA 235 (4), 398–401.
Stewart, R.D., Fisher, T.N., Hosko, M.J., Peterson, J.E., Baretta, E.D., Dodd, H.C., 1972. Experimental human exposure to methylene chloride. Arch. Environ. Health. 25 (5), 342–348.
Sun, T.K., Chen, Y.Y., Huang, S.H., Hsu, S.W., Lee, C.C., Chang, W.N., et al., 2018. Neurotoxicity of carbon monoxide targets caudate-mediated dopaminergic system. Neurotoxicology 65, 272–279.
Tamura, T., Sugihara, G., Takahashi, H., 2021. Memory impairment and hippocampal volume after carbon monoxide poisoning. Arch. Clin. Neuropsychol. 36 (1), 145–148.
Thom, S.R., Bhopale, V.M., Fisher, D., Zhang, J., Gimotty, P., 2004. Delayed neuropathology after carbon monoxide poisoning is immune-mediated. Proc. Natl. Acad. Sci. U.S.A. 101 (37), 13660–13665.
Thomas, G.W., Sousan, S., Tatum, M., Liu, X., Zuidema, C., Fitzpatrick, M., et al., 2018. Low-cost, distributed environmental monitors for factory worker health. Sensors (Basel) 18 (5), 1411.
Tsai, P.H., Chou, M.C., Chiang, S.W., Chung, H.W., Liu, H.S., Kao, H.W., Chen, C.Y., 2017. Early white matter injuries in patients with acute carbon monoxide intoxication: a tract-specific diffusion kurtosis imaging study and STROBE compliant article. Medicine 96 (5), e5982.
Uemura, K., Harada, K., Sadamitsu, D., Tsuruta, R., Takahashi, M., Aki, T., et al., 2001. Apoptotic and necrotic brain lesions in a fatal case of carbon monoxide poisoning. Forensic Sci. Int. 116 (2-3), 213–219.
Uemura, K., Hoshino, S., Uchida, K., Tsuruta, R., Maekawa, T., Yoshida, K., 2003. Hypothermia attenuates delayed cortical cell death and ROS generation following CO inhalation. Toxicol. Lett. 145 (2), 101–106.
United States Department of Labor, 2000. Directive Number: CPL 2-0.125. Subject: Home-Based Worksites. Effective Date February, 25 https://www.osha.gov/enforcement/directives/cpl-02-00-125#references.
Watt, S., Prado, C.E., Crowe, S.F., 2018. Immediate and delayed neuropsychological effects of carbon monoxide poisoning: a meta-analysis. J. Int. Neuropsychol. Soc. 24 (4), 405–415.

Weaver, L.K., 2009. Clinical practice. Carbon monoxide poisoning. N. Engl. J. Med. 360 (12), 1217–1225.

Weaver, L.K., Howe, S., Hopkins, R., Chan, K.J., 2000. Carboxyhemoglobin half-life in carbon monoxide-poisoned patients treated with 100% oxygen at atmospheric pressure. Chest 117 (3), 801–808.

Weaver, L.K., Hopkins, R.O., Chan, K.J., Churchill, S., Elliott, C.G., Clemmer, T.P., et al., 2002. Hyperbaric oxygen for acute carbon monoxide poisoning. N. Engl. J. Med. 347 (14), 1057–1067.

Weaver, L.K., Valentine, K.J., Hopkins, R.O., 2007. Carbon monoxide poisoning: risk factors for cognitive sequelae and the role of hyperbaric oxygen. Am. J. Respir. Crit. Care Med. 176 (5), 491–497.

Weaver, L.K., Oliver, L.C., Deru, K., Stemmer-Rachamimov, A.O., 2019. Myositis associated with carbon monoxide poisoning. Undersea Hyperb. 46 (1), 63–67.

Wong, C.S., Lin, Y.C., Hong, L.Y., Chen, T.T., Ma, H.P., Hsu, Y.H., et al., 2016. Increased long-term risk of dementia in patients with carbon monoxide poisoning: a population-based study. Medicine 95 (3), e2549.

Worsham, C.M., Woo, J., Kearney, M.J., Bray, C.F., Jena, A.B., 2022. Carbon monoxide poisoning during major U.S. power outages. N. Engl. J. Med. 386 (2), 191–192.

Xu, S.Y., Li, C.X., Li, L.Y., Song, Y., Sui, Y., 2020. Wallerian degeneration of bilateral cerebral peduncles after acute carbon monoxide poisoning. BMC Neurol. 20 (1), 96.

Yang, H.J., Jeon, B.S., 2011. Selective cerebellar degeneration following carbon monoxide poisoning. Am. J. Neuroradiol. 32 (2), E39.

Yang, K.C., Yang, B.H., Lirng, J.F., Liu, M.N., Hu, L.Y., Liou, Y.J., et al., 2021. Interaction of dopamine transporter and metabolite ratios underpinning the cognitive dysfunction in patients with carbon monoxide poisoning: a combined SPECT and MRS study. Neurotoxicology 82, 26–34.

Yoon, S.S., Macdonald, S.C., Parrish, R.G., 1998. Deaths from unintentional carbon monoxide poisoning and potential for prevention with carbon monoxide detectors. JAMA 279 (9), 685–687.

Zhou, L., Yuan, L., Bahrami, D., Thomas, R.A., Rowland, J.H., 2018. Numerical and experimental investigation of carbon monoxide spread in underground mine fires. J. Fire Sci. 36 (5), 406–418.

Further reading

Henn, S.A., Bell, J.L., Sussell, A.L., Konda, S., 2013. Occupational carbon monoxide fatalities in the US from unintentional non-fire related exposures, 1992-2008. Am. J. Ind. Med. 56 (11), 1280–1289.

Occupational Safety and Health Administration (OSHA), 2022a. United States Department of Labor, Occupational Chemical Data Base, Carbon Monoxide. https://www.osha.gov/chemicaldata/462.

Occupational Safety and Health Administration (OSHA), 2022b. United States Department of Labor, Occupational Chemical Data Base, Methylene Chloride. https://www.osha.gov/chemicaldata/572.

CHAPTER EIGHT

Mercury neurotoxicity in gold miners

Maria Elena Crespo-Lopez[a,*], Marcus Augusto-Oliveira[a], Amanda Lopes-Araújo[a], Letícia Santos-Sacramento[a], José Rogério Souza-Monteiro[b], Felipe Farias da Rocha[a], and Gabriela de Paula Arrifano[a]

[a]Laboratório de Farmacologia Molecular, Instituto de Ciências Biológicas, Universidade Federal do Pará, Belém, Brazil
[b]Faculdade de Medicina, Campus de Altamira, Universidade Federal do Pará, Altamira, Brazil
*Corresponding author: e-mail address: ecrespo@ufpa.br

Contents

1. Introduction

Mercury is a metal with remarkable physicochemical properties that can be found in the air, water and soil (O'connor et al., 2019; UNEP, 2019). It exists in three species reflecting different properties and usage:

(I) Elemental mercury (Hg^0), a silver-colored liquid (at room temperature) also known as metallic mercury. This form of mercury is used in many manufacturing processes including thermometers, barometers, blood pressure gauges, dental amalgams and in gold isolation process in both small-scale and industrial mining (WHO, 2021a,b). It is the most volatile specie of mercury, and once released into the air, it remains in the atmosphere for long periods with high potential

Advances in Neurotoxicology, Volume 7
ISSN 2468-7480
https://doi.org/10.1016/bs.ant.2022.04.003

to be transported and deposited globally (Crespo-Lopez et al., 2021; UNEP, 2019).

(II) Inorganic mercury (IHg), which occurs when the metal ions (Hg^+ and Hg^{2+}) combines with other elements such as sulfur, chloride or oxygen to form mercury salts. IHg has been largely used to make other chemicals and in cosmetic skin products (UNEP, 2019; WHO, 2021b).

(III) Organic mercury, which is formed when mercury combines with carbon, resulting in different organic compounds including methylmercury (MeHg), ethylmercury, and phenylmercuric acetate, with distinct uses and toxicities (UNEP, 2019; WHO, 2021a,b).

One of the most interesting aspects of mercury dynamics is the biotransformation, a process by which one specie transforms in other. For instance, once in the atmosphere, Hg^0 is oxidized to IHg and subsequently carried to the waters and sediments through the rain (Crespo-Lopez et al., 2021; O'connor et al., 2019). Then, some microorganisms (methanogenic bacteria) can transform both Hg^0 and IHg into organic mercury, mainly MeHg, which represents the most toxic form of mercury (Crespo-Lopez et al., 2021). Thus, it accumulates in the food chain reaching piscivorous fish and contributes to the human exposure, especially riverine populations and gold miners, who has fish ingestion as the main alimentary habit (Crespo-Lopez et al., 2021; Machado et al., 2021; Rodriguez Martin-Doimeadios et al., 2014).

Notably, a large number of people around the world are exposed to different species of mercury, subjected to acute and/or chronic intoxication and its associated harmful effects. Among these, gold miners seem to represent a particular group, as they are exposed to all three forms of mercury through food consumption (fish contaminated with MeHg and even IHg), and an intense occupational exposure to Hg^0, which is, as previously stated, used in the gold purification process.

Occupational exposure presents at least two aggravating factors. The first one is the steep increase in number of artisanal and small-scale gold mining (ASGM) activities around the world and particularly in the Amazon region (Crespo-Lopez et al., 2021; Veit and Vallejos, 2020). ASGM is the main source of high mercury indices in the Amazon (Crespo-Lopez et al., 2021), as well as affects countries such as Zimbabwe (Becker et al., 2021), Ghana (Ovadje et al., 2021) and Mongolia (Steckling et al., 2011).

It should be noted that the term "small-scale" does not refer to the reduced presence of this activity or the low impact of these processes, but to the fact of the rudimentary practices and absence of industrial large-scale machinery. The term ASGM was initially used because of the localized craft activity carried out without the presence of large equipment. This term continues to be used today (e.g., in international reports) when refer to the mining responsible for mercury emissions in regions such as the Amazon. However, it should be emphasized that gold mining in the Amazon currently occupies extensive regions of the territory, often using machines such as excavators, rafts and dredgers, among others, which require teams of people working together for processing of the ore (Fig. 1). At the same time, there is no requirement of formal qualification or knowledge in safety practices to work in this activity, which may have possibly contributed to the maintenance of the term ASGM considering the characteristics of this informal activity.

Fig. 1 Biogeochemical cycle of mercury in the environment. Artisanal small-scale mining (ASGM) causes mercury vapor (Hg^0) emissions to the atmosphere. This activity is exerted both manually and using machines such excavators, rafts and dredgers, usually without protection equipment. In the clouds, Hg^0 can be oxidized to inorganic mercury (Hg^{2+}), contaminating the soil and water bodies through the rain. In the watershed, the Hg^{2+} is biotransformed into methylmercury (MeHg) by methanogenic bacteria. MeHg enter the food chain reaching humans through contaminated fish. Created with BioRender.com.

The latter fact, however, does not mean that ASGM is not a very extensive and high-impact economic activity.

Despite the extension of this activity, ASGM, as the name suggests, is many times characterized by a rudimentary gold extraction based on a relatively small group of people working without large equipment in remote/undeveloped areas, usually without a legal license for extraction (WHO, 2010). Such activity results in intense use of Hg^0 (liquid form) to separate the gold from its impurities, in a process called "amalgamation" (WHO, 2021b) (Fig. 1). Subsequently, the amalgam is heated to isolate gold, thus releasing mercury vapor into the environment (UNEP, 2019; USEPA, 2021a,b). Ultimately, this process releases around hundreds of tons of mercury a year into the atmosphere, representing the main source of mercury emissions (Crespo-Lopez et al., 2021; UNEP, 2019). For example, in 2015, ASGM was responsible for around 1.220 tons of mercury to terrestrial and aquatic ecosystems, and they stand for about 70–80% of the emissions in the South America and Sub-Saharan Africa (UNEP, 2019). These two regions, in addition to East and South-east Asia, are responsible for 73% of all global emissions of the metal (UNEP, 2019).

The second aggravating factor of the occupational exposure in mining using mercury is the way this metal is handled in ASGM; as highlighted before, there is neither formal requirement for adequate equipment nor formal qualification and basic knowledge in safety practices to work in the activities, which reflect a mercury mishandling and consequent more intense exposure (Harianja et al., 2020). Furthermore, scientific evidence alerts to the fact that current mercury levels may be increasing due to anthropic actions in regions such as the Amazon, with global consequences (Crespo-Lopez et al., 2021). This is a serious problem that will not disappear, requiring the improvement of our knowledge about the effects on human health for proper management.

2. Human exposure and symptomatology of mercury intoxication

In humans, chronic mercury poisoning, also known as mercurialism or erethism, is the oldest occupational disease; described by Pope in 1665, although Paracelso had already mentioned it in 1567 (Azevedo, 2003). First found in miners, it was later detected in other workers, such as those in the hat industry due to its use as a smoothing agent (even naming the symptomatology as Mad Hatter's disease, supposedly portrayed by Lewis Carroll in the book Alice in Wonderland) (Azevedo, 2003; Park and Zheng, 2012).

Nowadays, about 19 million people (including men, women and children) work in ASGM in underdeveloped/developing countries, being exposed to mercury (Esdaile and Chalker, 2018), and up to 6.5 million suffer from chronic poisoning at a moderate level (Steckling et al., 2017). Moreover, according to WHO, all human beings are currently exposed to certain quantity of mercury (WHO, 2018), and all species of the metal are toxic to humans (O'donoghue et al., 2020; WHO, 2018), eventually leading to systemic toxicity (ATSDR, 2015, 2019).

The signs and symptoms of mercury intoxication depend on several factors, including, but not limited to, the specie of mercury, the dose and the type of exposure (Bernhoft, 2012; WHO, 2016, 2017). Frequently, in acute exposure, the signs may appear within hours or take weeks, depending on the concentration (USEPA, 2021b). Occupational exposure in mining is usually chronic and, therefore, even exposure to low concentrations has the potential to cause deleterious consequences.

Traditionally, mercury vapor (Hg^0), emitted to the environment during the amalgam burning, has been considered as the main specie of mercury affecting the miners and their families (WHO, 2008). Approximately 80% of this vapor is absorbed by the lungs, and thus distributed to the tissues and bloodstream (Park and Zheng, 2012). Damage to the respiratory system may cause dyspnea, pulmonary inflammation, pneumonia, chest pain, cough, bloody sputum, bronchitis, corrosive bronchiolitis and in severe cases pulmonary edema (Fisher et al., 2003) (Fig. 2). Once in the body, Hg^0 accumulates in kidneys leading to proteinuria and nephrotic syndrome (WHO, 2008). However, due to its high liposolubility, Hg^0 easily crosses biological barriers such as the blood-brain barrier, also accumulating in the central nervous system (CNS) and producing neurotoxic effects (Crespo-Lopez et al., 2005; Fisher et al., 2003; Park and Zheng, 2012). The exposure to the vapor causes symptoms such as blurred vision, dysarthria, memory loss, muscle weakness, headaches, insomnia, tremors, unsteady gait and slow mental response, dizziness, lack of attention, emotional instability, numbness, fatigue and cognitive deficits (Fisher et al., 2003; Park and Zheng, 2012) (Fig. 2).

Although the WHO recommended an annual average of 1 $\mu g/m^3$ of mercury in the air of workplaces and estimated a tolerable concentration

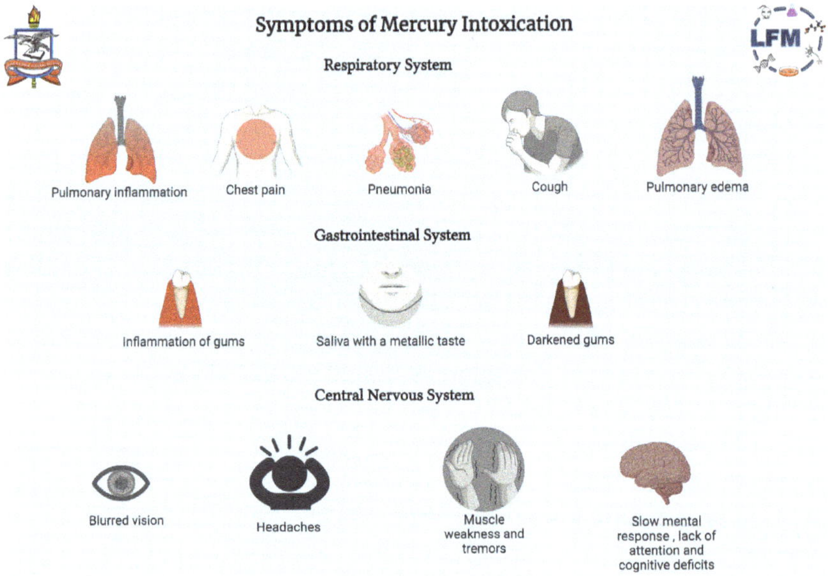

Fig. 2 Mercury intoxication symptomatology. Mercury poisoning may produce over 150 different symptoms depending on dose, time of exposure, among other factors. The most frequent signs and symptoms are related to the respiratory, gastrointestinal and central nervous systems. Created with BioRender.com.

of 0.2 µg/m^3 for long-term inhalation exposure to Hg^0 (WHO, 2021a), it should be emphasized that technologies for quantitation of air Hg^0 are hardly available in the most affected regions by ASGM, such as the Amazon (Santos-Sacramento et al., 2021). In fact, literature on mercury quantitation in the air in gold mining areas is extremely scarce.

Miners are also exposed to Hg^0 by dermal absorption, due to the direct skin contact related to manipulation of this metal using no gloves nor other safety equipment. Although dermal absorption is thought to be minimal, this can be an alternative pathway of exposure considering the frequency and amount of mercury. Additionally, mercury deposition on surrounding surfaces also originates sediment and water contamination (Afrifa et al., 2019), leading to the contamination of the food chain, mainly fish, as described above. Consequently, the consumption of fish from contaminated water bodies occurs concomitantly (being MeHg, the main specie found in the fish muscles) (Crespo-Lopez et al., 2021; Machado et al., 2021; Rodriguez Martin-Doimeadios et al., 2014). Approximately 95% of the MeHg in fish ingested by humans is absorbed in the gastrointestinal tract, directly entering the bloodstream, and rapidly distributed to all the organs of the body.

Oral exposure can also cause inflammation of the nasal mucosa and gums, saliva with a metallic taste, darkened gums, and excessive salivation (Fisher et al., 2003) (Fig. 2). Metabolic and cardiovascular alterations have been also described in populations chronically exposed to MeHg, especially when long-term consequences are considered (Arrifano et al., 2018b, 2021c; Hu et al., 2021). MeHg presents high affinity for the CNS and it is considered the most toxic specie of mercury for humans (WHO, 2017).

Thus, as mentioned before, miners are frequently exposed to both elemental and methylmercury (Calao-Ramos et al., 2021) and both mercury species can easily cross biological barriers, such as blood-placenta barrier and blood-brain barrier, reaching the CNS. A recent scoping review on human neurotoxicity of mercury in the Amazon described symptoms such as cognitive alterations, visual and motor impairments, and somatosensory and emotional deficits in the exposed populations (Santos-Sacramento et al., 2021).

Developmental CNS is particularly sensitive to mercury poisoning (WHO, 2010, 2021a). Children can be exposed to the metal in its various chemical species and through different pathways (WHO, 2010). Mercury can reach the fetus or babies via direct maternal transfer through placenta barrier and breastfeeding, respectively; children and adolescents consume contaminated fish and a significant part of them directly works in ASGM, compelled by poverty and lack of alternative economic opportunities (USEPA, 2021a,b; WHO, 2010). Even exposure to low doses of the metal can cause serious health consequences in these individuals (WHO, 2010). Intellectual disability, seizures, vision and hearing loss, delayed development, language disorders and memory loss are among the main symptoms described in intoxicated children (WHO, 2010).

Interestingly, after absorption, in vivo evidence suggests some interconversion between the various species of mercury (Bernhoft, 2012). For instance, all mercury species generate IHg (Hg^{+} or Hg^{2+}). In the case of Hg^{0}, the oxidation is facilitated by catalase in the erythrocytes; this process can take several minutes and occurs throughout the human body. Noteworthy, Hg^{0} crosses the blood-brain barrier and the placental barrier, before oxidation. Also, methylation and demethylation are known do happen in the human body (Bernhoft, 2012; Berzas Nevado et al., 2009; Li et al., 2019). The exact mechanisms involved in MeHg demethylation, however, as well as the rate of demethylation remains unknown. In the gastrointestinal tract, the intestinal microbiota is able to convert MeHg to IHg

(Li et al., 2019). In cells of CNS origin, MeHg is demethylated to both Hg^0 and IHg (Bernhoft, 2012). Consequently, the idea of neurological symptoms and signs of mercury intoxication due to a single specie of the metal would be an excessive oversimplification.

In the body, the half-life of Hg^0 and MeHg is 30–60 and 70–80 days, respectively. Urine and feces are the main excretion pathway for Hg^0 (mostly as IHg after oxidation), and bile and feces for MeHg (WHO, 2008). An important toxicokinetic aspect of MeHg is that its excretion in hair (about 10%) correlates with mercury content in brain (Crespo-Lopez et al., 2021; WHO, 2008). Therefore, urinary mercury is widely used to assess individual exposure to Hg^0 while hair mercury is an excellent marker for MeHg exposure (Crespo-Lopez et al., 2021). The latter biomarker is especially interesting in human biomonitoring since allows a historical registry of exposure with the hair growing about 1 cm/month (Crespo-Lopez et al., 2021).

Notably, the toxicokinetic of mercury is modulated by the genetic background (Llop et al., 2015). A number of genes, such as glutathione S-transferase family genes and metallothioneins genes, can modify mercury uptake, absorption, distribution and/or excretion (Llop et al., 2015). Moreover, the influence of apolipoprotein E genotypes (APOE) has been recently described (Arrifano et al., 2018c). In exposed populations, individuals carrying APOE4 genotype accumulated more mercury (Arrifano et al., 2018a), in addition to other possible deleterious consequences (Arrifano et al., 2018c). Furthermore, it has been recently discovered the relationship between APOE4-carriers and a high severity of disease caused by SARS-CoV-2 (Al-Jaf et al., 2021; Gkouskou et al., 2021; Goldstein et al., 2021; Kuo et al., 2020). Although the mechanism is not yet clear, the neurological consequences caused by the virus, especially in children, could be adding to those caused by exposure to the metal, being more severe in patients with APOE4. However, there is no study to date that has studied the role of APOE4 in this possible synergism.

It has been already demonstrated that neurological signs and sequalae are prevalent in COVID-19 patients (Chou et al., 2021). At the same time, a new "gold rush" is happening due to the rising gold price and the economic crisis, acting as driving forces that compel to expand ASGM (Veit and Vallejos, 2020). The possible consequences of the combination of these two factors (COVID-19 infection and chronic exposure to mercury) for the exposed brain are totally unknown and the affected populations will have to deal with this problem from now on.

3. Cellular and molecular mechanisms of mercury neurotoxicity

Considering that there are compelling evidence supporting the transformation of one mercury specie into another inside de CNS (Burbacher et al., 2005; Mailloux et al., 2015; Rooney, 2014; Shapiro and Chan, 2008; Vahter et al., 1995), and that miners are also exposed to the other species (IHg and MeHg), we shall approach cellular and molecular mechanisms involved in mercury neurotoxicity caused by the exposure to the different species of the metal, in neurons and glia.

Mercury neurotoxicity arises from neuronal and glial damage, disrupting the complex glia-neuron relationship. Once in the CNS, mercury triggers numerous deleterious events including oxidative stress (Teixeira et al., 2018; Yang et al., 2020a), calcium, GABA and glutamate unbalance (Ayensu et al., 2009; Colón-Rodríguez et al., 2020; Farina et al., 2003; Yang et al., 2020a), mitochondrial dysfunction (Liu et al., 2019), and cytokine release (Algarve et al., 2019), leading to defects in cellular signaling, cellular energetics and neuroinflammation, ultimately leading to behavioral and cognitive dysfunction, as we will see below. Interestingly, a recent study provided the first evidence on a possible novel mechanism underlying mercury neurotoxicity: the reduction in the number of circulating plasma exosomes, the smallest extracellular vesicles involved in cell communication and cellular regeneration (Arrifano et al., 2021b). Although not fully understood, a growing body of evidence has improved our understanding about the mechanisms involved in mercury neurotoxicity, and how this metal affects neurons and glia.

3.1 Mercury and neurons

After crossing the blood-brain barrier, mercury can severely damage neuronal structure and physiology (Chang, 1977). Neuronal structure fundamentally relies on cytoskeletal stability. Along with actin and intermediate filaments, microtubules constitute the cytoskeleton of eukaryotic organisms, and provide tracks for the transport of organelles, secretory vesicles, besides involvement in neurite growth, cell migration, proliferation and differentiation (Rolls et al., 2021). Thus, disruption on microtubules assembly and stability may affect neuronal viability, which occurs in mercury intoxication. Mercury binds to free sulfhydryl groups distributed uniquely on the surface and on the ends of microtubules, inhibiting microtubule assembly (Vogel et al., 1985), causing cytoskeleton breakdown and disturbing their function such as in cellular division (Castoldi et al., 2000; Crespo-Lopez et al., 2007, 2009, 2016; Stoiber et al., 2004).

In primary neuronal cultures obtained from cerebral cortex of neonatal rats from dams intoxicated with 8 mg/kg of methylmercury on gestational day 15, methylmercury reduced cell viability, neurite outgrowth (with retraction or collapse of some neurites) caused by dissolution of microtubules. Interestingly, the neuronal damage correlated with long term memory impairment, since adult rats from these dams presented dysfunctional performance when subjected to a passive avoidance task (Ferraro et al., 2009).

In vitro exposure to IHg (in which MeHg, EtHg and Hg^0 are transformed once in the brain) led to a disruption on neurite growth and decreased expression of microtubule associate protein 2 (MAP2), tubulin-β3 and tau, impairing cell differentiation (Chan et al., 2017). Additionally, another in vitro study suggests the cytoskeleton as a main neuronal target for MeHg neurotoxicity; cultured cortical neurons exposed to this mercury specie showed cytotoxic effects related to reduced glutathione (GSH) levels leading to disruption of cytoskeleton and consequent cell death (Pierozan et al., 2017). Besides reducing neuronal viability, MeHg induced caspase-3-dependent apoptosis along with downregulated PI3K/Akt pathway, which is directly involved in cell proliferation (Pierozan et al., 2017).

Also, MeHg in vitro interferes in the tropomyosin receptor kinase A (TrKA) pathway and eukaryotic elongation factor 1A1 (eEF1A1), impairing neurite outgrowth and synaptic protein expression, similar to what found in the cerebellum (Fujimura et al., 2016).

Synaptic transmission dysfunction is one of the major neuronal hallmarks of mercury intoxication. Indeed, mercury is known affecting dopaminergic, serotoninergic (Farina et al., 2017; Ke et al., 2020), GABAergic (Ayensu et al., 2009; Narahashi et al., 1994; Yuan and Atchison, 2003), and glutamatergic (Farina et al., 2003; Juárez et al., 2002) transmissions. Regarding glutamate, the most investigated neurotransmitter involved in mercury intoxication, there has been evidence for a long time of its involvement in mercury neurotoxicity (Albrecht and Matyja, 1996; Brookes, 1992). MeHg increases extracellular levels of glutamate by inhibiting its uptake and stimulating its release into the synaptic cleft resulting in excitotoxic events, which is exacerbated with over-activation of N-Methyl-D-Aspartate receptors (NMDA-R) (Farina et al., 2013). Indeed, NMDA-R are associated with neurobehavioral deficits in adult rats caused by early postnatal mercury intoxication (Liu et al., 2009). Data using 0.025–25 μM of mercuric chloride (IHg) showed neuronal excitability, with increased amplitude and frequency of synaptic potentials followed by sustained depolarization, and a two- to fivefold rises in intracellular calcium concentration ($[Ca^{2+}]_i$), which was reduced by the application of MK801, a non-competitive antagonist of NMDA-R (Xu et al., 2012). Similar results were found using a human neuroblastoma cell line (SHSY5Y); low dose of MeHg induced toxicity related to increased cellular response to glutamate through NMDA-R, suggesting NMDA-R antagonists as potential treatment for MeHg-associated neurological diseases (Petroni et al., 2013). Besides NMDA-R, α-amino-3-hydroxy-5-methyl-4-isoxazolepropionic acid receptors (AMPA-R) have a contribution to mercury-mediated alteration of intracellular Ca^{2+} concentration, hyper-excitability and cytotoxicity with cell death, as recently seen in human induced pluripotent stem cell motor neurons (Colón-Rodríguez et al., 2020) and in mouse lumbar spinal motor neurons (Sceniak et al., 2020).

This glutamate dyshomeostasis along with NMDA-R disorder elicited by MeHg intoxication may be accelerated by oxidative stress. Rats exposed to 12 μmol/kg of MeHg showed neuronal oxidative stress, with reactive oxygen species (ROS) accumulation and dysfunctional cellular antioxidant system, leading to disruption of Glu-Gln cycle, overactivation of NMDA-R, Ca^{2+} overload and accelerated degradation of NMDA-R (Yang et al., 2020a). Oxidative stress occurs in conditions which redox network is disrupted, and it is a major cause of MeHg-induced neurotoxicity triggered by over-activation of NMDA-R. Exposure to other species of mercury such as IHg also cause memory, motor coordination and cognitive impairment via neurotoxicity associated to oxidative stress in hippocampus and cortex (Aragao et al., 2018; Teixeira et al., 2014, 2018). This may explain, at least in part, why potent antioxidants are able to totally prevent neurological alterations such as motor incoordination, even when no decrease in mercury levels in the brain is detected (Crespo-Lopez et al., 2019). Of note, the low expression of antioxidative enzymes in cerebellar granular cells might explain, at least in part, the higher vulnerability of rat cerebellar granular cells to MeHg as compared to Purkinje cells and molecular layer neurons. In fact, expression of manganese-superoxide dismutase, glutathione peroxidase and thioredoxin reductase was significantly higher in Purkinje cells and molecular layer cells than in cerebellar granular cells (Fujimura and Usuki, 2014).

Although most of literature investigating neuronal damage induced by mercury intoxication focus on glutamatergic excitotoxicity and oxidative stress, recent evidence suggests that mercury exposure induces neuroinflammation and cell death signaling through neuronal release of pro-inflammatory cytokines and caspases, respectively. Using SHSY5Y neuron-like cell line, it was found that MeHg induced an increase of pro-inflammatory cytokines levels including IL-1β, IL-6, TNF-α and IFN-γ, and decreased anti-inflammatory cytokine IL-10. Furthermore, MeHg exposure increased levels of caspase-1, caspase-3 and caspase-8, indicating intracellular activation of apoptotic pathways (Algarve et al., 2019).

Mercury also induces neuronal apoptosis through ROS-mediated endoplasmic reticulum stress and mitochondrial apoptosis pathways activation. Significant apoptosis has been already detected in hippocampus and cortex of animals exposed to IHg (Aragao et al., 2018; Teixeira et al., 2018). Particularly, MeHg-induced mitochondrial dysfunction was supported by activation of caspase-3, caspase-9, and high levels of cytochrome C and apoptosis induce factor (Liu et al., 2019). Furthermore, MeHg causes neuronal apoptosis through suppression of the TrKA pathway, which is known to regulate neuronal differentiation and viability. MeHg disrupted phosphorylation of TrKA and its downstream effectors. Interestingly, GM1 ganglioside and MMC-257, which enhance TrKA phosphorylation, prevented that effect of MeHg in neurons, supporting the TrKA involvement in MeHg-induced neuronal damage (Fujimura and Usuki, 2015). Besides apoptosis, MeHg in higher concentrations may also trigger necrosis-related neuronal death (Yang et al., 2020b).

Finally, mercury intoxication may disrupt both embryonic/perinatal and adult neurogenesis, interfering in cell proliferation, migration and leading to acute cell death, with long-lasting effects on behavior and cognition (Abbott and Nigussie, 2021; Falluel-Morel et al., 2007; Sokolowski et al., 2013; Tian et al., 2016).

3.2 Mercury and astrocytes

Astrocytes comprise the most heterogeneous glial cells, being instrumental for numerous functions throughout life including water and potassium buffering, energy metabolism, blood-brain barrier maintenance, neurotransmitter recycling, oxidative balance, cytokine release, and synaptic behavior ultimately modulating animal behavior and cognition (Augusto-Oliveira et al., 2020). Astrocytes respond to extracellular environment through fluctuation of cytosolic ions organized both in space and time, and in cooperation with neurons and other glial cells, perform critical roles in the pathophysiology of CNS (Escartin et al., 2021; Sofroniew and Vinters, 2010), including MeHg intoxication, reviewed by Arrifano et al. (2021a).

Most of literature investigating astroglial involvement in mercury neurotoxicity focus on MeHg, and astrocytes seem to present a predisposition for MeHg uptake, placing them as an important target for this metal specie in the CNS (Aschner and Kimelberg, 1991; Aschner et al., 1990; Berzas Nevado et al., 2009). In fact, astrocytes uptake MeHg through neutral amino acid transporter system (LAT1) present in their membranes, being considered as MeHg deposits in the brain (Tiffany-Castiglion and Qian, 2001; Yin et al., 2008). Additionally, astrocytes present some characteristics that favor the accumulation of MeHg including: (i) astrocytic endfeet covering the neurovasculature (first contact with MeHg that crosses the blood-brain barrier), (ii) high levels of GSH and cytosolic metallothioneins with high affinity for MeHg (Tiffany-Castiglion and Qian, 2001), and (iii) higher resistance to mercury as compared to neurons, supporting higher concentration of metal before cell death (Berzas Nevado et al., 2009; Crespo-Lopez et al., 2007; Pierozan et al., 2017).

MeHg reduces GSH levels in astrocytes and increases intracellular ROS generation. Also, the synthesis of critical glutathione precursors (cystine and cysteine) is inhibited by MeHg, disrupting neuronal antioxidant defense, ultimately increasing neuronal vulnerability to MeHg intoxication (Allen et al., 2002; Shanker and Aschner, 2001). To cope with this effect, astrocytes increase their content of NF-E2-related factor 2 (Nrf-2) following MeHg intoxication, a key protective protein against oxidative stress (Ni et al., 2011). This mechanism has potential to protect astrocytes and neurons from cell death- induced neurotoxicity caused by oxidative stress; defective Nrf-2 response resulted in increased cell death following MeHg intoxication (Ni et al., 2011).

Furthermore, MeHg inhibits astroglial glutamate uptake resulting in increased glutamate concentration in the synaptic cleft and consequent excitotoxicity (Yin et al., 2007). MeHg can disrupt the uptake of glutamate by glutamate transporter 1 (GLT-1) and glutamate/aspartate transporters (GLAST) (Mutkus et al., 2005). Noteworthy, the reduced astroglial glutamate uptake caused by MeHg directly interferes in the synthesis of glutamate and glutathione (Arrifano et al., 2018c, 2021a; Augusto-Oliveira et al., 2020), which rather contribute to neurotoxicity (Mutkus et al., 2005). Additionally, accumulation of MeHg in cultured astrocytes leads to cellular swelling and consequent disruption of clearance of glutamate (Aschner et al., 1993), eventually resulting in neurotoxicity.

On the other hand, as the cells predominately responsible for glutamate buffering from synaptic cleft through GLT-1 and GLAST, astrocytes prevent MeHg-induced excitotoxicity, as shown in a human co-culture model. When cultured with astrocytes, neurons present higher resistance to MeHg as compared with neuronal monoculture (De Simone et al., 2017; Yin et al., 2009). Also, MeHg-induced decrease of glutamate in cultured neurons was recovered when co-cultured with astrocytes, suggesting that astrocyte may attenuate the neuronal glutamatergic imbalance induced by MeHg (Yin et al., 2009).

Additionally, astrocytes can protect neurons against MeHg exposure through release of growth and neurotrophic factors. It was found, in vitro, that astrocytes protect neurons through release of nerve growth factor (NGF) and brain-derived neurotrophic factor (BDNF) in response to MeHg (Takemoto et al., 2015). Similar finding was reported in vivo; in the inferior colliculus, BDNF release, increased in astrocytes, protected the auditory system following MeHg exposure (Ishihara et al., 2019). Moreover, astrocytes express vascular endothelial growth factor (VEGF), which can disrupt blood-brain barrier due to MeHg intoxication contributing to neurotoxicity (Takahashi et al., 2017).

Considering IHg, scarce literature shows astroglial involvement in mercury-induced neurotoxicity. Five types of human astrocytes uptake this specie of the metal including gray matter subpial, interlaminar, protoplasmic and varicose astrocytes, white matter fibrous astrocytes (Pamphlett and Kum Jew, 2018). In chronic exposure to IHg, decreased number of astrocytes in the motor cortex and hippocampus is associated to neurotoxicity and oxidative stress, and it seems to play roles in cognitive impairment and loss of balance and fine motor coordination observed in rats (Aragao et al., 2018; Teixeira et al., 2018).

Interestingly, an in vitro study in differentiated human neurons and astrocytes shows that IHg leads to neurons and astrocytes death, proposing that astrocytes are more resistant than neurons (Lohren et al., 2015). The inhibition of glutamate uptake by astrocytes through selective impairment of this transport represents a critical feature of IHg-induced neurotoxicity (Kim and Choi, 1995), as also observed for MeHg. Additionally, genotoxicity and DNA-damage response in astrocytoma and glioma cell lines were also suggested as mechanisms of neurotoxicity induced by both IHg (Pieper et al., 2014) and MeHg (Crespo-Lopez et al., 2016).

3.3 Mercury and microglia

Microglia represent the major immune cells of the CNS. Unlike other resident cells of nervous tissue, microglia derive from primitive macrophage from yolk sac and migrate to the CNS differentiating into microglia early in development. Microglia represent approximately 6% of all CNS cells and are involved in homeostasis and disease, being essentials for cellular network formation and refinement during CNS development, and for CNS immunosurveillance and defense (Augusto-Oliveira et al., 2019). To do so, microglia express numerous receptors which allow them to sense the CNS environment and rapidly responding to any change in homeostasis, including wide variety of molecules present in pathogens and released from damaged cell (Augusto-Oliveira et al., 2021).

Thus, microglia undergo a shifting from surveilling to a more reactive phenotype, accompanied by morphological remodeling and release of plenty of inflammatory mediators, which can either promoting tissue repair or aggravating detrimental conditions caused by numerous pathological events, including environmental pollutants such as mercury in its different species (Eanes et al., 2018; Minami et al., 2009; Ni et al., 2012; Toyama et al., 2021).

Most of literature about microglial involvement in mercury neurotoxicity, as well as for astrocytes, focus on MeHg. Microglia and astrocytes share common features in MeHg intoxication. For example, MeHg is deposited in both cell types producing concentration-dependent oxidative stress, with a direct consumption of intracellular GSH, eventually resulting in cell death (Ni et al., 2011). As a defense against MeHg-induced oxidative stress, both cells share an increase of Nrf-2 and its downstream genes heme oxygenase 1 (Ho1), NAD(*P*)H dehydrogenase quinone 1 (Nqo1), and x- C-type transporter (xCT), leading to MeHg detoxification (Ni et al., 2011).

Despite these similarities, microglia respond to MeHg faster than astrocytes, possibly due to higher metal uptake and increased ROS generation combined with lower basal GSH levels in microglia (Ni et al., 2011). Indeed, 1 min after MeHg exposure, there is an increased ROS generation, a decrease of GSH levels and an increased Nrf-2 with subsequent nuclear translocation in microglia, whereas these events are absent in astrocytes at this time point (Ni et al., 2012).

Besides being more responsive, microglia also present some remarkable features that seem to be more intense than in astrocytes, and may be responsible, at least in part, for the mercury neurotoxicity. After MeHg exposure, microglia display morphological remodeling, altered motility and increased inflammatory mediators release, which represent well characterized features of microglial reactivity. Regarding morphological remodeling, MeHg induced hypertrophic and rounded cell body, thickening and retraction of processes, and reduced number of processes both in vitro and in animal models of intoxication (Eskes et al., 2002; Hoshi et al., 2019; Sakamoto et al., 2008; Shinozaki et al., 2019). This morphological remodeling toward a less complex morphology, besides to indicate microglial reactivity, indicates a disruption of microglial survey activity, which is essential for brain homeostasis (Augusto-Oliveira et al., 2021).

In addition to changes in microglial morphology, MeHg induces changes in microglial motility. Chronic exposure to very low concentrations of MeHg caused substantial neuronal damage associated with microglial reactivity, revealed by a reduced motility associated with hypertrophy of cell body and process retraction (Shinozaki et al., 2019), which may implicate in dysfunctional cell surveillance and break of neural homeostasis (Madry and Attwell, 2015; Nimmerjahn et al., 2005).

Furthermore, MeHg causes neurotoxicity by inducing release of microglial inflammatory mediators. In optimized organotypic cerebral slice cultures, microglia of postnatal day 7 (similar to microglia from adult mice brains) express increased levels of CD16, CD32, TNFα and IL-1β after MeHg exposure (Hoshi et al., 2019).

The upregulation of pro-inflammatory molecules by MeHg in microglia is associated with neuronal damage and neurodegeneration, which are dependent on Rho-kinase (ROCK) signaling that regulates cellular motility, morphology, and gene expression (Shinozaki et al., 2019). Fasudil, a specific ROCK inhibitor, had a restorative effect on axonal degeneration by changing microglia toward an anti-inflammatory phenotype by suppressing TNF-α, iNOS, IL-1β and IL-6 and increasing the levels of anti-inflammatory molecules such as IL-10 and arginase-1 (Fujimura et al., 2019).

On the other hand, microglia can provide neuroprotection against MeHg. In 3D-cultures, microglial reactivity induced by noncytotoxic MeHg treatment resulted in neuroprotection via IL-6 release and interactions with astrocytes (Eskes et al., 2002). Later, it was shown that this mechanism of neuroprotection via IL-6 release by microglia/astrocyte interaction occurs via purinergic signaling. When in contact with MeHg, microglia release ATP via p38 MAPK- and vesicular nucleotide transporter-dependent mechanisms, which is sensed by astrocytes via P2Y1 receptors. It causes astrocytes release IL-6, protecting the surrounding neurons (Shinozaki et al., 2014).

As occurs with astrocytes, microglial response to MeHg seems to obey a complex dose- and time-dependent mechanisms. Despite the neuroprotective role exerted by microglia against low doses of MeHg, chronic exposure, even at low doses, resulted in neuronal damage-associated microglial reactivity in situ and in vitro (Shinozaki et al., 2019). In the microglial cell line N9, low dose (2 ng/mL) of MeHg conjugated with human serum albumin (HSA), a biomolecule with free thiol, induced cell proliferation, increased nitric oxide and intracellular calcium levels, and suppressed TNF-α and IL-1β release, with no cytotoxic effects. However, high dose (2 μg/mL) of MeHg-HSA exhibited a cytotoxic effect on N9 cells, including increased release of TNF-α and IL-1β and cell death (Tan et al., 2019).

The effects of mercury on microglia are not limited to MeHg as it can also involve other species such as IHg. Despite the extremely scarce literature, first evidence showed that both acute and chronic exposures to IHg (a single dose of 6 mg/kg $HgCl_2$ and 3 months infusion with low dose, respectively) were capable of inducing microglial proliferation (Charleston et al., 1995; Gajkowska et al., 1992). Recent data have demonstrated that, as observed for MeHg, IHg seems to exert a biphasic effect on microglia, depending on concentration of the metal. The neurotoxicant IHg in low dose (15 ng/mL) is not cytotoxic and decreases TNF-α and IL-1β secretion in N9 cell line, but a high dose (15 μg/mL) increases the secretion of the same cytokines, suggesting the dose-dependent effect of the microglia inflammatory response to this specie of the metal (Tan et al., 2019). Although the deleterious effects of the chronic exposure to IHg on neurons and astrocytes have been already described in vivo (Aragao et al., 2018; Teixeira et al., 2018), no data are available for microglia. To expand our knowledge on the behavior of these important cells against both IHg and Hg^0 exposures, in addition to MeHg, is necessary to better understand how the intoxicated brain responds to this metal.

3.4 Mercury and oligodendrocytes

Oligodendrocytes are glial cells responsible for myelin production in the CNS (Chesnut et al., 2021). There are few studies assessing the effects of mercury on oligodendrocytes and, consequently, the knowledge about the neurotoxic effects of mercury on myelination is extremely scarce. However, severe axonal loss and demyelination has been recently demonstrated in the exposed brain (Minj et al., 2021).

The evaluation of the number of oligodendrocytes and the intracellular accumulation of mercury in these cells reveals no change in the total number of cells in adult monkeys exposed to MeHg (Charleston et al., 1994). However, oligodendrocytes can accumulate both IHg and MeHg

(Charleston et al., 1995; Pamphlett and Kum Jew, 2018; Pamphlett et al., 2020), suggesting that these cells participate of the neurotoxic effects of Hg in the CNS in any protective or harmful manner, which remains poorly understood.

In addition to this uptake, oligodendrocytes are also vulnerable to both MeHg and IHg (Issa et al., 2003, 2008). Both mercury species highly inhibit the activity of 2′,3′-cyclic-nucleotide 3′-phosphodiesterase (Cnp), an enzyme essential for the performance of myelination in oligodendrocytes, which could explain, at least in part, the cytotoxic effects of the metal (Monnet-Tschudi et al., 1996; Raasakka and Kursula, 2014). In vitro data from a human oligodendrocytic cell line (MO3.13) exposed to IHg revealed morphological alterations corresponding to cell death by both apoptosis and necrosis. Cell death by apoptosis was induced by low concentrations of IHg, via activation of caspases. Otherwise, necrosis was detected with exposure to high concentrations of IHg (Issa et al., 2003).

Another important aspect to be highlighted is the important influence that MeHg can exert on proteins and genes associated with myelin production. In an interesting model of MeHg-induced amyotrophic lateral sclerosis (dose of 5 mg/kg), this pollutant considerably reduced the level of myelin basic protein (MBP), a main myelin protein (Minj et al., 2021). Also, in oligodendrocytes, MeHg reduced the transcription levels of the MBP gene (Bradl and Lassmann, 2010; Padhi and Pelletier, 2012) and the expression levels of Plp1 and Cnp (Lee and Fields, 2009; Padhi et al., 2015). Consequently, considering that the production of myelin in the CNS depends on oligodendrocytes and that mercury has a cytotoxic effect on these cells and exerts a negative influence on myelin-associated genes, it is reasonable to assume that mercury can affect the myelination process and, therefore, the conduction of the nerve impulse through of the CNS axons.

4. Current challenges in mercury neurotoxicology

Despite all the knowledge described above, the research on mercury neurotoxicology due to its use in mining is still challenging. As described at the beginning of this chapter, mercury is currently considered one of the most dangerous pollutants for human health (currently ranked third on the "ATSDR 2019 Substance Priority List" of the US Agency for Toxic Substances and Disease Registry (ATSDR, 2019)). Its chemical characteristics make it extremely useful and accessible in both industrial and traditional economics, with the greatest negative impact found on vulnerable populations in underdeveloped/developing countries. We need, for example, more data on the consequences for the brain of women in fertile age, older individuals, and children, among other vulnerable individuals. In addition to this worrying fact, scientific evidence alerts about the possible increase of current mercury levels due to anthropic actions in regions such as the Amazon, and particularly during the COVID-19 pandemic, with global consequences. Miners and their families are frequently exposed throughout their lifetime, both occupationally and through food from the contaminated environment of mining sites. The often remote/isolated location of these mining regions as well as the lack of political will and awareness of the problem, hamper research development. Moreover, consequences of the use of mercury in mining are felt beyond the mining regions. This is a serious problem that will not disappear soon, requiring the improvement of our knowledge about the effects of chronic exposure on human health for proper management.

Traditionally, elemental mercury has been considered as the main responsible for the harmful consequences of occupational exposure, with the kidneys and lungs being considered the main target organs. However, it would be to oversimplify the problem in an excessive and mistaken way. The ability of mercury to transform into different chemical species both in the environment and in the living beings means that the consequences of intoxication cannot be attributed to a single species, even when we talk about occupational exposure. Furthermore, recent evidence on the consumption of inorganic mercury reaching and damaging the CNS breaks the paradigm of any limitation in the distribution of mercury species. In this context, and especially considering chronic exposure, the CNS arises as the most important organ affected by exposure to mercury, with the most complex consequences.

However, when we look at the current knowledge on mercury neurotoxicity, two main obstacles emerge: difficulties in normalization/translation of the results when using mercury intoxication models, and the absence of a better knowledge on the dialogue between brain and periphery, especially considering that mercury is a multitarget pollutant. The first one is due to the lack of standardization in testing and analyzing data (required for direct comparison among in vivo studies), and adequate contextualization of the mercury dose (required for physiological/translational relevance). To standardize the studies, methodological suggestions have been recently recommended to surpass this issue (Arrifano et al., 2021a,b). Importantly, the simple and direct extrapolation of the dose between in vivo/in vitro models and epidemiological/clinical studies have to be avoided because of the large toxicokinetic differences. It is highly recommended selecting the dose based on physiological effects previously described (Aschner, 2012) and/or using an allometric approach that considers body surface area in addition to body weight (Nair and Jacob, 2016; Reagan-Shaw et al., 2007). Both together (body surface area and body weight) are well-correlated with critical biological parameters including basal metabolism, blood volume, oxygen consume, caloric expenditure, circulating plasma proteins and renal function (Nair and Jacob, 2016; Reagan-Shaw et al., 2007). Moreover, the type of exposure (acute or chronic), the via of exposure (oral, inhaled, injected, etc.) and the individual age should be taken in account.

The second problem (knowledge on the dialogue between brain and periphery) arises from the scarce literature on mercury neurotoxicity, which encompasses peripheral factors or consequences, such as dyslipidemia or inflammation, which may compose a bidirectional communication with cells and structures of SNC origin. Future studies should join efforts to overcome both obstacles and allow a more realistic and complete view of what really happens inside the intoxicated brain.

In addition, it is noteworthy that exposure markers (mercury levels in tissues) may change over time, with values that are dynamic and depend on several factors, such as the period of time after exposure in which the samples were collected. Conversely, damage to the CNS is unidirectional, especially in older individuals, and it eventually triggers neurodegeneration. In chronically exposed populations such as those of the Amazon, it would not be unusual to find individuals, especially adults, displaying neurological damage with current low levels of mercury, after possible past episodes of high exposure. Therefore, it is essential the pursuit of biomarkers (that can be detected peripherally) on mercury neurotoxicity for the diagnosis as soon as possible, even before the onset of subtle signs of cognitive dysfunction. In this sense, the use of matrixes such as saliva or the annual biomonitoring have been already proposed as important strategies (Augusto-Oliveira et al., 2021; Santos-Sacramento et al., 2021). The ultimate aim of collecting these data is to prevent the cases from evolving to neurodegenerative conditions and ultimately death.

In summary, considering that mercury emissions are a global problem, and despite the economic crisis that pandemic left us, all countries (and specially those significantly affected by ASGM) should join efforts and dedicate resources to surpass these challenges (Fig. 3) and minimize the problem in near future.

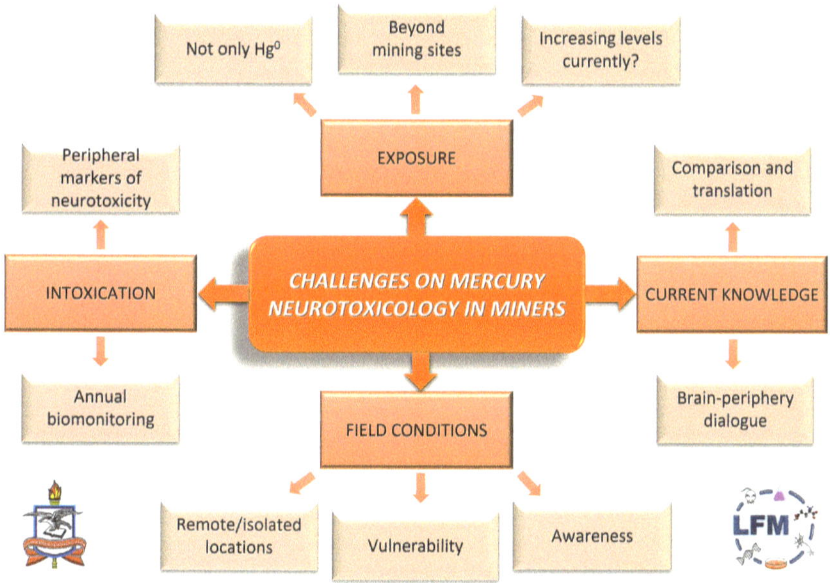

Fig. 3 Summary of the current challenges for the research on mercury neurotoxicology in miners.

Acknowledgments

This work was funded by the Conselho Nacional de Desenvolvimento Científico e Tecnológico (CNPq, grant numbers 427784/2018-2 and 313406/2021-9) and Porticus Fund. M.A.-O. thanks Porticus Fund for his post-doctoral fellowship and L.S.-S. and A.L.-A. thanks Fundação Amazônia de Amparo a Estudos e Pesquisas (FAPESPA) for their PhD fellowships. F.F.R. thanks CNPq for his undergraduate fellowship. Further, M.E.C.-L. thanks CNPq for her recognition as a highly productive researcher.

References

Abbott, L.C., Nigussie, F., 2021. Mercury toxicity and neurogenesis in the mammalian brain. Int. J. Mol. Sci. 22, 7520. https://doi.org/10.3390/Ijms22147520.

Afrifa, J., Opoku, Y.K., Gyamerah, E.O., Ashiagbor, G., Sorkpor, R.D., 2019. The clinical importance of the mercury problem in artisanal small-scale gold mining. Front. Public Health 7, 131. https://doi.org/10.3389/Fpubh.2019.00131.

Albrecht, J., Matyja, E., 1996. Glutamate: a potential mediator of inorganic mercury neurotoxicity. Metab. Brain Dis. 11, 175–184. https://doi.org/10.1007/Bf02069504.

Algarve, T.D., Assmann, C.E., Cadona, F.C., Machado, A.K., Manica-Cattani, M.F., Sato-Miyata, Y., Asano, T., Duarte, M., Ribeiro, E.E., Aigaki, T., Da Cruz, I.B.M., 2019. Guarana improves behavior and inflammatory alterations triggered by methylmercury exposure: an in vivo fruit fly and in vitro neural cells study. Environ. Sci. Pollut. Res. Int. 26, 15069–15083. https://doi.org/10.1007/S11356-019-04881-0.

Al-Jaf, S.M.A., Niranji, S.S., Ali, H.N., Mohammed, O.A., 2021. Association of apolipoprotein E polymorphism with SARS-CoV-2 infection. Infect. Genet. Evol. 95, 105043. https://doi.org/10.1016/J.Meegid.2021.105043.

Allen, J.W., Shanker, G., Tan, K.H., Aschner, M., 2002. The consequences of methylmercury exposure on interactive functions between astrocytes and neurons. Neurotoxicology 23, 755–759. https://doi.org/10.1016/S0161-813x(01)00076-6.

Aragao, W., Teixeira, F., Fagundes, N., Fernandes, R., Fernandes, L., Da Silva, M., Amado, L., Sagica, F., Oliveira, E., Crespo-Lopez, M., Maia, C., Lima, R., 2018. Hippocampal dysfunction provoked by mercury chloride exposure: evaluation of cognitive impairment, oxidative stress, tissue injury and nature of cell death. Oxid. Med. Cell. Longev. 2018, 7878050. https://doi.org/10.1155/2018/7878050.

Arrifano, G.P.F., Martin-Doimeadios, R.C.R., Jimenez-Moreno, M., Fernandez-Trujillo, S., Augusto-Oliveira, M., Souza-Monteiro, J.R., Macchi, B.M., Alvarez-Leite, J.I., Do Nascimento, J.L.M., Amador, M.T., Santos, S., Ribeiro-Dos-Santos, A., Silva-Pereira, L.C., Oria, R.B., Crespo-Lopez, M.E., 2018a. Genetic susceptibility to neurodegeneration in Amazon: apolipoprotein E genotyping in vulnerable populations exposed to mercury. Front. Genet. 9, 285. https://doi.org/10.3389/fgene.2018.00285.

Arrifano, G.P.F., Alvarez-Leite, J.I., Souza-Monteiro, J.R., Augusto-Oliveira, M., Paraense, R., Macchi, B.M., Pinto, A., Oria, R.B., Do Nascimento, J.L.M., Crespo-Lopez, M.E., 2018b. In the heart of the Amazon: noncommunicable diseases and apolipoprotein E4 genotype in the riverine population. Int. J. Environ. Res. Public Health 15, 1957. https://doi.org/10.3390/ijerph15091957.

Arrifano, G., De Oliveira, M., Souza-Monteiro, J., Paraense, R., Ribeiro-Dos-Santos, A., Vieira, J., Silva, A., Macchi, B., Do Nascimento, J., Burbano, R., Crespo-Lopez, M., 2018c. Role for apolipoprotein E in neurodegeneration and mercury intoxication. Front. Biosci. (Elite Ed.) 10, 229–241. https://doi.org/10.2741/e819.

Arrifano, G.D.P., Augusto-Oliveira, M., Souza-Monteiro, J.R., Macchi, B.D.M., Lima, R.R., Suñol, C., Do Nascimento, J.L.M., Crespo-Lopez, M.E., 2021a. Revisiting astrocytic roles in methylmercury intoxication. Mol. Neurobiol. 58, 4293–4308. https://doi.org/10.1007/s12035-021-02420-y.

Arrifano, G.P., Augusto-Oliveira, M., Sealey-Bright, M., Zainal, J., Imbiriba, L., Fernandes, L.M.P., Maia, C.S.F., Anthony, D., Crespo-Lopez, M.E., 2021b. Contributing to understand the crosstalk between brain and periphery in methylmercury intoxication: neurotoxicity and extracellular vesicles. Int. J. Mol. Sci. 22, 10855. https://doi.org/10.3390/ijms221910855.

Arrifano, G.P., Alvarez-Leite, J.I., Macchi, B.M., Campos, N., Augusto-Oliveira, M., Santos-Sacramento, L., Lopes-Araujo, A., Souza-Monteiro, J.R., Alburquerque-Santos, R., Do Nascimento, J.L.M., Santos, S., Ribeiro-Dos-Santos, A., Oria, R.B., Crespo-Lopez, M.E., 2021c. Living in the southern hemisphere: metabolic syndrome and its components in Amazonian riverine populations. J. Clin. Med. 10, 3603. https://doi.org/10.3390/jcm10163630.

Aschner, M., 2012. Considerations on methylmercury (MeHg) treatments in in vitro studies. Neurotoxicology 33 (3), 512–513. https://doi.org/10.1016/J.Neuro.2012.05.002.

Aschner, M., Kimelberg, H.K., 1991. The use of astrocytes in culture as model systems for evaluating neurotoxic-induced-injury. Neurotoxicology 12, 505–517.

Aschner, M., Eberle, N.B., Goderie, S., Kimelberg, H.K., 1990. Methylmercury uptake in rat primary astrocyte cultures: the role of the neutral amino acid transport system. Brain Res. 521, 221–228. https://doi.org/10.1016/0006-8993(90)91546-S.

Aschner, M., Du, Y.L., Gannon, M., Kimelberg, H.K., 1993. Methylmercury-induced alterations in excitatory amino acid transport in rat primary astrocyte cultures. Brain Res. 602, 181–186. https://doi.org/10.1016/0006-8993(93)90680-L.

ATSDR, Agency for Toxic Substances and Disease Registry, 2015. *Mercury Overview*. [Online]. Available at: https://www.atsdr.cdc.gov/sites/toxzine/mercury_toxzine.html [Accessed 21th Oct 2021].

ATSDR, Agency for Toxic Substances and Disease Registry, 2019. *Substance Priority List Agency for Toxic Substances and Disease Registry*. [Online]. Available: https://www.atsdr.cdc.gov/spl/index.html [Accessed 17th Oct 2021].
Augusto-Oliveira, M., Arrifano, G.P., Lopes-Araujo, A., Santos-Sacramento, L., Takeda, P.Y., Anthony, D.C., Malva, J.O., Crespo-Lopez, M.E., 2019. What do microglia really do in healthy adult brain? Cell 8, 1293. https://doi.org/10.3390/cells8101293.
Augusto-Oliveira, M., Arrifano, G.P., Takeda, P.Y., Lopes-Araujo, A., Santos-Sacramento, L., Anthony, D.C., Verkhratsky, A., Crespo-Lopez, M.E., 2020. Astroglia-specific contributions to the regulation of synapses, cognition and behaviour. Neurosci. Biobehav. Rev. 118, 331–357. https://doi.org/10.1016/J.Neubiorev.2020.07.039.
Augusto-Oliveira, M., Arrifano, G.P., Delage, C.I., Tremblay, M., Crespo-Lopez, M.E., Verkhratsky, A., 2021. Plasticity of microglia. Biol. Rev. Camb. Philos. Soc. 97, 217–250. https://doi.org/10.1111/brv.12797.
Ayensu, W.K., Isokpehi, R.D., Cohly, H.H., Murray, J.M., Webb, D.J., Tchounwou, P.B., 2009. Altered gabaa receptor expression as biomarker of mercury toxicity in embryonic neurogenesis. In: 2009 First Annual ORNL Biomedical Science & Engineering Conference, 18–19 March 2009, pp. 1–3. Available at: https://ieeexplore.ieee.org/document/5090491 [Accessed 20th Oct 2021].
Azevedo, F.A., 2003. Toxicologia Do Mercúrio. Rima, Sao Paulo.
Becker, J., Furu, P., Singo, J., Shoko, D., Elbel, J., Bose-O'reilly, S., Steckling-Muschack, N., 2021. Determinants of health and health needs assessment of artisanal and small-scale gold miners in Kadoma, Zimbabwe: a mixed method approach. Environ. Res. 197, 111081. https://doi.org/10.1016/J.Envres.2021.111081.
Bernhoft, R.A., 2012. Mercury toxicity and treatment: a review of the literature. J. Environ. Public Health 2012, 460508. https://doi.org/10.1155/2012/460508.
Berzas Nevado, J.J., Rodríguez Martín-Doimeadios, R.C., Jiménez Moreno, M., Do Nascimento, J.L.M., Crespo-López, M.E., 2009. Mercury speciation analysis on cell lines of the human central nervous system to explain genotoxic effects. Microchem. J. 93, 12–16. https://doi.org/10.1016/J.Microc.2009.03.008.
Bradl, M., Lassmann, H., 2010. Oligodendrocytes: biology and pathology. Acta Neuropathol. 119, 37–53. https://doi.org/10.1007/S00401-009-0601-5.
Brookes, N., 1992. In vitro evidence for the role of glutamate in the CNS toxicity of mercury. Toxicology 76, 245–256. https://doi.org/10.1016/0300-483x(92)90193-I.
Burbacher, T.M., Shen, D.D., Liberato, N., Grant, K.S., Cernichiari, E., Clarkson, T., 2005. Comparison of blood and brain mercury levels in infant monkeys exposed to methylmercury or vaccines containing thimerosal. Environ. Health Perspect. 113, 1015–1021. https://doi.org/10.1289/Ehp.7712.
Calao-Ramos, C., Bravo, A.G., Paternina-Uribe, R., Marrugo-Negrete, J., Diez, S., 2021. Occupational human exposure to mercury in artisanal small-scale gold mining communities of Colombia. Environ. Int. 146, 106216. https://doi.org/10.1016/J.Envint.2020.106216.
Castoldi, A.F., Barni, S., Turin, I., Gandini, C., Manzo, L., 2000. Early acute necrosis, delayed apoptosis and cytoskeletal breakdown in cultured cerebellar granule neurons exposed to methylmercury. J. Neurosci. Res. 59, 775–787. https://doi.org/10.1002/(Sici)1097-4547(20000315)59:6<775::Aid-Jnr10>3.0.Co;2-T.
Chan, M.C., Bautista, E., Alvarado-Cruz, I., Quintanilla-Vega, B., Segovia, J., 2017. Inorganic mercury prevents the differentiation of SH-SY5Y cells: amyloid precursor protein, microtubule associated proteins and ROS as potential targets. J. Trace Elem. Med. Biol. 41, 119–128. https://doi.org/10.1016/J.Jtemb.2017.02.002.
Chang, L.W., 1977. Neurotoxic effects of mercury—a review. Environ. Res. 14, 329–373. https://doi.org/10.1016/0013-9351(77)90044-5.

Charleston, J.S., Bolender, R.P., Mottet, N.K., Body, R.L., Vahter, M.E., Burbacher, T.M., 1994. Increases in the number of reactive glia in the visual cortex of Macaca fascicularis following subclinical long-term methyl mercury exposure. Toxicol. Appl. Pharmacol. 129, 196–206. https://doi.org/10.1006/Taap.1994.1244.

Charleston, J.S., Body, R.L., Mottet, N.K., Vahter, M.E., Burbacher, T.M., 1995. Autometallographic determination of inorganic mercury distribution in the cortex of the calcarine sulcus of the monkey Macaca fascicularis following long-term subclinical exposure to methylmercury and mercuric chloride. Toxicol. Appl. Pharmacol. 132, 325–333. https://doi.org/10.1006/Taap.1995.1114.

Chesnut, M., Paschoud, H., Repond, C., Smirnova, L., Hartung, T., Zurich, M.G., Hogberg, H.T., Pamies, D., 2021. Human IPSC-derived model to study myelin disruption. Int. J. Mol. Sci. 22, 9473. https://doi.org/10.3390/Ijms22179473.

Chou, S.H.-Y., Beghi, E., Helbok, R., Moro, E., Sampson, J., Altamirano, V., Mainali, S., Bassetti, C., Suarez, J.I., Mcnett, M., Consortium, G.-N, Consortium, E, 2021. Global incidence of neurological manifestations among patients hospitalized with COVID-19—a report for the GCS-NeuroCOVID consortium and the energy consortium. JAMA Netw. Open 4, e2112131. https://doi.org/10.1001/Jamanetworkopen.2021.12131.

Colón-Rodríguez, A., Colón-Carrión, N.M., Atchison, W.D., 2020. Ampa receptor contribution to methylmercury-mediated alteration of intracellular Ca2+ concentration in human induced pluripotent stem cell motor neurons. Neurotoxicology 81, 116–126. https://doi.org/10.1016/J.Neuro.2020.09.037.

Crespo-Lopez, M.E., Herculano, A.M., Corvelo, T.C., Do Nascimento, J.L., 2005. Mercury and neurotoxicity. Rev. Neurol. 40, 441–447. https://doi.org/10.33588/Rn.4007.2004499.

Crespo-Lopez, M.E., Lima De Sa, A., Herculano, A.M., Rodriguez Burbano, R., Do Nascimento, M., J. L., 2007. Methylmercury genotoxicity: a novel effect in human cell lines of the central nervous system. Environ. Int. 33, 141–146. https://doi.org/10.1016/J.Envint.2006.08.005.

Crespo-Lopez, M.E., Macedo, G.L., Pereira, S.I., Arrifano, G.P., Picanco-Diniz, D.L., Do Nascimento, J.L., Herculano, A.M., 2009. Mercury and human genotoxicity: critical considerations and possible molecular mechanisms. Pharmacol. Res. 60, 212–220. https://doi.org/10.1016/J.Phrs.2009.02.011.

Crespo-Lopez, M.E., Costa-Malaquias, A., Oliveira, E.H., Miranda, M.S., Arrifano, G.P., Souza-Monteiro, J.R., Sagica, F.E., Fontes-Junior, E.A., Maia, C.S., Macchi, B.M., Do Nascimento, J.L., 2016. Is low non-lethal concentration of methylmercury really safe? A report on genotoxicity with delayed cell proliferation. Plos One 11, E0162822. https://doi.org/10.1371/Journal.Pone.0162822.

Crespo-Lopez, M.E., Soares, E.S., Macchi, B.M., Santos-Sacramento, L., Takeda, P.Y., Lopes-Araujo, A., Paraense, R.S.O., Souza-Monteiro, J.R., Augusto-Oliveira, M., Luz, D.A., Maia, C., Rogez, H., Lima, M.O., Pereira, J.P., Oliveira, D.C., Burbano, R.R., Lima, R.R., Do Nascimento, J.L.M., Arrifano, G.P., 2019. Towards therapeutic alternatives for mercury neurotoxicity in the Amazon: unraveling the pre-clinical effects of the Superfruit acai (Euterpe Oleracea, Mart.) as juice for human consumption. Nutrients 11, 2585. https://doi.org/10.3390/Nu11112585.

Crespo-Lopez, M.E., Augusto-Oliveira, M., Lopes-Araujo, A., Santos-Sacramento, L., Yuki Takeda, P., Macchi, B.M., Do Nascimento, J.L.M., Maia, C.S.F., Lima, R.R., Arrifano, G.P., 2021. Mercury: what can we learn from the Amazon? Environ. Int. 146, 106223. https://doi.org/10.1016/J.Envint.2020.106223.

De Simone, U., Caloni, F., Gribaldo, L., Coccini, T., 2017. Human co-culture model of neurons and astrocytes to test acute cytotoxicity of neurotoxic compounds. Int. J. Toxicol. 36, 463–477. https://doi.org/10.1177/1091581817739428.

Eanes, T., Carvalho, C., Branco, V., 2018. P-128—modulation of NF-kB signaling by mercury compounds in mouse microglia cells and its relationship with glutathione levels. Free Rad. Biol. Med. 120, S83. https://doi.org/10.1016/J.Freeradbiomed.2018.04.275.

Escartin, C., Galea, E., Lakatos, A., O'callaghan, J.P., Petzold, G.C., Serrano-Pozo, A., Steinhäuser, C., Volterra, A., Carmignoto, G., Agarwal, A., Allen, N.J., Araque, A., Barbeito, L., Barzilai, A., Bergles, D.E., Bonvento, G., Butt, A.M., Chen, W.-T., Cohen-Salmon, M., Cunningham, C., Deneen, B., De Strooper, B., Díaz-Castro, B., Farina, C., Freeman, M., Gallo, V., Goldman, J.E., Goldman, S.A., Götz, M., Gutiérrez, A., Haydon, P.G., Heiland, D.H., Hol, E.M., Holt, M.G., Iino, M., Kastanenka, K.V., Kettenmann, H., Khakh, B.S., Koizumi, S., Lee, C.J., Liddelow, S.A., Macvicar, B.A., Magistretti, P., Messing, A., Mishra, A., Molofsky, A.V., Murai, K.K., Norris, C.M., Okada, S., Oliet, S.H.R., Oliveira, J.F., Panatier, A., Parpura, V., Pekna, M., Pekny, M., Pellerin, L., Perea, G., Pérez-Nievas, B.G., Pfrieger, F.W., Poskanzer, K.E., Quintana, F.J., Ransohoff, R.M., Riquelme-Perez, M., Robel, S., Rose, C.R., Rothstein, J.D., Rouach, N., Rowitch, D.H., Semyanov, A., Sirko, S., Sontheimer, H., Swanson, R.A., Vitorica, J., Wanner, I.-B., Wood, L.B., Wu, J., Zheng, B., Zimmer, E.R., Zorec, R., Sofroniew, M.V., Verkhratsky, A., 2021. Reactive astrocyte nomenclature, definitions, and future directions. Nat. Neurosci. 24, 312–325. https://doi.org/10.1038/S41593-020-00783-4.

Esdaile, L.J., Chalker, J.M., 2018. The mercury problem in artisanal and small-scale gold mining. Chem. A Eur. J. 24, 6905–6916. https://doi.org/10.1002/Chem.201704840.

Eskes, C., Honegger, P., Juillerat-Jeanneret, L., Monnet-Tschudi, F., 2002. Microglial reaction induced by noncytotoxic methylmercury treatment leads to neuroprotection via interactions with astrocytes and IL-6 release. Glia 37, 43–52. https://doi.org/10.1002/Glia.10019.

Falluel-Morel, A., Sokolowski, K., Sisti, H.M., Zhou, X., Shors, T.J., Dicicco-Bloom, E., 2007. Developmental mercury exposure elicits acute hippocampal cell death, reductions in neurogenesis, and severe learning deficits during puberty. J. Neurochem. 103, 1968–1981. https://doi.org/10.1111/J.1471-4159.2007.04882.X.

Farina, M., Dahm, K.C.S., Schwalm, F.D., Brusque, A.M., Frizzo, M.E.S., Zeni, G., Souza, D.O., Rocha, J.B.T., 2003. Methylmercury increases glutamate release from brain synaptosomes and glutamate uptake by cortical slices from suckling rat pups: modulatory effect of Ebselen. Toxicol. Sci. 73, 135–140. https://doi.org/10.1093/Toxsci/Kfg058.

Farina, M., Avila, D.S., Da Rocha, J.B., Aschner, M., 2013. Metals, oxidative stress and neurodegeneration: a focus on iron, manganese and mercury. Neurochem. Int. 62, 575–594. https://doi.org/10.1016/J.Neuint.2012.12.006.

Farina, M., Aschner, M., Da Rocha, J.B.T., 2017. Chapter Two—the catecholaminergic neurotransmitter system in methylmercury-induced neurotoxicity. In: Aschner, M., Costa, L.G. (Eds.), Advances in Neurotoxicology. Academic Press, pp. 47–81.

Ferraro, L., Tomasini, M.C., Tanganelli, S., Mazza, R., Coluccia, A., Carratù, M.R., Gaetani, S., Cuomo, V., Antonelli, T., 2009. Developmental exposure to methylmercury elicits early cell death in the cerebral cortex and long-term memory deficits in the rat. Int. J. Dev. Neurosci. 27, 165–174. https://doi.org/10.1016/J.Ijdevneu.2008.11.004.

Fisher, J.F., World Health Organization & International Programme On Chemical Safety, 2003. Elemental Mercury and Inorganic Mercury Compounds: Human Health Aspects. World Health Organization, Geneva. Available at: https://www.who.int/ipcs/publications/cicad/en/cicad50.pdf.[Accessed 20th Oct 2021].

Fujimura, M., Usuki, F., 2014. Low in situ expression of antioxidative enzymes in rat cerebellar granular cells susceptible to methylmercury. Arch. Toxicol. 88, 109–113. https://doi.org/10.1007/S00204-013-1089-2.

Fujimura, M., Usuki, F., 2015. Methylmercury causes neuronal cell death through the suppression of the TrkA pathway: in vitro and in vivo effects of TrkA pathway activators. Toxicol. Appl. Pharmacol. 282, 259–266. https://doi.org/10.1016/J.Taap.2014.12.008.
Fujimura, M., Usuki, F., Cheng, J., Zhao, W., 2016. Prenatal low-dose methylmercury exposure impairs neurite outgrowth and synaptic protein expression and suppresses TrkA pathway activity and eEF1A1 expression in the rat cerebellum. Toxicol. Appl. Pharmacol. 298, 1–8. https://doi.org/10.1016/J.Taap.2016.03.002.
Fujimura, M., Usuki, F., Nakamura, A., 2019. Fasudil, a Rho-associated coiled coil-forming protein kinase inhibitor, recovers methylmercury-induced axonal degeneration by changing microglial phenotype in rats. Toxicol. Sci. 168, 126–136. https://doi.org/10.1093/Toxsci/Kfy281.
Gajkowska, B., Szumanska, G., Gadamski, R., 1992. Ultrastructural alterations of brain cortex in rat following intraperitoneal administration of mercuric chloride. J. Hirnforsch. 33, 471–476. 1479187.
Gkouskou, K., Vasilogiannakopoulou, T., Andreakos, E., Davanos, N., Gazouli, M., Sanoudou, D., Eliopoulos, A.G., 2021. Covid-19 enters the expanding network of apolipoprotein E4-related pathologies. Redox Biol. 41, 101938. https://doi.org/10.1016/J.Redox.2021.101938.
Goldstein, M.R., Poland, G.A., Graeber, C.W., 2021. Coronavirus disease 2019, multisystem inflammatory syndrome in children, apolipoprotein E4, and race. J. Pediatr. 229, 313–314. https://doi.org/10.1016/J.Jpeds.2020.10.072.
Harianja, A.H., Saragih, G.S., Fauzi, R., Hidayat, M.Y., Syofyan, Y., Tapriziah, E.R., Kartiningsih, S.E., 2020. Mercury exposure in artisanal and small-scale gold mining communities in Sukabumi, Indonesia. J. Health Pollut. 10. https://doi.org/10.5696/2156-9614-10.28.201209, 201209.
Hoshi, T., Toyama, T., Shinozaki, Y., Koizumi, S., Lee, J.Y., Naganuma, A., Hwang, G.W., 2019. Evaluation of M1-microglial activation by neurotoxic metals using optimized organotypic cerebral slice cultures. J. Toxicol. Sci. 44, 471–479. https://doi.org/10.2131/Jts.44.471.
Hu, X.F., Lowe, M., Chan, H.M., 2021. Mercury exposure, cardiovascular disease, and mortality: a systematic review and dose-response meta-analysis. Environ. Res. 193, 110538. https://doi.org/10.1016/J.Envres.2020.110538.
Ishihara, Y., Itoh, K., Oguro, A., Chiba, Y., Ueno, M., Tsuji, M., Vogel, C.F.A., Yamazaki, T., 2019. Neuroprotective activation of astrocytes by methylmercury exposure in the inferior colliculus. Sci. Rep. 9, 13899. https://doi.org/10.1038/S41598-019-50377-9.
Issa, Y., Watts, D.C., Duxbury, A.J., Brunton, P.A., Watson, M.B., Waters, C.M., 2003. Mercuric chloride: toxicity and apoptosis in a human oligodendroglial cell line MO3.13. Biomaterials 24, 981–987. https://doi.org/10.1016/S0142-9612(02)00436-2.
Issa, Y., Brunton, P., Waters, C.M., Watts, D.C., 2008. Cytotoxicity of metal ions to human oligodendroglial cells and human gingival fibroblasts assessed by mitochondrial dehydrogenase activity. Dent. Mater. 24, 281–287. https://doi.org/10.1016/J.Dental.2007.09.010.
Juárez, B.I., Martínez, M.L., Montante, M., Dufour, L., García, E., Jiménez-Capdeville, M.E., 2002. Methylmercury increases glutamate extracellular levels in frontal cortex of awake rats. Neurotoxicol. Teratol. 24, 767–771. https://doi.org/10.1016/S0892-0362(02)00270-2.
Ke, T., Tsatsakis, A., Santamaría, A., Antunes Soare, F.A., Tinkov, A.A., Docea, A.O., Skalny, A., Bowman, A.B., Aschner, M., 2020. Chronic exposure to methylmercury induces puncta formation in cephalic dopaminergic neurons in Caenorhabditis Elegans. Neurotoxicology 77, 105–113. https://doi.org/10.1016/J.Neuro.2020.01.003.

Kim, P., Choi, B.H., 1995. Selective inhibition of glutamate uptake by mercury in cultured mouse astrocytes. Yonsei Med. J. 36, 299–305. https://doi.org/10.3349/Ymj.1995.36.3.299.

Kuo, C.L., Pilling, L.C., Atkins, J.L., Masoli, J.A.H., Delgado, J., Kuchel, G.A., Melzer, D., 2020. APOE E4 genotype predicts severe COVID-19 in the UK biobank community cohort. J. Gerontol. A Biol. Sci. Med. Sci. 75, 2231–2232. https://doi.org/10.1093/Gerona/Glaa131.

Lee, P.R., Fields, R.D., 2009. Regulation of myelin genes implicated in psychiatric disorders by functional activity in axons. Front. Neuroanat. 3, 4. https://doi.org/10.3389/Neuro.05.004.2009.

Li, H., Lin, X., Zhao, J., Cui, L., Wang, L., Gao, Y., Li, B., Chen, C., Li, Y.F., 2019. Intestinal methylation and demethylation of mercury. Bull. Environ. Contam. Toxicol. 102, 597–604. https://doi.org/10.1007/S00128-018-2512-4.

Liu, W., Wang, X., Zhang, R., Zhou, Y., 2009. Effects of postnatal exposure to methylmercury on spatial learning and memory and brain NMDA receptor mRNA expression in rats. Toxicol. Lett. 188, 230–235. https://doi.org/10.1016/J.Toxlet.2009.04.021.

Liu, W., Yang, T., Xu, Z., Xu, B., Deng, Y., 2019. Methyl-mercury induces apoptosis through ROS-mediated endoplasmic reticulum stress and mitochondrial apoptosis pathways activation in rat cortical neurons. Free Radic. Res. 53, 26–44. https://doi.org/10.1080/10715762.2018.1546852.

Llop, S., Ballester, F., Broberg, K., 2015. Effect of gene-mercury interactions on mercury toxicokinetics and neurotoxicity. Curr. Environ. Health Rep. 2, 179–194. https://doi.org/10.1007/S40572-015-0047-Y.

Lohren, H., Blagojevic, L., Fitkau, R., Ebert, F., Schildknecht, S., Leist, M., Schwerdtle, T., 2015. Toxicity of organic and inorganic mercury species in differentiated human neurons and human astrocytes. J. Trace Elem. Med. Biol. 32, 200–208. https://doi.org/10.1016/J.Jtemb.2015.06.008.

Machado, C.L.R., Crespo-Lopez, M.E., Augusto-Oliveira, M., Arrifano, G.D.P., Macchi, B.D.M., Lopes-Araújo, A., Santos-Sacramento, L., Souza-Monteiro, J.R., Alvarez-Leite, J.I., Souza, C.B.A.D., 2021. Eating in the Amazon: nutritional status of the riverine populations and possible nudge interventions. Foods 10, 1015. https://doi.org/10.3390/Foods10051015.

Madry, C., Attwell, D., 2015. Receptors, ion channels, and signaling mechanisms underlying microglial dynamics. J. Biol. Chem. 290, 12443–12450. https://doi.org/10.1074/Jbc.R115.637157.

Mailloux, R.J., Yumvihoze, E., Chan, H.M., 2015. Superoxide anion radical (O2(−)) degrades methylmercury to inorganic mercury in human astrocytoma cell line (CCF-STTG1). Chem. Biol. Interact. 239, 46–55. https://doi.org/10.1016/J.Cbi.2015.06.028.

Minami, T., Miyata, E., Sakamoto, Y., Kohama, A., Yamazaki, H., Ichida, S., 2009. Expression of metallothionein mRNAs on mouse cerebellum microglia cells by thimerosal and its metabolites. Toxicology 261, 25–32. https://doi.org/10.1016/J.Tox.2009.04.037.

Minj, E., Upadhayay, S., Mehan, S., 2021. Nrf2/HO-1 signaling activator acetyl-11-keto-beta boswellic acid (AKBA)-mediated neuroprotection in methyl mercury-induced experimental model of ALS. Neurochem. Res. 46, 2867–2884. https://doi.org/10.1007/S11064-021-03366-2.

Monnet-Tschudi, F., Zurich, M.G., Honegger, P., 1996. Comparison of the developmental effects of two mercury compounds on glial cells and neurons in aggregate cultures of rat telencephalon. Brain Res. 741, 52–59. https://doi.org/10.1016/S0006-8993(96)00895-5.

Mutkus, L., Aschner, J.L., Syversen, T., Aschner, M., 2005. Methylmercury alters the in vitro uptake of glutamate in GLAST- and GLT-1-transfected mutant CHO-K1 cells. Biol. Trace Elem. Res. 107, 231–245. https://doi.org/10.1385/Bter:107:3:231.

Nair, A.B., Jacob, S., 2016. A simple practice guide for dose conversion between animals and human. J. Basic Clin. Pharm. 7 (2), 27–31. https://doi.org/10.4103/0976-0105.177703.
Narahashi, T., Ma, J.Y., Arakawa, O., Reuveny, E., Nakahiro, M., 1994. Gaba receptor-channel complex as a target site of mercury, copper, zinc, and lanthanides. Cell. Mol. Neurobiol. 14, 599–621. https://doi.org/10.1007/Bf02088671.
Ni, M., Li, X., Yin, Z., Sidoryk-Wegrzynowicz, M., Jiang, H., Farina, M., Rocha, J.B., Syversen, T., Aschner, M., 2011. Comparative study on the response of rat primary astrocytes and microglia to methylmercury toxicity. Glia 59, 810–820. https://doi.org/10.1002/Glia.21153.
Ni, M., Li, X., Rocha, J.B., Farina, M., Aschner, M., 2012. Glia and methylmercury neurotoxicity. J. Toxicol. Environ. Health A 75, 1091–1101. https://doi.org/10.1080/15287394.2012.697840.
Nimmerjahn, A., Kirchhoff, F., Helmchen, F., 2005. Resting microglial cells are highly dynamic surveillants of brain parenchyma in vivo. Science 308, 1314–1318. https://doi.org/10.1126/Science.1110647.
O'connor, D., Hou, D., Ok, Y.S., Mulder, J., Duan, L., Wu, Q., Wang, S., Tack, F.M.G., Rinklebe, J., 2019. Mercury speciation, transformation, and transportation in soils, atmospheric flux, and implications for risk management: a critical review. Environ. Int. 126, 747–761. https://doi.org/10.1016/J.Envint.2019.03.019.
O'donoghue, J.L., Watson, G.E., Brewer, R., Zareba, G., Eto, K., Takahashi, H., Marumoto, M., Love, T., Harrington, D., Myers, G.J., 2020. Neuropathology associated with exposure to different concentrations and species of mercury: a review of autopsy cases and the literature. Neurotoxicology 78, 88–98. https://doi.org/10.1016/J.Neuro.2020.02.011.
Ovadje, L., Calys-Tagoe, B.N., Clarke, E., Basu, N., 2021. Registration status, mercury exposure biomarkers, and neuropsychological assessment of artisanal and small-scale gold miners (ASGM) from the Western region of Ghana. Environ. Res. 201, 111639. https://doi.org/10.1016/J.Envres.2021.111639.
Padhi, B.K., Pelletier, G., 2012. Perturbation of myelin basic protein (Mbp) splice variant expression in developing rat cerebellum following perinatal exposure to methylmercury. Toxicol. Lett. 213, 374–380. https://doi.org/10.1016/J.Toxlet.2012.07.011.
Padhi, B.K., Rosales, M., Pelletier, G., 2015. Perinatal methylmercury exposure perturbs the expression of Plp1 and Cnp splice variants in cerebellum of rat pups. Neurotoxicology 48, 223–230. https://doi.org/10.1016/J.Neuro.2015.04.006.
Pamphlett, R., Kum Jew, S., 2018. Inorganic mercury in human astrocytes, oligodendrocytes, corticomotoneurons and the locus ceruleus: implications for multiple sclerosis, neurodegenerative disorders and gliomas. Biometals 31, 807–819. https://doi.org/10.1007/S10534-018-0124-4.
Pamphlett, R., Kum Jew, S., Doble, P.A., Bishop, D.P., 2020. Elemental imaging shows mercury in cells of the human lateral and medial geniculate nuclei. Plos One 15, E0231870. https://doi.org/10.1371/Journal.Pone.0231870.
Park, J.D., Zheng, W., 2012. Human exposure and health effects of inorganic and elemental mercury. J. Prev. Med. Public Health 45, 344–352. https://doi.org/10.3961/Jpmph.2012.45.6.344.
Petroni, D., Tsai, J., Mondal, D., George, W., 2013. Attenuation of low dose methylmercury and glutamate induced-cytotoxicity and tau phosphorylation by an N-methyl-D-aspartate antagonist in human neuroblastoma (SHSY5Y) cells. Environ. Toxicol. 28, 700–706. https://doi.org/10.1002/Tox.20765.
Pieper, I., Wehe, C.A., Bornhorst, J., Ebert, F., Leffers, L., Holtkamp, M., Hoseler, P., Weber, T., Mangerich, A., Burkle, A., Karst, U., Schwerdtle, T., 2014. Mechanisms of Hg species induced toxicity in cultured human astrocytes: genotoxicity and DNA-damage response. Metallomics 6, 662–671. https://doi.org/10.1039/C3mt00337j.

Pierozan, P., Biasibetti, H., Schmitz, F., Avila, H., Fernandes, C.G., Pessoa-Pureur, R., Wyse, A.T.S., 2017. Neurotoxicity of methylmercury in isolated astrocytes and neurons: the cytoskeleton as a main target. Mol. Neurobiol. 54, 5752–5767. https://doi.org/10.1007/S12035-016-0101-2.

Raasakka, A., Kursula, P., 2014. The myelin membrane-associated enzyme 2′,3'-cyclic nucleotide 3'-phosphodiesterase: on a highway to structure and function. Neurosci. Bull. 30, 956–966. https://doi.org/10.1007/S12264-013-1437-5.

Reagan-Shaw, S., Nihal, M., Ahmad, N., 2007. Dose translation from animal to human studies revisited. FASEB J. 22, 659–661. https://doi.org/10.1096/Fj.07-9574lsf.

Rodriguez Martin-Doimeadios, R.C., Berzas Nevado, J.J., Guzman Bernardo, F.J., Jimenez Moreno, M., Arrifano, G.P., Herculano, A.M., Do Nascimento, J.L., Crespo-Lopez, M.E., 2014. Comparative study of mercury speciation in commercial fishes of the Brazilian Amazon. Environ. Sci. Pollut. Res. Int. 21, 7466–7479. https://doi.org/10.1007/S11356-014-2680-7.

Rolls, M.M., Thyagarajan, P., Feng, C., 2021. Microtubule dynamics in healthy and injured neurons. Dev. Neurobiol. 81, 321–332. https://doi.org/10.1002/Dneu.22746.

Rooney, J.P., 2014. The retention time of inorganic mercury in the brain—a systematic review of the evidence. Toxicol. Appl. Pharmacol. 274, 425–435. https://doi.org/10.1016/J.Taap.2013.12.011.

Sakamoto, M., Miyamoto, K.-I., Wu, Z., Nakanishi, H., 2008. Possible involvement of cathepsin B released by microglia in methylmercury-induced cerebellar pathological changes in the adult rat. Neurosci. Lett. 442, 292–296. https://doi.org/10.1016/J.Neulet.2008.07.019.

Santos-Sacramento, L., Arrifano, G.P., Lopes-Araujo, A., Augusto-Oliveira, M., Albuquerque-Santos, R., Takeda, P.Y., Souza-Monteiro, J.R., Macchi, B.M., Do Nascimento, J.L.M., Lima, R.R., Crespo-Lopez, M.E., 2021. Human neurotoxicity of mercury in the Amazon: a scoping review with insights and critical considerations. Ecotoxicol. Environ. Saf. 208, 111686. https://doi.org/10.1016/J.Ecoenv.2020.111686.

Sceniak, M.P., Spitsbergen, J.B., Sabo, S.L., Yuan, Y., Atchison, W.D., 2020. Acute neurotoxicant exposure induces hyperexcitability in mouse lumbar spinal motor neurons. J. Neurophysiol. 123, 1448–1459. https://doi.org/10.1152/Jn.00775.2019.

Shanker, G., Aschner, M., 2001. Identification and characterization of uptake systems for cystine and cysteine in cultured astrocytes and neurons: evidence for methylmercury-targeted disruption of astrocyte transport. J. Neurosci. Res. 66, 998–1002. https://doi.org/10.1002/Jnr.10066.

Shapiro, A.M., Chan, H.M., 2008. Characterization of demethylation of methylmercury in cultured astrocytes. Chemosphere 74, 112–118. https://doi.org/10.1016/J.Chemosphere.2008.09.019.

Shinozaki, Y., Nomura, M., Iwatsuki, K., Moriyama, Y., Gachet, C., Koizumi, S., 2014. Microglia trigger astrocyte-mediated neuroprotection via purinergic gliotransmission. Sci. Rep. 4, 4329. https://doi.org/10.1038/Srep04329.

Shinozaki, Y., Danjo, Y., Koizumi, S., 2019. Microglial rock is essential for chronic methylmercury-induced neurodegeneration. J. Neurochem. 151, 64–78. https://doi.org/10.1111/Jnc.14817.

Sofroniew, M.V., Vinters, H.V., 2010. Astrocytes: biology and pathology. Acta Neuropathol. 119, 7–35. https://doi.org/10.1007/S00401-009-0619-8.

Sokolowski, K., Obiorah, M., Robinson, K., Mccandlish, E., Buckley, B., Dicicco-Bloom, E., 2013. Neural stem cell apoptosis after low-methylmercury exposures in postnatal hippocampus produce persistent cell loss and adolescent memory deficits. Dev. Neurobiol. 73, 936–949. https://doi.org/10.1002/Dneu.22119.

Steckling, N., Boese-O'reilly, S., Gradel, C., Gutschmidt, K., Shinee, E., Altangerel, E., Badrakh, B., Bonduush, I., Surenjav, U., Ferstl, P., Roider, G., Sakamoto, M.,

Sepai, O., Drasch, G., Lettmeier, B., Morton, J., Jones, K., Siebert, U., Hornberg, C., 2011. Mercury exposure in female artisanal small-scale gold miners (ASGM) in Mongolia: an analysis of human biomonitoring (HBM) data from 2008. Sci. Total Environ. 409, 994–1000. https://doi.org/10.1016/J.Scitotenv.2010.11.029.

Steckling, N., Tobollik, M., Plass, D., Hornberg, C., Ericson, B., Fuller, R., Bose-O'reilly, S., 2017. Global burden of disease of mercury used in artisanal small-scale gold mining. Ann. Glob. Health 83, 234–247. https://doi.org/10.1016/J.Aogh.2016.12.005.

Stoiber, T., Bonacker, D., Bohm, K.J., Bolt, H.M., Thier, R., Degen, G.H., Unger, E., 2004. Disturbed microtubule function and induction of micronuclei by chelate complexes of mercury(II). Mutat. Res. 563, 97–106. https://doi.org/10.1016/J.Mrgentox.2004.06.009.

Takahashi, T., Fujimura, M., Koyama, M., Kanazawa, M., Usuki, F., Nishizawa, M., Shimohata, T., 2017. Methylmercury causes blood-brain barrier damage in rats via upregulation of vascular endothelial growth factor expression. Plos One 12, E0170623. https://doi.org/10.1371/Journal.Pone.0170623.

Takemoto, T., Ishihara, Y., Ishida, A., Yamazaki, T., 2015. Neuroprotection elicited by nerve growth factor and brain-derived neurotrophic factor released from astrocytes in response to methylmercury. Environ. Toxicol. Pharmacol. 40, 199–205. https://doi.org/10.1016/J.Etap.2015.06.010.

Tan, Q., Zhang, M., Geng, L., Xia, Z., Li, C., Usman, M., Du, Y., Wei, L., Bi, H., 2019. Hormesis of methylmercury-human serum albumin conjugate on N9 microglia via ERK/MAPKs and STAT3 signaling pathways. Toxicol. Appl. Pharmacol. 362, 59–66. https://doi.org/10.1016/J.Taap.2018.10.017.

Teixeira, F.B., Santana, L.N., Bezerra, F.R., De Carvalho, S., Fontes-Junior, E.A., Prediger, R.D., Crespo-Lopez, M.E., Maia, C.S., Lima, R.R., 2014. Chronic ethanol exposure during adolescence in rats induces motor impairments and cerebral cortex damage associated with oxidative stress. Plos One 9, E101074. https://doi.org/10.1371/Journal.Pone.0101074.

Teixeira, F.B., De Oliveira, A.C.A., Leao, L.K.R., Fagundes, N.C.F., Fernandes, R.M., Fernandes, L.M.P., Da Silva, M.C.F., Amado, L.L., Sagica, F.E.S., De Oliveira, E.H.C., Crespo-Lopez, M.E., Maia, C.S.F., Lima, R.R., 2018. Exposure to inorganic mercury causes oxidative stress, cell death, and functional deficits in the motor cortex. Front. Mol. Neurosci. 11, 125. https://doi.org/10.3389/Fnmol.2018.00125.

Tian, J., Luo, Y., Chen, W., Yang, S., Wang, H., Cui, J., Lu, Z., Lin, Y., Bi, Y., 2016. Mehg suppressed neuronal potency of hippocampal NSCs contributing to the puberal spatial memory deficits. Biol. Trace Elem. Res. 172, 424–436. https://doi.org/10.1007/S12011-015-0609-8.

Tiffany-Castiglion, E., Qian, Y., 2001. Astroglia as metal depots: molecular mechanisms for metal accumulation, storage and release. Neurotoxicology 22, 577–592. https://doi.org/10.1016/S0161-813x(01)00050-X.

Toyama, T., Hoshi, T., Noguchi, T., Saito, Y., Matsuzawa, A., Naganuma, A., Hwang, G.-W., 2021. Methylmercury induces neuronal cell death by inducing TNF-α expression through the ASK1/p38 signaling pathway in microglia. Sci. Rep. 11, 9832. https://doi.org/10.1038/S41598-021-89210-7.

UNEP, United Nations Environment Programme, 2019. *Global Mercury Assessment 2018* [Online]. Available at: https://www.unenvironment.org/resources/publication/global-mercury-assessment-2018 [Accessed 17th June 2020].

USEPA, United States Environmental Protection Agency, 2021a. *How People are Exposed to Mercury*. [Online]. Available at: https://www-epa-gov.translate.goog/mercury/how-people-are-exposed-mercury?_x_tr_sl=en&_x_tr_tl=pt&_x_tr_hl=pt-br&_x_tr_pto=nui,sc#metallicmercury [Accessed 18th Oct 2021].

USEPA, United States Environmental Protection Agency, 2021b. *Reducing Mercury Pollution from Artisanal and Small-Scale Gold Mining*. [Online]. Available at: https://www-epa-gov.translate.goog/international-cooperation/reducing-mercury-pollution-artisanal-and-small-scale-gold-mining?_x_tr_sl=en&_x_tr_tl=pt&_x_tr_hl=pt-br&_x_tr_pto=nui,op,sc [Accessed 17th Oct 2021].

Vahter, M.E., Mottet, N.K., Friberg, L.T., Lind, S.B., Charleston, J.S., Burbacher, T.M., 1995. Demethylation of methyl mercury in different brain sites of Macaca fascicularis monkeys during long-term subclinical methyl mercury exposure. Toxicol. Appl. Pharmacol. 134, 273–284. https://doi.org/10.1006/Taap.1995.1193.

Veit, P., Vallejos, P.Q., 2020. *COVID-19, Rising Gold Prices and Illegal Mining Threaten Indigenous Lands in the Amazon* [Online]. World Research Institute. Available at: https://www.wri.org/insights/covid-19-rising-gold-prices-and-illegal-mining-threaten-indigenous-lands-amazon [Accessed 21th Oct 2021].

Vogel, D.G., Margolis, R.L., Mottet, N.K., 1985. The effects of methyl mercury binding to microtubules. Toxicol. Appl. Pharmacol. 80, 473–486. https://doi.org/10.1016/0041-008x(85)90392-8.

WHO, World Health Organization, 2008. *Guidance for Identifying Populations at Risk From Mercury Exposure* [Online]. Switzerland. Available at: http://www.who.int/foodsafety/publications/risk-mercury-exposure/en/ [Accessed 17th Sep 2021].

WHO, World Health Organization, 2010. *Children's Exposure to Mercury Compounds* [Online]. Environment, Climate Change and Health, WHO Headquarters (HQ). Available at: https://www.who.int/publications/i/item/9789241500456 [Accessed 21th oct 2021).

WHO, World Health Organization, 2016. Environmental and Occupational Health Hazards Associated With Artisanal and Small-Scale Gold Mining [Online]. Available at: https://apps.who.int/iris/handle/10665/247195 [Accessed 12th oct 2021.

WHO, World Health Organization, 2017. Mercury and Health [Online]. Available at: https://www.whoint/news-room/fact-sheets/detail/mercury-and-health [Accessed 20th August 2021].

WHO, World Health Organization, 2018. *Health Sector Involvement in the Minamata Convention on Mercury: Outcomes of World Health Organization Regional Workshops for Ministries of Health* [Online]. World Health Organization, Geneva. Available at: https://apps.who.int/iris/handle/10665/275938 [Accessed 21th Oct 2021].

WHO, World Health Organization, 2021a. Exposure to mercury: a major public health concern. In: *Preventing Disease Through Healthy Environments*. [Online], *second ed.* Available at: https://www.who.int/publications/i/item/9789240023567 [Accessed 17th Oct 2021].

WHO, World Health Organization, 2021b. *Mercury and Human Health: Educational Course* [Online]. World Health Organization, Copenhagen. Regional Office for Europe. Available at: https://apps.who.int/iris/handle/10665/345443 [Accessed 20th Aug 2021].

Xu, B., Xu, Z.F., Deng, Y., Liu, W., Yang, H.B., Wei, Y.G., 2012. Protective effects of MK-801 on methylmercury-induced neuronal injury in rat cerebral cortex: involvement of oxidative stress and glutamate metabolism dysfunction. Toxicology 300, 112–120. https://doi.org/10.1016/J.Tox.2012.06.006.

Yang, T., Xu, Z., Liu, W., Xu, B., Deng, Y., 2020a. Oxidative stress accelerates synaptic glutamate dyshomeostasis and NMDARs disorder during methylmercury-induced neuronal apoptosis in rat cerebral cortex. Environ. Toxicol. 35, 683–696. https://doi.org/10.1002/tox.22904.

Yang, L., Zhang, Y., Wang, F., Luo, Z., Guo, S., Strähle, U., 2020b. Toxicity of mercury: molecular evidence. Chemosphere 245, 125586. https://doi.org/10.1016/j.chemosphere.2019.125586.

Yin, Z., Milatovic, D., Aschner, J.L., Syversen, T., Rocha, J.B., Souza, D.O., Sidoryk, M., Albrecht, J., Aschner, M., 2007. Methylmercury induces oxidative injury, alterations in permeability and glutamine transport in cultured astrocytes. Brain Res. 1131, 1–10. https://doi.org/10.1016/j.brainres.2006.10.070.

Yin, Z., Jiang, H., Syversen, T., Rocha, J.B., Farina, M., Aschner, M., 2008. The methylmercury-L-cysteine conjugate is a substrate for the L-type large neutral amino acid transporter. J. Neurochem. 107, 1083–1090. https://doi.org/10.1111/J.1471-4159.2008.05683.X.

Yin, Z., Albrecht, J., Syversen, T., Jiang, H., Summar, M., Rocha, J.B., Farina, M., Aschner, M., 2009. Comparison of alterations in amino acids content in cultured astrocytes or neurons exposed to methylmercury separately or in co-culture. Neurochem. Int. 55, 136–142. https://doi.org/10.1016/J.Neuint.2009.01.015.

Yuan, Y., Atchison, W.D., 2003. Methylmercury differentially affects GABA(A) receptor-mediated spontaneous IPSCs in Purkinje and granule cells of rat cerebellar slices. J. Physiol. 550, 191–204. https://doi.org/10.1113/Jphysiol.2003.040543.

FSC
www.fsc.org
MIX
Papier | Fördert gute Waldnutzung
FSC® C083411

Kolibri 360 GmbH
Ferdinand-Jühlke-Straße 7
99095 Erfurt, Deutschland
produktsicherheit@kolibri360.de